SUSTAINABLE CITY AND CREATIVITY

Sustainable City and Creativity
Promoting Creative Urban Initiatives

Edited by

LUIGI FUSCO GIRARD
Università degli Studi di Napoli Federico II, Italy

TÜZİN BAYCAN
Istanbul Technical University, Turkey

PETER NIJKAMP
VU University Amsterdam, The Netherlands

Routledge
Taylor & Francis Group

LONDON AND NEW YORK

First published 2011 by Ashgate Publishing

2 Park Square, Milton Park, Abingdon, Oxon OX14 4RN
711 Third Avenue, New York, NY 10017, USA

Routledge is an imprint of the Taylor & Francis Group, an informa business

First issued in paperback 2016

British Library Cataloguing in Publication Data
Sustainable city and creativity: promoting creative urban initiatives.
    1. City planning–Social aspects. 2. City planning–Economic aspects. 3. Urban ecology (Sociology) 4. Cultural industries–Social aspects. 5. Cultural industries–Economic aspects. 6. Sustainable urban development.
    I. Fusco Girard, Luigi. II.Baycan, Tüzin
    III. Nijkamp, Peter.
    307.7'6-dc22

Library of Congress Cataloging-in-Publication Data
Fusco Girard, Luigi.
    Sustainable city and creativity : promoting creative urban initiatives / by Luigi Fusco Girard, Tüzin Baycan, and Peter Nijkamp.
        p. cm.
    Includes bibliographical references and index.
    ISBN 978-1-4094-2001-9 (hardback) – ISBN 978-1-4094-2002-6 (ebook)
    1. Urban ecology (Sociology) 2. Sustainable development. 3. City planning.
    I. Baycan, Tüzin. II. Nijkamp, Peter. III. Title.
    HT241.F87 2011
    307.76–dc22

                                                                2011015339

ISBN 978-1-4094-2001-9 (hbk)
ISBN 978-1-138-24893-9 (pbk)

Contents

List of Figures

List of Tables

Notes on Contributors

Aliye Ahu Akgün is researcher at the Istanbul Technical University Department of Urban and Regional Planning. Her main research interests cover urban and regional planning, sustainable rural development, suburbanization, gated communities and creativity. In recent years, she has focused her research in particular on sustainable rural development. She is a member of the Regional Science Association International and the Regional Studies Association. She is also one of the reviewers of the *Journal of Civil Engineering and Architecture*.

Robert U. Ayres, physicist and economist, noted for his work on thermodynamics in the economic process; on material flows and transformations (industrial ecology or industrial metabolism) – and, most recently, on the role of energy in economic growth. He is Emeritus Professor of Economics and Technology at the international business school INSEAD. Ayres was trained as a physicist at the University of Chicago, University of Maryland, and King's College London (PhD in Mathematical Physics). He was Professor of Engineering and Public Policy at Carnegie-Mellon University in Pittsburgh from 1979 until 1992, when he was appointed Professor of Environment and Management at INSEAD. He is also an Institute Scholar at IIASA. He is author and co-author of 18 books and more than 200 journal articles and book chapters. His books range from *A Handbook of Industrial Ecology*, with Leslie Ayres (Edward Elgar 2002), to *The Economic Growth Engine: How Energy and Work Drive Material Prosperity*, with Benjamin Warr (Edward Elgar 2009), to *Crossing the Energy Divide*, with Edward H. Ayres (Wharton School 2010).

Tüzin Baycan is Professor of Urban and Regional Planning at the Istanbul Technical University (ITU) Department of Urban and Regional Planning. Her main research interests cover urban and regional development and planning, urban systems, sustainable development, environmental issues, creativity, innovation and entrepreneurship, and diversity and multiculturalism. In the past years she has focused her research in particular on urban economics and creativity, innovation and entrepreneurship, knowledge commercialization and valorization. She has published widely in these fields: many book chapters and numerous articles in international journals including *Annals of Regional Science, European Planning Studies, European Urban and Regional Studies, Regional Studies, Journal of Urban Planning and Development, Entrepreneurship and Regional Development, International Journal of Social Economics, Papers in Regional Science, Studies in Regional Science, International Journal of Service*

Technology and Management, International Journal of Entrepreneurship and Innovation Management, International Journal of Foresight and Innovation Policy, Greener Management International, Tijdschrift voor Economische en Sociale Geografie (*Journal of Economic & Social Geography*). She is currently an Editorial Board member of *International Journal of Sustainable Society, Romanian Journal of Regional Science* and *A/Z ITU Journal of Faculty of Architecture* and an Advisory Board member of *Studies in Regional Science*. She is a member of Regional Science Association International, Turkish National Committee of Regional Science and International Society of City and Regional Planners (ISoCaRP). She is also a member of Turkish National Delegation of ISoCaRP and Administrative Board of Turkish National Committee of Regional Science. Recently, she has served as ERC Advanced Grants Panel Member of Social Science and Humanities: SH3 Environment and Society (2008 and 2010).

Giorgio Tavano Blessi, PhD, is Adjoin Professor at Free University of Bozen and University of Trento, Italy. He is post-doctorate fellowship at IULM University, Milan – Italy, and Honorary Associate at UTS – University Technology, and Sydney. His research focuses on cultural economics, cultural policies, urban areas and local development process, culture and social well-being. He has written articles published in *Urban Studies, Cultural Trends, Journal of Art Management Law and Society, Cities, Journal of Happiness Studies*, and *Health and Quality of Life*.

Roberto Camagni is Full Professor of Urban Economics and of Economic assessment of urban transformations, Politecnico di Milano; President of ERSA – European Regional Science Association 2003–2005 and President of GREMI – Groupe de Recherche Européen sur les Milieux Innovateurs, Paris, since 1987. Head of the Department for Urban Affairs at the Presidency of the Council of Ministers, Rome, under the Prodi Government, 1997–1998. In 2010 he received the ERSA – European Investment Bank Prize for his scientific contribution to regional and urban economics.

Lata R. Chatterjee is Professor Emeritus at the Department of Geography and Environment at Boston University, USA. Professor T.R. Lakshmanan and Professor Lata Chatterjee are currently completing a book on 'The Rise and Evolution of Entrepreneurial Cities'.

Peter W. Daniels gained a BSc (Hons) and later a PhD in Geography at University College London. He embarked on his academic career at the University of Liverpool where he was promoted to Reader in Geography before moving to the University of Portsmouth as Professor of Geography and Head of Department. He then joined Birmingham as Professor of Geography and Head of the School of Geography and was subsequently Deputy Dean of Science and later Physical Science and Engineering. From July 2005 he was Dean of Physical Science and Engineering. Since 2008 he assists the Provost and Vice-Principal as Deputy

Pro-Vice-Chancellor. Peter's research interests centre on service industries and economic development. Most of his work explores the way in which advanced services are key drivers of metropolitan and regional economic restructuring. He has a portfolio of publications spanning single authored books, journal papers, reports and working papers. He is a co-investigator on a four-year EU FP7 Collaborative Project (URBACHINA) commencing on 1 January 2011.

Eduardo Dias was an Assistant Professor at the VU University Amsterdam at the time of this research. Currently he works full time for a geo-spatial company based in Southeast Asia, Geodan SEA, developing projects for rain forest monitoring and conservation. He received his PhD in Spatial Economics from the VU University in 2007. His research Interests Include Geo-spatial science, natural interfaces for spatial planning, geo-spatial information in environmental conservation, crowd sourcing and human computation in Geo-analysis.

Guido Ferilli, PhD at the Edinburgh Napier University, is post doctorate researcher at IULM University in Cultural and Creative Industries in the Emerging countries and the Far East. He is Adjoint Professor of economic of design at the IUAV University of Venice and economic thought at University San Raffaele of Milan, Italy. His research focuses on cultural economics, cultural and creative industries, and regional development. He has reached long experience in International research projects, and European funding projects in culture-led local development.

Francesco Forte, graduated in architecture in 1964 at the Faculty of Architecture, University of Naples. In 1969 he was Assistant Professor of Urban Planning. He was Associated Professor in 1982, Extraordinary Town Planning from 1988 to 1990 and Professor of Urban Planning at the University of Naples Federico II since 1991. He directed the seminary of urban planning 'Alberto Calza Bini' from 1994 to 1998 and the Interdepartmental Centre for Urban Research 'Alberto Calza Bini' from 1999 to 2004. He was Director of the Department of Conservation, University of Naples Federico II from 2005 to 2009. He established and directed the annual course of specialization in 'Urban Planning and Sustainable Development'. He has focused his research in particular on urban design, landscape and environmental planning, enhancement of the historical sites, regional and metropolitan developing.

Luigi Fusco Girard is full professor in 'Appraisal' (ICAR/22 SSD) to the Department of Conservation of Architectural and Environmental Heritage, University of Naples 'Federico II' from 1979. He is Coordinator of the PhD School of Architecture Faculty of the University of Naples, Director of the PhD in 'Evaluation Methods for the Integrated Conservation, Development, Maintenance and Management of Architectural Urban and Environmental Heritage' and Director of the Interdepartmental Research Center 'Alberto Calza Bini'. President of the International Scientific Committee for Economics of Conservation, International Council of Monuments and Sites (ICOMOS), Co-president of 'GUD Program

Committee on Celebrating Our Urban Heritage', The Prague Institute for Global Urban Development, Barcellona, Pechino, Londra, Praga, Washington D.C. He is author of several publications nationally and internationally. His interests includes sustainable cities and the methods and assessment tools for sustainability.

Mario Giampietro is ICREA Research Professor at the Institute of Environmental Science and Technology (ICTA) of the Universitat Autònoma de Barcelona (UAB). His research focuses on multi-criteria analysis of sustainability; integrated assessment of scenarios and technological changes; sustainability indicators; multi-scale integrated analysis of societal and ecological metabolism; and science for governance. He published more than 150 papers and book chapters, as well as four books: *Multi-Scale Integrated Analysis of Agro-ecosystems* (2003), *The Jevons Paradox and the Myth of Resource Efficiency Improvements* (2008), *The Biofuel Delusion* (2009), and *The Metabolic Pattern of Societies: Where Economists Fall Short* (2011).

Gamboa Gonzalo holds a PhD in Environmental Sciences, Ecological Economics option, from the Universitat Autònoma de Barcelona . He holds a Bachelor degree in Mechanical Engineering from the Catholic University of Valparaiso, Chile. His previous research has focused on the combination of participatory approaches and multi-criteria evaluation within the public policy domain. Specifically, applying the Social Multi-criteria Evaluation framework in order to support decisions on renewable energies implementation (MCDA-RES project) and costal management (MESSINA project) (2004–2005). He coordinated a research project on local water management carried out by FLACSO-Guatemala (2006). He has also collaborated with the University of the Basque Country in the research project EKO-LURRALDEA (2008).He is part of the following research groups at UAB: Integrated Assessment: Sociology, Technology and the Environment and Agriculture, Livestock and Food in the Globalization (ARAG-UAB).

Xavier Greffe is Professor of Economics at the University Paris I – Sorbonne where he manages the PhD programme in Economics, after having taught in Algiers, Los Angeles (UCLA), Poitiers and Orléans where he was Rector. He is Visiting Professor at the Graduate Research Institute for Policy Studies, in Tokyo. Professor Xavier Greffe has specialised in the fields of local development, economic policy and the economics of arts and culture. His major publications include: *Managing our Cultural Heritage* (Paris, 1999, New Delhi, 2003 and Milano, 2004), *Arts and Artists from an Economic Perspective* (Unesco Publishing & Brookings, 2002), *Le Développement Local* (Paris, 2003), *Création et diversité au miroir des industries culturelles* (La documentation française, 2006), *French Cultural Policy* (in Japanese), (Tokyo : Bookdom, 2007), *Artistes et marchés* (Paris, la documentation française, 2008), *CultureWeb* (Paris : Dalloz).

Emiko Kakiuchi obtained her doctoral degree in Urban Planning from the University of Tokyo. She is professor and Director of the Cultural Policy Program of the National Graduate Institute for Policy Studies (GRIPS). She previously held faculty positions at Hitotsubashi University and Shiga University, and has held short-term lectureships at several universities in Japan and overseas. She is a member of the National Land Council of Japan and has conducted consultancy activities for the central and local governments of Japan and international institutions including the World Bank. Most of her publications have been in the field of cultural policy. She received the City Planning Institute of Japan Award in 2002, and the Japan Association for Planning Administration Prize for Excellent Papers in 2009. She published a paper in *Tourism and Community Development: Asian Practices* (UNWTO, 2008).

T.R. Lakshmanan is Professor Emeritus at the Department of Geography and Environment at Boston University, USA. Professor T.R. Lakshmanan and Professor Lata Chatterjee are currently completing a book on 'The Rise and Evolution of Entrepreneurial Cities'.

Faroek Lazrak studied economics at VU University, Amsterdam. Currently, he is working at the VU University at the spatial economics department writing a dissertation on the economic valuation of cultural heritage. He is focusing on the effect which cultural amenities have on the transaction prices of houses.

Agustin Lobo (Institut de Ciències de la Terra 'Jaume Almera', CSIC) specializes in applications of remote sensing and spatial analysis for the study of the environment. His work deals with the analysis of vegetation dynamics and land cover change based on time series of satellite images. At a more detailed scale, he works on the study of landscapes from satellite and airborne imagery, with applications such as the assessment of environmental impact and the design and management of natural areas. Agustin Lobo is also interested in theoretical aspects regarding the spatial structure of ecosystems, with increasing attention to the role of human activity.

Peter Nijkamp is professor in regional and urban economics and in economic geography at the VU University, Amsterdam. His main research interests cover quantitative plan evaluation, regional and urban modelling, multi-criteria analysis, transport systems analysis, mathematical systems modelling, technological innovation, entrepreneurship, environmental and resource management, and sustainable development. In the past years he has focused his research, in particular on new quantitative methods for policy analysis, as well as on spatial-behavioural analysis of economic agents. He has a broad expertise in the area of public policy, services planning, infrastructure management and environmental protection. In all these fields he has published many books and numerous articles. He is a member of editorial/advisory boards of more than 30 journals. He has been visiting professor in many universities all over the world. According to the RePec list he belongs to the top 30 of well-known economists world-wide. He is past president of the European Regional Science Association and of the Regional Science Association International. He is also fellow of the Royal Netherlands Academy of Sciences, and past vice-president of this organization. From 2002 to 2009 he served as president of the governing board of the Netherlands Research Council (NWO). In addition, he is past president of the European Heads of Research Councils (EUROHORCs). He is also fellow of the Academia Europaea, and member of many international scientific organizations. He has acted regularly as advisor to (inter)national bodies and (local and national) governments. In 1996, he was awarded the most prestigious scientific prize in the Netherlands, the Spinoza award. Detailed information can be found on http://staff.feweb.vu.nl/pnijkamp.

Haifeng Qian joined the Levin College of Urban Affairs in 2010 as an assistant professor. His areas of expertise are economic development, economic geography, innovation and entrepreneurship, and public policy. As a consultant at The World Bank, Dr. Qian also worked on energy policy in the context of developing countries. His research has been accepted or published in *Annals of Regional Science, Economic Development Quarterly, Environment and Planning A, Small Business Economics*, among others.

Waldemar Ratajczak holds a PhD in Spatial Management and is Director of the Institute of Socio-Economic Geography and Spatial Management established in 1984 at Adam Mickiewicz University in Poznan, Poland. He has also studied in the USA and his research is devoted to urban renewal and development and the methodology of spatial economy.

Joe Ravetz is Co-Director of the Centre for Urban and Regional Ecology at the University of Manchester. He is a researcher/designer of sustainable city-regions and the systems which create them: he has a wide range of interests in environment policy, urban development, new economics and governance, innovation and futures studies. With a background as an architect and development manager, he is also an active graphic facilitator and foresight trainer. His main publications include *City-Region 2020, Environment and City* (co-authored), and *Synergicity: pathways to Shared Intelligence in the Urban Century* (forthcoming).

Arda Riedijk was a researcher at the VU University Amsterdam at the time of this research. Her interests include improving operational processes (such as spatial planning, disaster management) through use of (geo)information and communication. Currently she works as a consultant for Geodan.

Piet Rietveld studied econometrics at Erasmus University, Rotterdam (cum laude degree) and received his PhD in economics at Vrije Universiteit, Amsterdam. He worked at the International Institute of Applied Systems Analysis (Austria) and was research co-ordinator at Universitas Kristen Satya Wacana in Salatiga, Indonesia. Since 1990 he has been professor in Transport Economics at the Faculty of Economics, VU University, Amsterdam. He is a fellow at the Tinbergen Institute. His research interests concern: transport and regional development, valuing quality of transport services, economics of public transport, pricing in transport, modelling land use, methods for policy analysis, and valuation of cultural heritage.

Jan Rouwendal (1959) is Associate Professor at the Department of Spatial Economic of VU University. He studied spatial Economics at Erasmus Universiteit Rotterdam obtained his PhD in 1988 from VU university with a dissertation about discrete choice models and housing market analysis. He worked for at Wageningen University and MuConsult and returned to a full time position at VU University in 2005. His research interests are housing and labour markets, commuting behavior, cost-benefit analysis, economic evaluation of cultural heritage, transportation economics, discrete choice modelling. He is a research fellow of the Tinbergen Institute, academic partner of the Netherlands Bureau for Policy Analysis, and member of the editorial board of the Journal of Housing and the Built Environment and Growth and Change. He has published extensively in a wide variety of scientific journals. He is currently program coordinator of the MSc in Spatial, Transport and Environmental Economics. He has taught courses at both the graduate and undergraduate level in microeconomics, real state economics, urban economics, and applied spatial economics.

Pier Luigi Sacco, PhD, is Full Professor of Cultural Economics and Dean of the Faculty of Arts, Markets and Heritage at IULM University, Milan. His research focuses on pro-social game theory, cultural economics, cultural policies, cultural and creative industries, and local development processes. He has written more than 100 articles published in national and international peer-reviewed journals and authors of numerous books. He serves as consultant for national and international cultural agencies and public institutions.

Henk J. Scholten is professor in Spatial Informatics at the Free University in Amsterdam and Scientific Director of the SPINlab, Center for Research and Education on Spatial Information at the Vrije Universiteit Amsterdam. He is president of UNIGIS International, the association of 17 universities around the world that coordinates education and research in GIS. Professor Scholten has written various articles about GIS for publication in international books and journals. He is co-founder and CEO of Geodan in Amsterdam. In his role as director of Geodan, he has been supervisor on a large number of national and international GIS projects. He is advisor for several ministries in different countries. In 2005, Prof. Scholten received a Royal Decoration for his significant contribution to geo-information, both on a national and international level.

Roger Stough's education includes a BS in International Trade, Ohio State University; a MA in Economic Geography, University of South Carolina; Ph.D. in Geography and Environmental Engineering, Johns Hopkins University. He also received a Honoris Causa Doctoral degree from Jonkoping University (Jonkoping, Sweden) in 2006. His research specializations include leadership and entrepreneurship in regional economic development, regional economic modeling, and transport analysis and planning. During the past decade Dr. Stough was heavily involved in the development of research entrepreneurship training, and education programs including advising enterprise development and incubation centers in China and India. His publication record includes many scholarly and professional publications and more than 30 books with sponsored research and matching awards totaling more than $80,000,000 from a variety of sources in the U.S. and abroad. One of his recent books (with Robert Stimson and Maria Salazar) is entitled *Leadership and Institutions in Regional Endogenous Development* (Heidelberg: Springer 2009). He has supervised and/or participated on numerous Ph.D. dissertation committees and his students hold various positions around the world including professorial and leadership posts in universities, government agencies, corporations, think tanks and in international donor agencies. Dr. Stough served as President of the Regional Science Association International (RSAI) (2007–2008) and has been the joint editor-in-chief of the Annals of Regional Science (1984–2010). He holds a variety of other editorial positions. He is also the President of the Technopolicy Network (TPN), a global membership organization that promotes the use of science and technology in the development process. TPN is based in South Holland, the Netherlands. He is a Fellow of the Western Regional Science Association and a Fellow of the Regional Science Association International, the highest professional recognitions of these organizations. Before accepting the position of Vice President for Research & Economic Development in the summer of 2008, Dr. Stough was the Associate Dean for Research, External Relations and Development in the School of Public Policy at Mason where he also directed the Mason Enterprise Center, the National Center for ITS Deployment Research (until 2007) and the National Center for Transport and Regional Economic Development.

Rob van de Velde is currently director of Geonovum, the Dutch National SDI Executive Committee. Geonovum's mission is to realize better access to geo-information in the public sector and full integration of its services to public and industry. Geonovum has been mandated by the National GI-Council to implement the EU INSPIRE Directive in The Netherlands and to develop a National Georegister. Furthermore Geonovum develops and manages the national framework of geo-standards. He has a part-time position as lecturer in Spatial Informatics at the Vrije Universiteit Amsterdam, Department of Economics. He has contributed to more than 40 publications on spatial decision support systems, environmental impact assessments and e-Government services based on geospatial technology. He graduated as a human geographer (Vrije Universiteit, Amsterdam), and started his career in geospatial technology when joining the National Planning Agency in 1985, the first Governmental Agency in the Netherlands to acquire GIS technology from ESRI inc. Later he served the National Environmental Protection Agency, responsible for developing a corporate geographical information system. He directed a consortium that developed the 10 Minutes Pan-European Land Use Database. In 2000 he joined the Ministry of Agriculture, Nature Conservation and Food Quality, heading the GIS Competence Center. Here he experienced the key value of accurate spatial data and geospatial technology in managing the severe outbreaks of avian influenza and foot-and-mouth disease. Building on this he initiated the development of a National Spatial Data Infrastructure for Disaster management, a joint effort of several public agencies. He was managing partner of an international collaboration between governmental agencies and universities under the EU-INTERREG3 program, entitled 'Participatory Spatial Planning in Europe'.

Editorial Preface

Luigi Fusco Girard, Tüzin Baycan and Peter Nijkamp

In recent years, the concept of 'creative cities' has gained increasing interest among both academics and policy-makers. Cities and regions around the world are trying to develop, facilitate or promote concentrations of creative, innovative and/ or knowledge-intensive industries in order to become more competitive. The main interest of all stakeholders is to better understand creative cities – where cultural activities and creative and cultural industries play a crucial role in supporting urban creativity – and their contributions to the new creative economy. This challenge calls for a multidisciplinary orientation, as creativity presupposes an amalgation of different perspectives.

Undoubtedly, a main challenge of the modern creativeness fashion is to translate creative and cultural assets and expressions into commercial values (value added, employment, visitors, etc.), which means that private-sector initiatives are a sine qua non for effective and successful urban creativeness strategies. Consequently, an orientation towards local identity and local roots ('the sense of place'), a prominent commitment of economic stakeholders (in particular, the private sector), and the creation of a balanced and appealing portfolio of mutually complementary urban activities are critical success conditions for a flourishing urban creativeness strategy. Cities offer through their agglomeration advantages a broad array of business opportunities for creative cultures, in which self-employment opportunities and small- and medium-sized enterprises (SMEs) in particular may play a central role in creating new urban vitality. Clearly, flanking and supporting urban conditions – for example, local identity, an open and attractive urban 'milieu' or atmosphere, usage of tacit knowledge, presence of urban embeddedness of new business initiatives, and access to social capital and networks – provide additional opportunities for a booming urban creativeness culture and an innovative, vital and open urban social ecology. Urban creativeness presupposes an open and multi-faceted culture and policy.

The collection of chapters in this volume provides a valuable overview and introduction to this fascinating field for academics, policy-makers, researchers and students who share a common concern about the sustainable city and creativity. The original feature of the volume is its aim to bring together the different approaches of various disciplines on sustainability and creativity in an interdisciplinary and multidisciplinary perspective. This mixed perspective of the volume considers contributions ranging from economic studies to cultural studies, from architecture to urban and regional planning, from theoretical approaches to practical experiences, from strategic visions to implementations, and from institutional tools to wider policies. Thus it offers refreshing contributions, in particular to academia and social scientists, planners and policy-makers, the business sector and society at large. An important conclusion from the chapters in this volume is that the research in the field of sustainability and creativity does significantly contribute to a better understanding of the complexity of the issue concerned and is able to develop relevant policies for a sustainable city.

August 2011

Introduction

Chapter 1

Creative and Sustainable Cities: A New Perspective

Tüzin Baycan, Luigi Fusco Girard and Peter Nijkamp

Creative Creatures in Cities

Our world is dominated by an urban culture. Cities not only represent economies of density, but also economics of interaction. They incorporate both quantity and quality. Cities are not just geographical settlements of people, they are also the 'home of man' (Ward 1976). They reflect the varied history of mankind and are at the same time contemporaneous expressions of the diversity of human responses to future challenges. A great example of the way urban architecture reflects and shapes the future can be found in Dubai, a city that has deliberately left behind its old history and has decided to shape a spectacular new urban design and lifestyle. In doing so, it tries to find a balance between economy, technology, society and culture by deploying urban space as an action platform for accelerated economic growth and by mobilizing all resources for elite lifestyles in the city. Dubai intends to become a symbol of creative architecture.

Dubai is not an exception. Actually, modern urban planning shows an avalanche of varying initiatives focused on creative urban development, in particular by centring on culture and arts as multifaceted cornerstones for innovative development of the city. Consequently, it has become fashionable to regard cultural expressions such as arts, festivals, exhibitions, media, communication and advertising, design, sports, digital expression and research as signposts for urban individuality and identity and as departures for a new urban cultural industry (see Florida 2002, Scott 2003). 'Old' cities such as London, Liverpool, Amsterdam, Berlin, Barcelona, New York, San Francisco, Sydney or Hong Kong witness a profound transformation based on creative cultures. This new orientation not only provides a new dynamism for the city, it also has a symbolic value by showing the historical strength of these places as foundation stones for a new and open future. Clearly, blueprint planning of the city has become outdated. Hence, the creative sector has become an important signpost for modern urban planning and architecture, with major implications for both the micro structures of the city and its macro image towards the outer world.

Since Florida's ideas on the creative class, the creative industry and the creative city (see for an overview Florida 2002), an avalanche of studies has been undertaken to study the features and success conditions of creative environments (see, e.g. Gabe 2006, Heilbrun and Gray 1993, Hesmondhalgh 2002, Landry 2003, Markusen 2006, Power and Scott 2004, Pratt 1997, Scott 2003, Vogel 2001). Despite several empirical studies, an operational conceptualization of creativity infrastructure and suprastructure has as yet not been developed and calls certainly for more profound applied research. This is once more important, as there is a growing awareness of and interest in the dynamics-enhancing impact of creative activities.

In an open world dictated by global competitiveness, it is clear that cities are no longer islands of stable development, but are instead dynamic agglomerations operating in a force field where growth and decline are both possible. Creativity and innovation may become a competitive asset to improve the socio-economic performance of cities. Which factors are decisive for a sustainable development of cities that is able to cope with both local and global forces?

The creative city is, in general, understood and used in four ways: (i) a creative city as a focal point of arts and cultural infrastructure, (ii) a creative city as the action place of a creative economy, (iii) a creative city as synonymous with a strong creative class, and (iv) a creative city as a place that fosters a culture of creativity. In creative cities, most strategies are concerned with strengthening the arts and cultural fabric. The arts and cultural heritage, the media and entertainment industries, and the creative business-to-business services are the drivers of innovation in the creative economy; there is competition to attract, keep or grow their own creative class, while the factors that contribute to this such as 'quality of place' are of high importance; and there is an integrated system of multiple organizations and an amalgam of cultures in the public, private and community sectors. A creative city is supposed to develop imaginative and innovative solutions to a range of social, economic and environmental problems: economic stagnancy, urban shrinkage, social segregation, global competition or more.

The cultural and socio-ethnic pluriformity of modern cities seems to challenge the sense of a common identity. Urban fragmentation seems to become a new trend. For example, in restaurants in Miami it is sometimes impossible to use English as a communication language. Cities not only show cultural and ethnographic diversity, but are also becoming multilingual meeting places (cities as a modern 'Babylon') (see Extra and Yagmur 2004). But even in countries with a generally common language (e.g. the Netherlands or Italy), we observe an increasing popularity of local dialects as a vehicle for showing a common identity ('connotational value'). In this context, the region or city tends to become a geographic platform for establishing and showing a spatial–social identity. This shows that global openness and accessibility may run parallel to closed and fragmented cultural niches.

Urban research calls for a broader orientation in the field of cultural dynamics, with a focus on the following: citizenship and identity, creative activities and innovation, intermediality, the impact of popular culture, and the interface between traditional societal perspectives and open attitudes regarding modern cultures. Against this background, cities have always been meeting places for people of different cultures, education and talents. The modern city is an open 'agora', where ideas from a diversity of cultures and nations come together. The major challenge for a modern city will be to turn possible tensions in such a multicultural 'agora' into positive synergetic energy.

Composition of this Volume

Against the background of increasing interest in creative cities, this volume offers an original set of chapters on sustainable and creative cities that address modern theories and concepts related to research on sustainability and creativity, and analyse principles and practices of the creative city for the formulation of policies and recommendations towards the sustainable city. The volume investigates the relationship between creativity and sustainability in order to provide society and policy-making with the instruments and tools for managing creative cities as a key element of a new strategy for sustainable urban development. It focuses in particular on the following issues: (i) what is a creative city, how does it work and what are the key features and critical elements that foster creativity and innovation in cities; (ii) what are the key assets, infrastructures and tools required to promote creative processes in cities towards competitive, sustainable and cohesive places; (iii) what are the relevant urban policies towards a creative city and society and how can public policies influence the creative city? While addressing the role of the city as an engine of creativity, the volume highlights the new challenges for creative cities and societies.

The volume consists of four main parts. The first part puts the debate on sustainable and creative cities into context and innovative perspectives. From a system point of view, this part highlights the principles of and approaches to urban sustainability and creativity. Part I consists of three chapters. The first chapter by Tüzin Baycan aims to put current debates on creative cities in a comprehensive context while addressing recent studies from a multidisciplinary perspective. The chapter offers a framework that is based on 10 most frequently asked questions (FAQs). In other words, the chapter highlights the 'Top 10 FAQs' in the creative cities debate, which are: (i) What is a creative city? (ii) What makes a creative city? (iii) What are the key assets and infrastructure required to foster the development of creative, competitive and cohesive places? (iv) Why are some places (cities and regions) more attractive than others for new and creative activities? (v) What is the interconnection between creativity and (urban) space and how are they mutually related? (vi) What drives innovation in creative cities? (vii) What are the benefits of creative cities? (viii) What are the opportunities and barriers in the development of creative cities? (ix) How can public policy influence creative cities? (x) What is required to build a creative city? On the basis of these 'Top 10 FAQs', the chapter investigates the general features of creative cities and highlights creative city strategies and challenges.

In the second chapter, Luigi Fusco Girard aims to discuss city creativity in relation to four key problems: economic competition, ecological/climate stability, social cohesion and self-government promotion, and to identify a set of principles that will guide actions and possible initiatives to better achieve in a systemic perspective the objectives of human sustainable development. The focus of the chapter is on complex interdependencies among creativity and resilience in implementing sustainability and on principles, approaches and tools for nurturing the city's economic, social, cultural and ecological resilience. Resilience is proposed by Fusco Girard as the notion that ties creativity to sustainability. On the basis of the new and complex relations in the economic and social system as well as in urban planning and design, Fusco Girard underlines the need for an innovative governance that is based on tests, simulations and experiments. He proposes an evaluation process as an operational tool to effective implementation that includes the criteria and indicators of creativity and urban resilience.

The third and last chapter of Part I by Waldemar Ratajcezak addresses the relations among the concepts of urban complex dynamic system, city vulnerability, sustainable city, liveability of cities, and the newly introduced XXQ of cities. The chapter aims to give a short description of those notions and to emphasize the role that they play in assessing a creative city. The chapter demonstrates the relation between city's vulnerability and its status defined as XXQ and underlines that all elements of this relation contribute significantly to establishing the city's position, but resilience is especially vital among them, as shown by many real-life examples.

The second part of the volume addresses the entrepreneurial and creative urban economy and investigates the creative and knowledge-based industries and their roles in the urban, regional and global economy. This part, which consists of three chapters, also discusses the role of skills, training and professional services in sustaining creative cities, the geographic distribution of talents among megacities and regions, and the factors that affect regional talent intensity. The first chapter by T.R. Lakshmanan and Lata Chatterjee addresses the issue of entrepreneurial creative clusters in the new global economy, offering a survey of the recent rise, evolution and defining characteristics of entrepreneurial cities, viewed as complex entities. It reviews the factors underlying the 'resurgence' of cities as sites of innovation-led growth in an era of globalization and an increasing knowledge-intensive economy, and presents and discusses the three models of such innovation-led urban/regional development ('new economic geography', 'new growth theory' and a 'spatial–relational interaction model of innovation-led agglomeration'), highlighting the new and evolving dimension of decision and management context in these dynamic entrepreneurial clusters. The chapter illustrates the entrepreneurial functions of urban public and social sector agents in the process of co-production of urban value, and offers some selected American experiences in the rise of entrepreneurial creative clusters.

Next, Peter Daniels discusses the role of skills, training, and business and professional services in sustaining creative cities. The chapter suggests that the process of enhancing the creative capacity of cities in ways that allow them to compete in an increasingly globalized world requires an endogenous approach that incorporates skills acquisition and renewal. It underlines that this is especially critical for cities reconfiguring the legacies of the Industrial Revolution and that are located in a semi-peripheral or peripheral region. On the basis of a study of business and professional service (BPS) firms in Birmingham, UK, the chapter confirms that the ability of these firms to induce the creative capacity of the city is derived from a combination of technical competence, the personality of individual professional and support staff, and a set of soft skills. The findings of the study show that academic excellence and world-class technical competence alone do not provide a platform for harnessing the contribution of BPS to creative cities; it is also crucially founded upon local relationship building, soft skills and personality.

In the next chapter, Roger R. Stough and Haifeng Qian address the geography of talents in China. Their chapter presents the geographic distributions of two types of talent – human capital and professional personnel – among Chinese megacities and regions, and examines geographically bounded factors that affect regional talent intensity. A panel data analysis covers both market factors (the wage level, high technology and research and development (R and D) activity) and non-market factors (the quality of life, social openness and the university) as explanatory variables. Empirical results of this study show different geographical patterns and different determinants of talent both between human capital and professional personnel, and between China and developed countries.

The third part of the volume, consisting of five chapters, focuses on urban cultural landscape and creative milieu and evaluates the relationship between creativity, culture and urban milieu, the role of cultural landscape, and more specifically, cultural heritage in urban development and culture-led regeneration and local development. The first chapter by Roberto Camagni analyses the relationships between culture, creativity and urban milieu. This chapter aims at contributing to the creative city debate in a twofold way. First of all, he positions the creative city phenomenon in a more general and complex theoretical framework with respect to many contributions that straightforwardly underline single explicative elements for urban creativity. The general argument of the chapter is that attention should be addressed not just to traditional functional elements (human capital, externalities, external linkages …), but mainly to symbolic and cognitive elements (codes, representations, languages, values) replicating the ways in which individuals, groups and communities fully exploit their creative potential through synergy, cooperation, and associative thinking. The second aim of the chapter is to present alternative strategies for relaunching the cities' development by enhancing the preconditions for innovation and creativity: reorienting traditional 'vocations' and competencies of the local context towards new and modern activities through the provision of interaction opportunities and places, the creative utilization of urban cultural heritage and atmosphere, the enhancement of knowledge-intensive functions and the development of new urban governance styles.

The next chapter by Xavier Greffe focuses on the landscape changes and argues that while, traditionally, a landscape is considered as an expanse of natural scenery that people come to see and enjoy, this romantic perspective has widened in a post-modern city towards looking for feelings and emotions, and the landscape has become an experience. In this new understanding landscape has a more subjective content and, according to Greffe, it may be better in such a case to use the terms 'atmosphere' or 'environment' instead of landscape. He argues that once the conservation of cultural landscapes is recognized as an important element for reinforcing the economic base of a territory rather than as a simple expression of an aesthetic need, the issues change, and changing the traditional view implies new quality assessment criteria and instruments. According to Greffe, the motivations – ecological, tourist or cultural – will determine the type of actors who will play a role in the formulation of long-term policies to take action for the area's sustainable economic development by avoiding useless and irreversible damage to natural, cultural and, therefore, human environments. Some will intervene in the name of safeguarding the quality of the living conditions of the local inhabitants. Others will intervene in the name of preserving culture as an intangible element, while still others may invoke the beauty and integrity of a landscape. In this context, new cultural assets such as retrofits, cultural districts and quarters deserve attention.

Next, the chapter by Faroek Lazrak, Peter Nijkamp, Piet Rietveld and Jan Rouwendal addresses cultural heritage and creative cities from an economic evaluation perspective. While focusing on cultural heritage as a resource of both historico-cultural and socio-economic significance in a modern society, the chapter highlights that a main challenge of the modern creativeness fashion is to translate creative and cultural assets and expressions into commercial values (value added, employment, visitors, etc.), which means that private-sector initiatives are a sine qua non for effective and successful urban creativeness strategies. The chapter offers a survey of methods to value cultural heritage, including the widely used stated preference methods, as well as a limited set of studies using hedonic price approaches. The findings of the survey demonstrate that given the orientation towards the local benefits of cultural heritage, the latter class of methods is a promising area of research on the valuation of amenities in creative cities.

The chapter by Guido Ferilli, Pier Luigi Sacco, and Giorgio Tavano Blessi discusses the recent literature on the role of creativity in local and city development models. The chapter states that the key issue underlying the debate is that of instrumentality. Only those initiatives and polices that may credibly rely upon the intrinsic motivations of cultural producers and of the local community turn out to be socially and economically sustainable in the long run. The chapter presents a model of culture-led local development – that is, system-wide cultural districts – that may effectively cope with the instrumentality issue and inspire effective, sustainable policies.

The final chapter of Part III by Aliye Ahu Akgün, Tüzin Baycan and Peter Nijkamp aims to pinpoint the key factors that stimulate rural areas to become creative milieus. While focusing on 60 creative rural areas from Belgium, France and Italy, the chapter, first, creates an index to identify the positive changes – the socio-economic progress – in each of the rural towns investigated by means of a standard multidimensional analysis technique, viz. Principal Component Analysis and, second, identifies the critical factors inducing these changes by applying a recent artificial intelligence method, viz. Rough Set Data analysis. The results of the analysis show that the creativity of rural areas can be explained on the basis of their distance to the nearest urban centre, their main economic activity and the socio-economic composition of their inhabitants. Among these factors, the main economic activity turns out to be the critical factor for investing a rural area into a creative milieu.

The last part of the volume, Part IV, highlights new methodological approaches and planning instruments from theoretical approaches to practices and from design to visualization techniques. This part consists of six chapters. The first chapter by Francesco Forte aims to explore creativity as an expression of a geopolitical atmosphere that forms a stimulus for creative places. The chapter proposes to relate the concepts of creativity and sustainability to the geopolitical context, as a background context of urban design. Through an interpretation of stages of development, the chapter examines theories that played a role in directing design choices, city plans and architecture for the city, reflects on the culture of city living in the reconstruction plan after the Second World War, interprets the epochal revolution that occurred in the post-industrial age as global and comprehensive, explores significant city signs related to a floating society of minorities that recognize mutual dignity and meaningfulness values, and finally addresses planning culture in the present age, described as the stage of production and financial disintegration, and the role of creativity in highlighting sustainable goals. Based on this interpretation, the chapter turns to future hopes for cities, to the threats that can dominate and overshadow these hopes, and to strategies and policies to deal with the threats.

The next chapter by Mario Giampietro, Gonzalo Gamboa and Agustin Lobo illustrates the concept of energy metabolism to analyse the development and sustainability of urban society. It analyses the effect of the Industrial Revolution, which brought about a transition from pre-industrial rural societies based on renewable energy sources to modern urban societies, totally dependent on fossil energy. The chapter emphasizes that this transition dramatically changed the role of the city in society: from a passive consumer of resources generated in the countryside, the city became the engine of change and development making available resources to the countryside. That is, this change dramatically boosted the function of the city's creativity for society. However, the extreme density of both population and metabolic flows (energy, food, wastes) generated in modern cities prompts a new challenge for scientists willing to study these structural and functional changes. The second part of the chapter presents an innovative approach capable of establishing a link between the metabolic pattern of the modern city (the characteristics of the whole) with the metabolic patterns expressed by the various household types in a city (the characteristics of the parts). In this way it becomes possible to establish a much-needed link between the analysis of demographic changes and the resulting changes in the pattern of production and consumption determining the overall urban energy metabolism. In particular, the chapter distinguishes between activities related to efficiency (doing better what is already done) and adaptability (exploring new behaviours and generating new functions). Finally, the chapter underlines that to analyse these changes in spatial terms is essential for understanding the sustainability of urban development.

The chapter by Joe Ravetz aims to explore the relationship between a creative city and deliberative visualization while demonstrating the new approaches in research and social learning, and addressing the challenges and opportunities emerging from different parts of society. The chapter focuses on some innovative examples to represent a broad field and looks at cities from the point of view of peri-urban areas, rather than conventional urbanist centres, the sustainability agenda from the point of view of the low carbon economy, and the concept of creativity linking personal and political creative process in cities. In order to bring this together, the chapter demonstrates an emerging body of theory and practice: an application of 'relational thinking' – a new approach to systems mapping and transition/evolutionary analysis – which has been developed to respond to such complex and multiple challenges, and 'relational visualization' – a new approach to multiple channels of dialogue and experiential process. The chapter offers visual examples, with a range of possibilities from concept mapping to figurative drawing, and from static images to an interactive process of 'relational visualization'.

The following chapter by Eduardo Dias, Henk J. Scholten, Arda Riedijk and Rob van de Velde introduces geospatial technology and especially the role of geospatial visualization as crucial tools in supporting the creative process of participatory and sustainable city planning. The chapter focuses on the factors that determine the successful integration of geospatial visualization tools and techniques for creative input into agencies responsible for participatory sustainable city planning. In order to determine these factors, it reviews the literature about the adoption and acceptance of innovative geospatial visualization tools and techniques from the perspective of urban agencies and the user. The chapter emphasizes the dynamic nature of information and communication technology (ICT) and defends the idea that organizations should follow these dynamic trends and keep their tools for participatory processes up to date and attractive to use, and describes how new technologies, and specifically Geo-ICT, are accepted and implemented. The chapter discusses several theories that explain how they are taken up within organizations and diffused through society, and the user's role in the diffusion process, compares several transition theories and discusses the mechanisms and operational requirements from different perspectives. While addressing the initiatives taken by governments to set up Spatial Data Infrastructures, the chapter offers a number of case studies to illustrate the role of Geo-ICT in public participation.

Next, the chapter by Robert U. Ayres addresses the future of transport in the light of certain mega-trends that seem irreversible: population growth, urbanization, the 'peak oil' (with the decreasing availability of liquid petroleum), the increasing greenhouse gases in the atmosphere, and the search of electric alternatives. The chapter argues that the enormous industrial superstructure built around the internal combustion engine over the last 150 years, and ultimately based on cheap liquid hydrocarbon fuels, is going to have to change radically. The chapter discusses the different approaches, scenarios and solutions for the urban transport system in 2050 in terms of their strengths and weaknesses from the medium and the long-term perspective.

Finally, the last chapter of Part IV and the volume by Emiko Kakiuchi discusses the theory and practices of sustainable and creative cities in Japan. The chapter states that the importance of 'creativity' has long been discussed in relation to education, culture, and science and technology in Japan. In the field of business management, creativity has been viewed as an important element for innovation, which enables companies to adapt to the ever-changing environment. Tacit knowledge and corporate culture shared by employees are also considered to play an important role. Therefore, the newly introduced 'creative city' concept seems basically to be an extension of this philosophy in Japan, with the slight difference of paying more attention to culture. The chapter highlights that the capital city, Tokyo, is the most creative city in Japan in terms of a competitive knowledge-based economy. However, for smaller cities the creative city concept serves better as an instrument to mobilize the community in problem-solving efforts rather than as a goal. The case studies offered by the chapter show that tacit knowledge and culture shared by the community could serve as an infrastructure for creativity and innovation, and provide a basis for attaining a balance between quality of life and economic vitality, ultimately increasing social welfare for all residents.

Conclusions

The city is a centre of action for a modern society. It is unparalleled in terms of agglomeration advantages (despite the existence of clear negative externalities). The modern urban creative 'melting pot' may offer a new opportunity to a balanced and maybe even accelerated urban development, provided the negative externalities involved (social stress, unemployment, etc.) remain lower than the positive opportunities that are the result of a pluriform socio-cultural urban ecosystem. Urban policy has to operate at a difficult edge between checks and balances, and has to seek out creative strategies to exploit the potential benefits of a pluriform urban climate. Fortunately, as illustrated in the present volume, we have many good examples that demonstrate that a pluriform urban culture may enhance welfare and well-being. Such role models are badly needed in a period where sometimes socio-cultural tensions may seem to overshadow the beauty of urban life.

The overall conclusion of this volume is that research in the field of creativity and sustainability does significantly contribute to a better understanding of the complexity of the issues concerned and is able to develop relevant policies for creative and innovative cities and regions. It suggests a form of change not only that generates productive improvements, enhanced performance, but also sustainable growth and development, and emphasizes the crucial role of cities and regions in innovative policies. The volume also provides important insights into the ongoing transformation of cities and regions to become core areas of a dynamic knowledge-based economy. All in all, this publication offers a wealth of refreshing studies with a high value for academia (in particular, social scientists), planners and policy-makers, the business sector and society at large.

Bibliography

Extra, G. and Yagmur, K. 2004. *Urban Multilingualism in Europe*. Clevedon: Multilingual Matters.

Florida, R. 2002. *The Rise of the Creative Class*. New York: Basic Books.

Gabe, T.M. 2006. Growth of Creative Occupations. *U.S. Metropolitan Areas, Growth and Change*, 37(3), 396–415.

Heilbrun, J. and Gray, C.M. 1993. *The Economics of Art and Culture*. New York: Cambridge University Press.

Hesmondhalgh, D. 2002. *The Cultural Industries*. London: Sage.

Landry, C. 2003. *The Creative City*. London: Earthscan.

Markusen, A. 2006. Urban Development and the Politics of a Creative Class. *Environment and Planning* A, 38(10), 1921–40.

Power, D. and Scott, A. 2004. *Cultural Industries and the Production of Culture*. London: Routledge.

Pratt, A. 1997. The Cultural Industries Production System. *Environment and Planning* A, 29, 1953–74.

Scott, A. 2003. *The Cultural Economy of Cities*. London: Sage.

Vogel, H. 2001. *Entertainment Industry Economics*. New York: Cambridge University Press.

Ward, B. 1976. *The Home of Man*. New York: Norton.

PART I
Creative and Sustainable Cities:
Principles and Perspectives

Chapter 2

Creative Cities: Context and Perspectives

Tüzin Baycan

Cities as Creative Milieux

In today's information economy, knowledge and creativity are increasingly recognized as key strategic assets and powerful engines driving economic growth. Cities have become the strategic sites, as they represent the ideal scale for the intensive, face-to-face interactions that generate the new ideas that power knowledge-based innovation (Bradford 2004a). While creativity constitutes a response to some of the economic challenges raised by globalization, paradoxically it requires initiative and organization at a local level (KEA 2006). Kalandides and Lange (2007) have called these 'globalization paradox' and 'identity paradox'. 'Globalization paradox' addresses the ambivalence between local-based creativity and transnational networks of production systems, whereas 'identity paradox' addresses the ambivalence between individual or collective careers, identities and reputations. While creativity is an essential parameter in global competition, it is fostered and nurtured by exchanges of information and experiences at a local level (KEA 2006). The high concentrations of heterogeneous social groups with different cultural background and different ways of life have made cities incubators of culture and creativity (Baycan-Levent 2010, Merkel 2008, UNCTAD 2008). Besides knowledge and innovation, culture and creativity have become the new key resources in urban competitiveness. Cultural production in itself has become a major economic sector and a source for the competitive advantage of cities (Florida 2002, Merkel 2008, Miles and Paddison 2005, Musterd and Ostendorf 2004, Zukin 1995). Cultural and creative activities have shaped the competitive character of a city by enhancing both its innovative capacity and the quality of place, which is crucial to attracting creative people (Gertler 2004).

Knowledge, culture and creativity have also become the new keywords in the understanding of new urban transformations (Hall 2004). While cities are the key drivers of economic change, culture plays a crucial role in this process not just as a condition to attract the creative people but also as a major economic sector (Florida 2002, Miles and Paddison 2005, Musterd and Ostendorf 2004, Zukin 1995). The existing literature shows that cultural and creative industries are deeply embedded in urban economies (Foord 2008, Pratt 2008, Scott 2000). The role of cultural production in the new economy has radically changed the patterns of cultural consumption (Quinn 2005), and cities have transformed from functioning as 'landscape of production' to 'landscape of consumption' (Zukin 1998). The new patterns of cultural consumptions offer big opportunities for local and regional development (Marcus 2005).

In parallel to this transformation or 'cultural turn' a corresponding movement of 'image production' has emerged, especially in the advanced industrial societies (Quinn 2005). 'Culture-led regeneration' and 'city marketing' have become the main strategies of cities in order to create a 'good image' and attractive and high-quality places. As mentioned also by (Peck 2005) the creative economy is not only about creative people and creative industries but also about marketing, consumption and real-estate development. The new economy pushes cities to search for new spatial organization through urban restructuring (Sassen 2001) and in the restructuring process art and creativity play an important role as the key growth resources (Sharp et al. 2005). In order to create visual attractions and appealing consumption spaces urban renewal programmes consider waterfront revitalization, museum quarters, and so on (Merkel 2008).

Today over 60 cities worldwide call themselves 'creative cities', from London to Toronto and from Brisbane to Yokohama (UNCTAD 2008). Creative cities use their creative potential in various ways: some function as nodes for generating cultural experiences through the performing and visual arts (UNCTAD 2008); some use festivals that shape the identity of the whole city as part of their urban regeneration and city marketing strategies (Quinn 2005); and others look to broader cultural industries to provide employment and incomes.

Creative cities have also become a global movement reflecting a new planning paradigm in recent decades. The general features of creative cities are described by many scholars; however, as also mentioned by Bradford (2004a), the studies describe the conditions that foster creativity, and the mechanisms, processes and resources that turn ideas into innovations remain very limited. From the point of view of this need, this study aims to put the current debate about creative cities in context and perspective while addressing the most recent studies from a multidisciplinary perspective. The study offers a framework that is based on the 10 most frequently asked questions (FAQs). In other words, the study aims to highlight the 'Top 10 FAQs' in the creative cities debate. The next section begins by describing creativity and its dimensions, and is followed by a discussion of the key concepts in urban creativity in the third section. Then, the fourth section investigates the general features of creative cities and highlights the Top 10 FAQs in the creative cities debate. The last section considers the policy roadmap to the creative city and the challenges for government.

Creativity: Dimensions and Perspectives

Creativity has been studied by different disciplines such as psychology, psychometrics, cognitive science, artificial intelligence, philosophy, history, economics, design research, business and management, and from different perspectives including artistic, scientific, economic and technological. These

Table 2.1   Different definitions of creativity

Hadamard (1939)	Invention or discovery, be it in mathematics or anywhere else, takes place by combining ideas
Snow (1986)	Creativity is not a light bulb in the mind, as most cartoons depict it. It is an accomplishment born of intensive study, long reflection, persistence, and interest
Rothenberg (1990)	Creativity is the production of something that is both new and truly valuable
Sternberg (2000)	Ability to produce work that is novel (i.e. original, unexpected) high in quality, and appropriate (i.e. useful, meets task constraints)
Torrance (1989)	The process of becoming sensitive to problems, deficiencies, gaps in knowledge, missing elements, disharmonies, and so on; identifying the difficult, searching for solutions, making guesses, or formulating hypotheses and possibly modifying them and retesting them; and finally communication the results
Simon (2001)	We judge thought to be creative when it produces something that is both novel and interesting and valuable
City of Ottawa (2003)	Creativity is about new ideas, and the discipline of developing, sharing and applying them
UNCTAD (2004)	Creativity is not a given resource but a resource that is deeply embedded in every country's social, cultural and historical context. As such, it is a ubiquitous asset
Smith (2005)	Creativity should be defined by the novelty of its products, not by their usefulness, value, profit-ability, beauty, and so on
KEA (2006)	Creativity is a complex process of innovation mixing several dimensions such as technology, science, management, and culture
UNCTAD (2008)	Creativity refers to the formulation of new ideas and to the application of these ideas to produce original works of art and cultural products, functional creations, scientific inventions and technological innovations
Boston's Creative Economy, BRA/ Research, USA (2009, cited in UNCTAD 2008)	Creativity can be defined as the process by which ideas are generated, connected and transformed into things that are valued
The Ride (2009, cited in Wikipedia 2009)	Creativity is the ability to illustrate what is outside the box from within the box
Wikipedia (2009)	Creativity is a mental and social process involving the generation of new ideas or concepts, or new associations of the creative mind between existing ideas or concepts

studies have covered everyday creativity, exceptional creativity and artificial creativity. However, there is no single or simple definition of creativity (the notion has more than 60 definitions (Sternberg 1999) that includes all the different dimensions of this phenomenon). On the one hand, there have been inconsistencies concerning the definition of creativity and, on the other, creativity definitions overlap and intertwine (Table 2.1). Similar to the existence of inconsistencies in the definition of creativity, there are also inconsistencies in the methodologies used to measure creativity. There is neither a standardization of methods or measurement techniques nor a systematic approach (Afolabi et al. 2006, Wikipedia 2009).

As can be seen from Table 2.1, creativity is a complex phenomenon and associated with originality, imagination, inspiration, ingenuity and inventiveness. A unifying definition of creativity is challenging but also difficult as it has been argued by various researchers that creativity is domain specific (Afolabi et al. 2006). However, in order to find a unifying definition of creativity Rhodes (1961) did an extensive search and found that the definitions are not mutually exclusive. According to Rhodes (1961), creativity definitions form four strands: 4P (PERSON – identification of the characteristics of the creative person, PROCESS – the components of creativity, PRODUCT – the outcome of creativity and PRESS – the qualities of the environment that nurture creativity) and each strand has a unique identity. Torrance (1976) similarly observed that creativity has usually been defined in terms of process, product, personality and environment. He redefined creativity as 'a successful step into the unknown, getting away from the main track, breaking out of the mold, being open to experience and permitting one thing to lead to another, recombining ideas or seeing new relationships among ideas'. Torrance's definition of creativity has been widely accepted by many researchers.

The characteristics of creativity in different areas have been suggested by UNCTAD (2008: 9) as 'artistic creativity', 'scientific creativity', 'economic creativity' and 'technological creativity':

- 'artistic creativity' involves imagination and a capacity to generate original ideas and novel ways of interpreting the world, expressed in text, sound and image;
- 'scientific creativity' involves curiosity and a willingness to experiment and make new connections in problem solving;
- 'economic creativity' is a dynamic process leading towards innovation in technology, business practices, marketing, etc., and is closely linked to gaining competitive advantages in the economy;
- all these characteristics of creativity in different areas are interrelated and involve 'technological creativity'.

Therefore, 'creativity' is defined in a cross-sector and multidisciplinary way, mixing elements of 'artistic creativity', 'economic creativity' or 'economic innovation', 'scientific creativity' as well as 'technological creativity' or 'technological

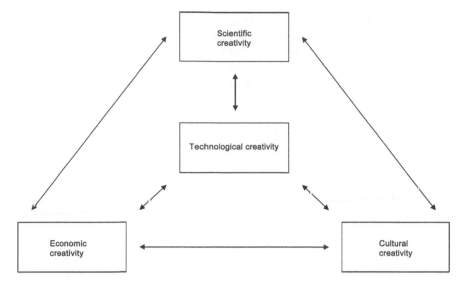

Figure 2.1   Characteristics of creativity

Source: KEA 2006: 42.

innovation' (Figure 2.1). Here creativity is considered as a process of interactions and spillover effects between different innovative processes (KEA 2006).

Creativity has various dimensions and attributes. In psychology and cognitive science, the concept refers to individual creativity and mental representations and processes underlying creative thought (Afolabi et al. 2006, Kaufman and Sternberg 2006, Sternberg 1999, Wikipedia 2009). However, there is no agreement whether creativity is an attribute of people or a process by which original ideas are generated. The psychological dimension of creativity includes the attributes such as 'convergent' and 'divergent' thinking, age and gender differences, and the role of education in creativity (Baer 1997, Fasko Jr 2001, Kaufman and Sternberg 2006, Reis 2002, Sternberg 1999). The formal starting point for the scientific study of creativity in psychology is generally considered to be the studies after the 1950s and in particular J.P. Guilford's scientific approach to conceptualizing creativity and measuring it psychometrically (Afolabi et al. 2006, Sternberg 1999, Wikipedia 2009). Guilford (1967) performed important work in the field of creativity, drawing a distinction between 'convergent' and 'divergent' thinking. Convergent thinking involves aiming for a single, correct solution to a problem, whereas divergent thinking involves the creative generation of multiple answers to a set problem. Divergent thinking is sometimes used as a synonym for creativity in the psychology literature.

The cultural dimension of creativity includes creativity in diverse cultures and different modes of artistic expression. Creativity in general seems to be enhanced by cultural diversity and cultural diversity provides sources for creative expression. The relationship between diversity and creativity has been investigated by many scholars in different disciplines from socio-economic, cultural and psychological perspectives (see for a comprehensive evaluation Baycan-Levent 2010). In these studies, diversity has been analysed in terms of demographic attributes (age, sex, ethnicity) and cognitive (knowledge, skills, abilities) aspects in order to explain whether it has a positive or negative effect on performance, creativity and innovation (Bechtoldt et al. 2007, Herring 2009). Many studies of collective creativity (teams, organizations) find that diversity fosters creativity. The results of research on heterogeneity in groups suggest that diversity offers a great opportunity for organizations and an enormous challenge. More diverse groups have the potential to consider a greater range of perspectives and to generate more high-quality and innovative solutions than do less diverse groups. In brief, while diversity leads to contestation of different ideas, more creativity, and superior solutions to problems, in contrast, homogeneity may lead to greater group cohesion but less adaptability and innovation.

Creativity is seen from an economic perspective as an important element in the recombination of elements to produce new technologies and products and, consequently, economic growth. In the early twentieth century, Joseph Schumpeter (1950) introduced the economic theory of 'creative destruction' to describe the way in which old ways of doing things are endegenously destroyed and replaced by the new. Today, the economic dimension of creativity can be so called because the 'creative economy' is an evolving and a holistic concept dealing with complex interactions between culture, economics and technology (KEA 2006, UNCTAD 2008). The creative economy is based on creative assets potentially generating economic growth and development. These creative assets consider 'creative class', 'creative industries/cultural industries', 'entrepreneurship and innovation' and 'cultural entrepreneurialism' (Boschma and Fritsch 2007, Ellmeier 2003, Florida 2002, Hartley 2005, KEA 2006, Fleming and NORDEN 2007, Musterd et al. 2007, UNCTAD 2008, Wu 2005). Creativity is increasingly recognized by economists as a powerful engine driving economic growth. It contributes to entrepreneurship, fosters innovation, enhances productivity and promotes economic growth.

The physical-spatial dimension of creativity refers to 'creative milieu/ innovative milieu', 'creative city' and 'quality of place' (Bradford 2004a and 2004b, Crane 2007, Drake 2003, Evans 2009, Florida 2002, Foord 2008, Hall 2000, Jones 2007, Landry 2000 and 2007, Musterd et al. 2007, UNCTAD 2008, Wu 2005, Yip 2007). Creativity is seen from an urban perspective as 'a collective process emerging from a broader crerative field characterized by place-based clusters of cultural industries' (Rantisi et al. 2006: 1790). In general, creative industries tend to cluster in large cities and regions that offer a variety of economic opportunities, a stimulating environment and amenities for different lifestyles. Creative industry development is often considered part of the inherent dynamic of urban spaces, and urban environments provide ideal conditions – a creative milieu – for cluster development (Landry 2000, Porter 1998, Porter and Stern 2001). Cities and regions around the world are trying to develop, facilitate or promote concentrations of creative, innovative and/or knowledge-intensive industries in order to become more competitive. These places are seeking new strategies to combine economic development with quality of place that will increase economic productivity and encourage growth.

Although the wealth of literature regarding the development of creativity indicates wide acceptance that creativity is desirable, it is not as highly rewarded in practice as it is supposed to be in theory (Kaufman and Sternberg 2006, McLaren 1999). There is a dark side to creativity (McLaren 1999) as it represents a 'quest for a radical autonomy apart from the constraints of social responsibility'. This means by encouraging creativity we may also encourage a departure from society's existing norms and values. Therefore, there is a paradox between the ability to 'think outside the box' and maintaining traditional, hierarchical organization structures. In these traditional structures individual creativity is not rewarded.

Urban Creativity

Creativity has found many reflections in urban and regional studies and has offered some new concepts such as 'creative class', 'creative industries', 'creative milieu' and 'creative city' to urban literature. The international debate in the past few years has been dominated by these concepts (Florida 2002, Hall 1998 and 2000, Helbrecht 2004, Howkins 2002, ISoCaRP 2005, Jones 2007, Kalandides and Lange 2007, Landry 2000 and 2007, Musterd et al. 2007, Scott 2006, Wu 2005).

The concept of 'creative class' was first developed by Richard Florida (2002). In his book *The Rise of Creative Class*, Florida described the 'creative class' as a fast growing, highly educated, and well-paid segment of the workforce. The 'creative class' consists of the 'super-creative core', which includes scientists and engineers, university professors, poets and novelists, artists, entertainers, actors, designers, architects and the 'thought leadership' – non-fiction writers, editors, cultural figures, think-tank researchers, analysts, and other opinion-makers – and the 'creative professionals' who work in a wide range of knowledge-intensive industries such as the high-tech sectors, financial services, the legal and healthcare professions, and business management. Florida has argued that creative people are a key driver of urban and regional growth and the 'creative class' is not evenly distributed among cities and regions. This class is especially attracted to places that are characterized by an urban climate of tolerance that is open to new ideas and new people. According to Florida, regions with a high share of creative people will perform economically better because they generate more innovations, have a higher level of entrepreneurship, and attract creative businesses. Analysing the role of creativity in economic development and urban and regional success, Florida has stated that 'talent', 'technology' and 'tolerance' (3Ts) are important conditions. In his 3T model he argues that growth is powered by creative people (talent), who prefer places that are culturally diverse and open to new ideas (tolerance), and the concentration of 'cultural capital' wedded to new products (technology). All these together result in business formation, job generation and economic growth.

Florida's 'creative class' has had a great influence on analysing the effects of creative class on employment growth and new business formation in different countries (Andersen and Lorenzen 2005, Boschma and Fritsch 2007). The empirical evidence of recent studies on mapping and analysis of the creative class supports the relevance of Florida's theory that the creative class has a positive and significant effect on employment growth and new business formation, and that it is attracted to a climate of tolerance and openness (Andersen and Lorenzen 2005, Boschma and Fritsch 2007). However, Florida's ideas have met also some criticism. Many geographers and economists (Glaeser 2004, Hall 2004, Markusen 2006, Musterd and Ostendorf 2004, Musterd et al. 2007) have argued that the existing research evidence is far from convincing. The critiques concern, on the one hand, the empirical issues such as how to distinguish which occupations are creative and which are not, and, on the other, whether the rise of the 'creative class' and the 'creative industries' is a long-term trend or the next 'hype' in the footsteps of the 'new economy' of the late 1990s.

The term 'creative industries' was first used in Australia in the early 1990s (Cunningham 2002, UNCTAD 2004) and was extended in the UK to highlight the economic contribution of cultural production and activities in the late 1990s when the Department of Culture, Media and Sport (DCMS) set up its Creative Industries Unit and Task Force (DCMS 1998). Creative industries are defined, in general, as a profit-oriented segment and thus cover all enterprises, entrepreneurs and self-employed persons producing, marketing, distributing, and trading profit-oriented cultural and symbolic goods (Kalandides and Lange 2007). Britain's Creative Task Force has defined the creative industries as 'those activities which have their origin in individual creativity, skill and talent and which have a potential for wealth and job creation through the generation and exploitation of intellectual property' (DCMS 2001: 3). On the basis of this definition, creative industries include the following activities: advertising, architecture, arts and antiques, crafts, design, designer fashion, film and video, interactive leisure software, television and radio, performing arts, music and software and related computer services. Creative industries are defined by UNCTAD (2008: 4) as 'the cycles of creation, production and distribution of goods and services that use creativity and intellectual capital as primary inputs'. They comprise a set of knowledge-based activities that produce tangible goods and intangible intellectual or artistic services with creative content, economic value and market objectives. Creativity is also seen as another labour distribution in an intelligent era, regarded as the 'fourth industry', firmly linked with the other three traditional industries and promoting integration among them (Jing and Rong 2007). However, the concepts 'creative industries', 'knowledge-intensive industries' and 'cultural industries' are often used interchangeably, which causes confusion. There is a big debate about what is and what is not included in the creative industries (Florida 2002, Markusen 2006). There is neither a precise definition nor a consensus yet about the concept.

In recent years, there has been a shift from a more traditional concept of culture and cultural industries as linked to the classical fine arts towards an understanding of creative industries that centres on the productive and innovative capacity of knowledge and information (Askerud 2007, Cunningham 2002, Cooke and Lazeretti 2008, Evans 2009, UNCTAD 2004). In this sense, creative industries are more open to trade and exchange and are positioned at the crossroads between the arts, business and technology (UNCTAD, 2004).

Today, creative industries are among the most dynamic sectors in world trade. Globally, creative industries are estimated to represent 7 per cent of employment, more than 7 per cent of the world's gross domestic product (GDP) and forecast to grow on average by 10 per cent annually (UNCTAD 2004 and 2008, UNESCO 2005, World Bank 2003, Wu 2005). This positive trend is observed in all regions and countries (see for a comprehensive evaluation Baycan-Levent 2010), and is expected to continue into the next decade, assuming that the global demand for creative goods and services continues to rise. Creative industries represent a leading sector in the Organisation for Economic Co-operation and Development (OECD) economies (EESC 2003), represent one of the leading assets and opportunity areas in the EU and other European countries (Andersen and Lorenzen 2005, Florida and Tinagli 2004, KEA 2006, MEA 2006, Fleming and NORDEN 2007, UNCTAD 2008), in East Asian countries such as Korea, Singapore, Hong Kong (China) and increasingly mainland China (Chang 2000, HKTDC 2002, Jing and Rong 2007, Xu and Chen 2007, UNCTAD 2004 and 2008), and in many developing countries (UNCTAD 2008). The creative economy in general and the creative industries in particular are opening up new opportunities for developing countries to increase their participation in global trade. However, the importance of creative industries is more remarkable when examined at city level (see for a comprehensive evaluation at city level Baycan-Levent 2010), and for some cities the stated level of creative employment is higher than national levels of creative employment.

In general, creative industries tend to cluster in large cities and regions that offer a variety of economic opportunities, a stimulating environment and amenities for different lifestyles. Creative industry development is often considered part of the inherent dynamic of urban spaces and urban environments provide ideal conditions – a creative milieu – for cluster development (Landry 2000, Porter 1998, Porter and Stern 2001). A 'creative milieu' can be defined as 'a locational hub combining hard and soft infrastructure, acting as a crucible for creative people and enterprises' (Landry 2000). Creative milieu is similar to what historians have termed as a 'moral temperature' allowing a particular kind of talent to develop in one place at one time (Hall 2000). A creative milieu, a notion similar to that of the 'innovative milieu' has four key features: information transmitted among people, knowledge or the storage of information, competence in certain activities, and creation of something new out of these three activities (Hall 2000, Törnqvist 1983, Wu 2005). A creative milieu and the characteristics of the social and economic networks are considered to be important in fostering creativity.

Creative industry development requires a creative milieu that is based on highly developed 'hard' and 'soft' infrastructures. Hard infrastructure refers to classic location factors and includes the labour force, rent levels, availability of office space, accessibility, local and national tax regimes, and other regulations and laws affecting the functioning of companies. Nearness to global financial centres, a major international airport, telecommunication services and other service suppliers and clients, and the availability of an international labour pool are also important considerations (Musterd et al. 2007, Sassen 2001). Soft infrastructure, on the other hand, includes a highly skilled and flexible labour force, a culture of entrepreneurship, a high-quality and attractive living environment, cultural richness and tolerance of alternative lifestyles and/or diversity, a lively cultural scene, the creation of meeting places for business and leisure purposes, education and social support systems, research resources and the support of networks and marketing (Evans 2009, Foord 2008, Musterd et al. 2007, UNCTAD 2008, Yip 2007).

The 'creative city' concept was first developed by Charles Landry in the late 1980s and his study *The Creative City: A Toolkit for Urban Innovators* in 2000 has become the main reference document on creative city. Landry's creative city philosophy is based on people's imagination and he has described the creative city as 'places where people think, plan and act with imagination'. Following Landry, many other scholars have focused on creative cities and they have developed different definitions from different perspectives. However, the main concern in these studies has been the role of creative people, and cultural and creative activities in the city development process. A further examination of creative cities is considered in the next section, but, on the basis of the above-mentioned concepts of creativity, the creative city can be defined as a place where a high proportion of 'creative people' exists, creative industries are the leading sectors in the urban economy, and a creative milieu is provided by high-quality 'hard' and 'soft' infrastructure.

Against this background, 'urban creativity' can be seen as an 'umbrella' concept that combines different dimensions of creativity from economic and social creativity to technological creativity or innovation. Therefore, 'urban creativity' refers to all other concepts of creativity including 'creative class', 'creative industries', 'creative milieu' and 'creative city'. However, urban creativity is focused more on the interconnection between creativity and (urban) space. From an urban creativity perspective the critical questions are: 'Why some places (cities and regions) are more attractive than some others for new and creative activities?' and 'What are the essential locational factors to attract the new and creative activities?'. The next section will focus on these questions and some others in creative cities debate.

Creative Cities: Top 10 FAQs

Creative cities have become a global movement reflecting a new planning paradigm in recent decades. The general features of creative cities are described by many scholars; however, as noted by Bradford (2004a), the studies describe the conditions that foster creativity, and the mechanisms, processes and resources that turn ideas into innovations remain very limited. From this need, this study aims to put current debate about creative cities in context and perspective while addressing the most recent studies from a multidisciplinary perspective. The study offers a framework that is based on the 10 most frequently asked questions (FAQs). In other words, the study aims to highlight the 'Top 10 FAQs' in the creative cities debate, which are:

1. What is a creative city?
2. What makes a creative city?
3. What are the key assets and infrastructure required to foster the development of creative, competitive and cohesive places?
4. Why some places (cities and regions) are more attractive than others for new and creative activities?
5. What is the interconnection between creativity and (urban) space and how are they mutually constituted?
6. What drives innovation in creative cities?
7. What are the benefits of creative cities?
8. What are the opportunities and barriers to the development of creative cities?
9. How can public policy influence creative cities?
10. What is required to build a creative city?

What is a Creative City?

The 'creative city' concept was first mentioned in a seminar organized by the Australia Council, the City of Melbourne, the Ministry of Planning and Environment of Victoria in September 1988 (Wikipedia 2009). In that seminar the focus was on how arts and cultural concerns could be better integrated into the planning process for city development. However, as mentioned above, the 'creative city' concept was first developed by Charles Landry in the late 1980s. His first detailed study of the concept was *Glasgow: The Creative City and its Cultural Economy* in 1990. This was followed in 1994 by a study on urban creativity called *The Creative City in Britain and Germany*. His next study, *The Creative City: A Toolkit for Urban Innovators*, in 2000 has become the main reference document on the creative city since its publication. Landry's creative city philosophy is based on people's imagination. According to Landry, the people's imagination is a city's greatest resource, creativity can come from any source (not only from artists and those involved in creative economy), and if the chance is given, ordinary people can make extra-ordinary things. Thus, the creative city philosophy assumes that there is always more creative potential in a place.

Table 2.2   Different definitions of creative cities

Landry (2000)	Creative cities are places where you can think, plan and act with imagination.
Hall (2000)	Creative cities are places of great social and intellectual turbulence: not comfortable places at all.
City of Ottawa (2003)	A creative city will therefore be a place where outsiders can enter and feel a certain state of ambiguity: they must neither be excluded from opportunity, nor must they be so warmly embraced that the creative drive is lost.
City of Toronto (2003)	A creative city must be able to sustain a concentration of artists, creative people, cultural organizations and creative industries within its boundaries. Creative cities are dense urban centers whose ecomies are dominated by ideas, and by people who bring new ideas to life ... These cities work with their minds.
Bradford (2004a)	Creative cities are dynamic locales of experimentation and innovation, where new ideas flourish and people from all walks of life come together to make their communities better places to live, work and play.
Bradford (2004b)	The creative city is home to diversity: different talents are recognized and represented.
Landry (2006)	The creativity of the creative city is about lateral and horizontal thinking, the capacity to see parts and the whole simultaneously as well as the woods and the trees at once.
Kalandides and Lange (2007)	Creative city concept implies a holistic, creative thinking process that can be applied to a range of social, economic and environmental problems.
Smith and Warfield (2008)	Creative city is a place with strong flourishing arts and culture, creative and diverse expressions, and inclusivity, artistry and imagination. Place of diverse and inclusive arts and culture (culture-centric definition). Place of economic innovation, creative talent, and creative industries (econo-centric definition).
UNCTAD (2008)	A creative city is an urban complex where cultural activities of various sorts are an integral component of the city's economic and social functioning.

As stated earlier, Landry (2000) has described creative cities as 'places where people think, plan and act with imagination'. Following Landry's (2002) study, many scholars and planning authorities in different cities have focused on what a creative city is and have described creative cities from different perspectives (Table 2.2). While some of the definitions have taken the key features of a creative city into consideration (Bradford 2004a and 2004b, Hall 2000, Smith and Warfield 2008, City of Toronto 2003, UNCTAD 2008), others have adressed the holistic and creative thinking process (Kalendides and Lange 2007, Landry 2000 and 2006).

Creative city is in general understood and used in four ways: (i) creative city as arts and cultural infrastructure; (ii) creative city as the creative economy; (iii) creative city as synonymous with a strong creative class; and (iv) creative city as a place that fosters a culture of creativity (UNCTAD, 2008). In creative cities: most of the strategies are concerned with strengthening the arts and cultural fabric; the arts and cultural heritage, the media and entertainment industries, and the creative business-to-business services are the drivers of innovation in the creative economy; there is a competition to attract, keep or grow their own creative class and the factors that contribute to this, such as 'quality of place', are of high importance; and there is an integrated system of multiple organizations and an amalgam of cultures in the public, private and community sectors (Baycan-Levent 2010).

What Makes a Creative City

A number of features and factors distinguish the creative city have been described in various studies (Bradford 2004b, Hall 1998 and 2000, Wu 2005). These studies have found that the key features of a creative city are 'uniqueness' and 'authenticity'. However, a creative city has also some other features, such as 'unsettled' and 'dynamic' structures.

'Uniqueness' and 'authenticity' refer to the city's own story, constitute the unique identity of the city and the community, and build the city's own niches of excellence on national and global stages (Bradford 2004b). Here, 'authenticity' can be seen as a kind of fixed thing from the past. With this feature 'authenticity' overlaps with the 'cultural heritage' of the city. According to Bradford, creative cities express their uniqueness and authenticity in three principle settings: the 'arts', 'commerce' and 'community' and the quality and intensity of the connections among them strongly influence the city's creative capacity.

'Unsettled' and 'dynamic' structures of a creative city stem from different visions, values and cultures (Bradford, 2004b, Hall, 2000). Creative cities are 'cosmopolitan' cities (Hall, 2000), and the different values and cultures intersect and lead to cross-fertilization of ideas. Here, contrary to the 'fixed' feature of 'authenticity', the need for innovation and doing things that are 'novel' or different represent the 'unsettled' and 'dynamic' features of a creative city. This 'dynamic' structure requires, on the one hand, a transition into new and unexplored modes of organization and, on the other, a transformation in social relationships and values. Therefore, a tension between a set of conservative forces and values, and a set of radical values emerges. In other words, in creative cities there is always a tension between 'authenticity' and 'novelty' (Bradford 2004b, Hall 2000), however this tension leads to a creative change. According to Hall (2000), the periods of structural instability, with great uncertainty about the future, offer a great potential for a creative change. When everything is uncertain a group of creative people can take the city or region to a new stable phase. A total breakdown of the established system or a kind of revolutionary movement turns creative cities into places of great and intellectual turbulence, and makes them uncomfortable and unstable.

Besides the above-mentioned key features, three main factors viz. 'people', 'place' and 'investment' play a crucial role in making a creative city (Bradford 2004b). A creative city brings together talented and diverse 'people' who bring ideas, inspiration and passion to a place; high-quality built and natural 'places' that nurture the creativity of residents and attract other creative people; and new 'investments' in the infrastructure of urban creativity from physical environment to social, cultural and institutional organizations. These factors together drive innovation in creative cities.

What are the Key Assets and Infrastructure Required to Foster the Development of Creative, Competitive and Cohesive Places?

The key asset and infrastructure for fostering the development of creative, competitive and cohesive places is a 'creative milieu'. As mentioned earlier, a 'creative milieu' is a locational hub combining 'hard' and 'soft' infrastructure. While 'hard' infrastructure includes classic location factors such as the labour force, availability of office space, accessibility, and transportation and telecommunication services, 'soft' infrastructure includes more qualitative location factors such as an attractive residential environment, tolerance of alternative lifestyles and diversity and a lively cultural scene. 'Hard' and 'soft' location factors are essential in developing creative places.

The key features of a creative milieu have been described by many scholars. The term 'creative milieu' was first developed by Törnqvist in 1983, and he has described the key features of a creative milieu as:

*   information transmitted among people;
*   knowledge or storage of information;
*   competence in certain activities;
*   creation of something new out of these three activities;

Following Törnqvist, Andersson (1985) has underlined the critical prerequisites for creative milieux as:

*   a sound financial basis, but without tight regulation;
*   basic original knowledge and competence;
*   an imbalance between experienced need and actual opportunities;
*   a diverse milieu;
*   good internal and external possibilities for personal transport and communication;
*   structural instability – uncertainty about the future – facilitating synergetic development.

Landry (2000) has highlighted the characteristics of the social and economic networks that are considered as to be important in fostering creativity to assist economic advantage. He has described the crucial factors for creative industry development as:

- personal qualities, including a motivation and capability to innovate;
- will and leadership, both moral and intellectual, to guide and mentor others;
- human diversity and access to varied talent, in age and outlook, from the available urban pool;
- organizational capacity, both to learn and also to follow through and deliver;
- local identity, an awareness of people and place;
- urban places and facilities, a combination of public spaces and more private venues;
- networking dynamics, embedded both within and between sectors.

Wu (2005) has emphasized that cities need to build institutional and political mechanisms that nurture creativity and channel innovation. He has described the crucial factors derived from the collective experience of successful creative centres as:

- outstanding university research and commercial linkages;
- the availability of venture capital;
- successful anchor firms and mediating organizations;
- an appropriate base of knowledge and skill;
- targeted public policies;
- quality of services and infrastructure;
- diversity and quality of place.

The above-mentioned factors are among the most important contributors to creative and dynamic cities. Today, the availability and quality of the local cultural resources determine whether an area is a good place to live. Culture and creativity is increasingly associated with quality of life, and while the 'hard' and more classic location factors are still very important in explaining the location patterns of people and companies, there is a growing emphasis on 'soft' location factors including quality of life components (Florida 2002, Gertler 2004, Landry 2000, Fleming and NORDEN 2007, Musterd et al. 2007, Wu 2005). The 'soft' infrastructure is of critical importance as it serves to shape both the social character and physical environment of cities (Gertler 2004). High availability and quality of the local cultural resources – in other words a 'cultural infrastructure' – is a real asset for the creative economy. The presence of a cultural infrastructure as well as a dynamic creative industries sector fosters the development of creative and competitive places. In this process, the coordination and collaboration among the actors as well as institutional and political mechanisms – or the presence of an 'institutional infrastructure' – also play a crucial role.

*Why are Some Places (Cities and Regions) More Attractive than Others for New
and Creative Activities?*

In the debate over why some places (cities and regions) are more attractive than
others for new and creative activities the concerns are related to 'path dependence'
in association with 'cluster formation' and urban and regional development, and
'soft location factors' often associated with the emergence of creative industries
and creative class (Bagwell 2008, Cumbers and MacKinnon 2004, Evans 2009,
Foord 2008, Hall 2004, Musterd et al. 2007, Porter 1998, Porter and Stern 2001,
Pratt 2008, Simmie 2005, Wu 2005, Yip 2007).

Path dependency means 'history matters' and refers to the historic development
paths of cities and regions and the consequences of these paths for recent and
future development. The logic of path dependence is that the chance of a city
or region specializing in creative and innovative activities and attracting the
talent needed are considerably larger where there is a long tradition of creativity
and innovation. It is difficult to generate a new and creative cluster where none
previously exists, as cluster development often is path dependent (Musterd et al.
2007, Wu 2005). Therefore, building a 'creative city' requires a strong social and
cultural infrastructure (Pratt 2008: 35): 'A creative city cannot be founded like
a cathedral in the desert; it needs to be linked and be part of an existing cultural
environment. We need to appreciate complex interdependencies, and not simply
use one to exploit the other'.

Creative activities often take place in clusters – geographic concentrations of
interconnected firms and institutions in a particular industry or sector (Porter 1998,
Porter and Stern 2001). Clustering leads to a number of advantages for both firms
and the regions in which they operate, including increased competitiveness, higher
productivity, new firm formation, growth, profitability, job growth and innovation
(Bagwell 2008, Cumbers and MacKinnon 2004, Porter 1998). Recent studies have
stressed the advantages of clusters for cities and regions seeking to compete in a
knowledge-driven global economy. Clustering can be particularly beneficial for
creative industries as they tend to have a large number of small firms. They can
benefit from competitive advantage that can be derived through efficiency gains
that a small firm may not manage on its own (Wu 2005). However, the results
of recent studies (Bagwell 2008, Evans 2009, Foord 2008) show that 'creative
clusters' are not conventional business clusters; they have distinct characteristics
that differentiate them from other types of business clusters and additional factors
are critical to their development and form, notably local area regeneration,
conservation/heritage, cultural tourism and related visitor economies. Creative
clusters differently from conventional business clusters have social objectives
such as goals of inclusion and cultural development.

Currently, 'creative clusters' are among the 'most wanted' targets of cities,
regions and countries, and 'cluster policy' is one of the most common instruments
to transform an urban or regional economy into a creative and knowledge-intensive
economy. Policy-makers have supported clusters as an economic development

strategy and clusters have become a prominent element of many national, regional and urban development strategies.

This comprehensive and multidimensional evaluation of creative cities shows that the more attractive cities for new and creative activities have some common characteristics: these cities are authentic and unique and have a local identity; they have human diversity as well as a diversity of cultural heritage; they have a 'history' and a long tradition of creativity, innovation and cluster development; they provide a creative milieu including highly developed 'hard' and 'soft' infrastructure; and they are 'open' and 'cool' to new ideas and different life styles. Therefore, 'creative cities' are 'open cities' and 'cool cities' at the same time.

What is the Interconnection between Creativity and (Urban) Space and How are they Mutually Constituted?

Clustering of certain new economic activities in some cities and within these cities at certain locations draw attention to the 'relevance of place' (Musterd and Ostendorf 2004). Relevance of place is linked to the intrinsic quality of the sites where creative activities settle and creative people want to live. Therefore, 'quality of place' is a crucial factor in the creation of the conditions with which the creative activities and creative people can be attracted. It is well known that creative people are attracted by diversity and a tolerant and open urban atmosphere (Florida 2002). They demand attractive places in which; (i) they can live; (ii) they can find opportunities for cultural and recreational activities and, therefore, they can consume; and (iii) perhaps also they can 'reload their creative battery' (Musterd and Ostendorf 2004). The built environment is crucial for establishing this atmosphere or milieu, and this milieu creates the mood of the city and its culture (UNCTAD 2008). At this point the historically developed identity, authenticity and uniqueness of cities, in other words its cultural heritage, and urban image enter the picture (Florida 2002, Musterd and Ostendorf 2004). Authenticity comes from several aspects of a community – historic buildings, established neighbourhood or specific cultural attributes – and an authentic place offers unique and original experiences (Florida 2002). Cultural heritage not only determines the image of the city, but is also an essential ingredient in establishing the context that stimulates creativity. Cultural heritage reflects the 'soul of the city' and contains the essential elements to build a sustainable future (Fusco Girard et al. 2003).

Drake (2003) provides empirical evidence for the links between place and creativity and argues that these links can be important in the creative process. According to Drake, locality is acting as an 'exhibition space' for social and cultural innovation and the attributes of locality can be a catalyst for individual creativity in three ways. First, locality can be a source of visual materials and stimuli. Second, locality-based intensive social and cultural activity may be a key source of inspiration. Third, locality as a brand based on reputation and tradition can be a catalyst for creativity.

In the interconnection between creativity and urban space three factors viz. 'spacemaking', 'placemaking' and 'building knowledge' are of crucial importance (Jones 2007). Jones has described 'spacemaking' as creating affordable space for artists, designer-makers and creative entrepreneurs; 'placemaking' as an integrated and transformative process that connects creative and cultural resources in a neighbourhood, district or city to build authentic, dynamic and resilient places; and 'building knowledge' as building and sharing knowledge in culture-led regeneration. The term 'regeneration' has been defined as 'the transformation of a place – residential, commercial or open space – that has displayed the symptoms of physical, social and/or economic decline' (Evans 2005: 967). Culture is a driver, a catalyst or a key player in the process of regeneration or renewal. Evans (2005) has classified culture's contributions to regeneration into three groups: (i) culture's contribution to sustainable development (physical regeneration); (ii) culture's contribution to competitiveness and growth (economic regeneration); and (iii) culture's contribution to social inclusion (social regeneration). Jones (2007: 195) has defined the specific form of regeneration, 'culture-led regeneration', as 'a multi-dimensional approach to the re-use, renewal, or revitalization of a place where art, culture, and creativity plays a leading or transformative role'. Culture-led urban regeneration has become a major force of change in European, North American and South-East Asian cities (Bagwell 2008, Jones 2007, Miles and Paddison 2005, Murray et al. 2007, Yeoh 2005).

'Culture-led regeneration' is directly linked to 'placemaking' and leads to regenerating of 'cultural quarters' or 'creative districts'. A widespread ambition to encourage cultural or creative quarters has emerged in recent years (Cooke and Lazzeretti 2008, Crane 2007, Jones 2007, Landry 2004, Murray et al. 2007). Cultural quarters have been widely developed over the last 15 to 20 years as mechanisms for synthesizing the cultivation of creative industries with urban regeneration objectives (Murray et al. 2007). Creative quarters provide quality of life amenities that complement the creative worker's lifestyle (Crane 2007). Workspace located in mixed-use facilities enables the merging of work, social culture and housing, to maximize the transference of creative ideas and the mixing of 'work' life and 'home' life for creative people (Crane 2007).

Abandoned workshops, warehouses, and other old commercial or residential buildings are the hottest real estate now, and are the ideal spaces for artists or other creative people (Xu and Chen 2007). Artists and other creative people play a key role in colonizing underused, neglected and devalorized urban neighbourhoods. Underutilized spaces and derelict neighbourhoods first become home to artists seeking affordable working spaces, then the areas for the vibrant cultural life with the influx of people followed by improved public transport and municipal services, and finally an environment of tolerance characterized by differences (Crane 2007, Gertler 2004, Jones 2007). Therefore, artists and other creative people play a leading role as 'dynamic agents of positive transformation' in communities (Gertler 2004). With this leading role they contribute enormously to culture-led regeneration.

'Culture-led regeneration' is also directly linked to 'city marketing'. Cultural and creative industries are often at the forefront of urban restructuring, place-based regeneration and marketing strategies. Through the creation of cultural quarters based on the idea of clustering or city rebranding campaigns, cultural and creative industries contribute to the regeneration and renewal of redundant buildings and depressed urban areas (Bagwell 2008, Porter 1995). After the 1980s, the decline in city centres impelled policy-makers and city authorities to find ways of rescuing city centres by locating creative industries in central locations (Evans 2005 and 2009). In the United States, clustering has been promoted as a way of encouraging the restructuring of deprived inner city areas (Porter 1995). This US-inspired model of business-led regeneration has led to many cultural strategy initiatives focusing on 'feeding' existing creative clusters in inner city areas (Bagwell 2008). Much research acknowledges the power of concentration of specialized industries in particular localities named as cultural districts. Santagata (2002) has stated that these localized cultural districts have become an example of sustainable and endogenous growth.

City marketing, on the other hand, is directly linked to notions of urban competitiveness and place identity (Kalandides and Lange 2007). Quality of place is widely used as one of the main instruments of city marketing besides events and advertising in order to attract creative people and creative activities. Some kind of quality, an 'air', 'atmosphere' or 'ambiance' makes one place more creative than another. Cities worldwide are using culture and creativity to brand themselves (Richards 2001). The production of culture has become central to many development strategies worldwide (McCann 2002). Creativity is increasingly used by cities and regions as a means of preserving their cultural identity and developing their 'socio-economic vibrancy' (Ray 1998). Montgomery (2007) suggests that successful cities of the new economy will be the ones that have invested heavily in their capacity for creativity and that understand the importance of locality and cultural heritage.

However, the culture-led urban regeneration and city marketing strategies are also questioning and criticized from different perspectives. First, urban regeneration may lead to gentrification, which is usually more associated with the negative effects than the positive ones. The negative effects include community resentment and conflict, loss of affordable housing, displacement of lower income households and loss of social diversity (Musterd et al. 2007). Second, city marketing strategies often attempt to mask social, ethnic, class and gender polarizations by mobilizing every aesthetic power of illusion and image and setting in motion a politics of 'forgetting' and 'remembering', of 'inclusion', 'exclusion' and 'revalorization' (Lee and Yeoh 2004, Yeoh 2005), and by-pass anything that does not fit the picture such as the spaces of migrants and the urban poor (Kalandides and Lange 2007). Third, there is a question mark where the promotional strategies of cities include cultural activities and international events about who participates in these activities and events and whether this includes the total population in general (Musterd et al. 2007). Fourth, another question is what the consequences will be of economic and urban change towards a creative economy on the population that is not highly skilled or skilled. Many people lacking the basic entry skills, experience and social networks might be excluded from the whole process (Musterd et al. 2007). Against these negative effects, Morrison (2003: 1629) suggests that a 'cultural justice' that involves 'a more fundamental revaluing of socially excluded people and places needs to be carried out in order to achieve greater inclusion'.

Another important interconnection between creativity and urban space appears in location decisions of creative industries and those who are working in that sector (Bayliss 2007, Bianchini and Parkinson 1994, Gornostaeva 2008, Helbrecht 2004, Mommaas 2004, Montgomery 2007, Musterd 2004, Musterd and Deurloo 2006, Musterd et al. 2007, Nachum and Keeble 2003, Newman and Smith 2000, Scott 2000, Yigitcanlar et al. 2008). Creative people prefer particular locations in the city. Helbrecht (2004) has argued that location choice is strongly based on 'look and feel' from the building to neighbourhood and city level. Helbrecht describes the urban landscape in the perception of creative people as a 'geographical capital'. The dilemma between investing in city centres or urban peripheries is one of the key cultural policy issues within the restructuring process (Bianchini and Parkinson 1994, Montgomery 2007). Creative and innovative industries can be found in large cities and, within them, tend to cluster in inner city areas. However, concentrations of creative and innovative companies can also be found at city edges or in (former) suburbs. Newman and Smith (2000) highlight the importance of the concentration of cultural production and location of creative industries within inner cities as co-location offers advantages. Hutton (2004) puts forward the importance of supporting inner-city investments to harness rapid growth in the new economy. Yigitcanlar et al. (2008) also emphasize the importance of centrality for creativity. On the other hand, Musterd (2004) and Musterd and Deurloo (2006) have emphasized that different types of creative professionals have different distribution patterns across the city and the region. While certain activities appear to show a distribution pattern in inner city, some others tend to be located in urban peripheries They have found that cultural creatives (such as artists, media and entertainment workers, scientists, teachers, designers and advertisers) tend to show a strong intercity orientation. In contrast, professional creatives (such as managers in commercial, financial and juridical services) are much more spread across the city and region. Therefore, there are clear differences in terms of spatial orientation of different types of firms and different types of workers (Musterd 2004). Although theory stresses the importance of centrality for creativity, creative industry companies tend to move towards periphery or to sub-centres because of the problems of city centres or the attractiveness of other locations (Gornostaeva 2008, Scott 2000). Nachum and Keeble (2003) highlight this paradox between theory and practice: between theories of clustering in city centres and tendencies for decentralization from city centres to the periphery. As mentioned also by Yigitcanlar et al. (2008), it is important to investigate the locational requirements of creative industries in order to respond to their specific needs, whether regenerating existing cultural quarters or developing new districts.

What Drives Innovation in Creative Cities?

An important distinction in the creativity debate has emerged between creativity and innovation. This distinction has been emphasized in many studies (Bradford 2004b, Duxbury 2004, Landry 2000, Fleming and NORDEN 2007, Wikipedia 2009). In general, it is thought that creativity is about generating 'new ideas' whereas innovation is the 'process' through which they are implemented. While creativity refers to producing new ideas, approaches and actions, innovation refers to the process of both generating and applying such creative ideas in some specific context (Wikipedia 2009). In other words, creativity is focusing on the origination of new ideas and innovation on their successful exploitation (Fleming and NORDEN 2007). In a similar way, Bradford (2004b) has mentioned that creativity is more 'utopian' and innovation is more 'pragmatic'. According to Bradford, creativity involves 'utopian thinking' whereas innovation is bringing 'discipline to imagination' for the practical application of new ideas. The distinctions among creativity, innovation and learning are also crucial in Landry's analysis (2000). According to Landry, learning connects creativity and innovation while testing the feasibility of ideas. There is a consensus in the literature that creativity is a necessary condition and a starting point for innovation, but it is not a sufficient condition for it.

Innovation is an economic and social phenomenon; however, the empirical studies have shown that there is a distinctive 'geography of innovation' (Simmie 2005). The relationships between innovation and space have been identified by Simmie (2005) in three periods characterized by three major shifts: (i) an early period that focused on 'growth pole' and 'agglomeration economies'; (ii) a second wave in the early 1980s that focused on a 'new industrial geography' that includes 'new industrial districts' and 'innovative milieu' as well as 'embeddedness' that emphasizes the significance of social relationships; and (iii) a final period that focused on 'modern evolutionary theory' in order to explain the developments through time of innovation and its relationships to space. The relationships between innovation and space offer a great potential to combine economic and spatial theories in order to explain the most significant forces behind economic growth and the significant role of localities in this process.

A critical question from a social perspective is what is 'innovative action' in a community. While emphasizing that innovation is relative to its context Duxbury (2004: 3) defines innovative action as 'doing something out of the norm, something new to that situation or context'. New ideas in social context, in other words 'social innovation', have begun to receive more attention in studies of creativity. Mumford (2002: 253) has defined social innovation as 'the generation and implementation of new ideas about people and their interactions within a social system'. Social innovation is a significant form of creativity, leading to the formation of new institutions, new industries, new policies and new forms of social interaction (Mumford 2002, Mumford and Moertl 2003).

The factors that motivate a city or community to innovate are complex. On the one hand, the global pressures for economic renewal drive innovation. Such economic renewal is required to develop a 'niche' in the global economy based on distinctive local assets including location, culture, skills and knowledge, and also a multifaceted approach that involves attracting creative people, investments and jobs; improving the 'quality of place'; and building local identity and pride (Duxbury 2004). On the other hand, to achieve their creative potential, cities and communities have also to manage a number of cross-pressures deriving from growing inequality, social exclusion and spatial segregation. The social sustainability of cities is required to reverse growing economic inequality, social exclusion, cultural tension and spatial segregation, and to generate new social perspectives and problem-solving capacities (Bradford 2004a) or, in other words, social innovation. Briefly, successful economic and social innovation requires the will and capacity to act in new and different ways.

To motivate a city or community to innovate is also required to provide a balance between local community roots and global influences, heritage and novelty, formal high culture and informal street scenes, non-profit arts activity and creative industry clusters, local knowledge and professional expertise, neighbourhood regeneration and social inclusion, rule-based accountability and grassroots experimentation, and holistic thinking and strategic action (Bradford 2004a). Balancing the synergies and the two-way innovation spillovers between the large and micro firm, the 'street' and corporate headquarters, and therefore consumption and production through new consumption and product modes and media and through city place-branding is also important (Evans 2009).

What are the critical factors for a relatively high urban innovation potential? Forte et al. (2005: 949) have classified and explained the preponderant factors for a high urban innovation potential into five different types of milieus: (i) *economic milieu* (i.e. the composition and spatial size distribution of economic sectors); (ii) *social milieu* (i.e. the demographic and population composition, including gender, age and ethnic diversity); (iii) *information milieu* (i.e. the interaction and telecommunication facilities, as well as educational and research infrastructure); (iv) *physical milieu* (i.e. the availability of physical infastructure and locations for entrepreneurial activities); (v) *institutional milieu* (i.e. the organized support infrastructure based on regulatory and decision-making arrangements). A blend of all these factors will increase the urban innovation potential. However, to increase the urban innovation potential requires also necessary public investments in public infrastructure including physical network infrastructure, environmental infrastructure and knowledge infrastructure, and an urban innovative policy that combines both public and private initiatives.

What are the Benefits of Creative Cities?

Creativity not only leads to economic and social innovation but also to artistic, cultural, civic and governance innovation. A combination of these factors generates successful places and 'innovative cities'. Creative and innovative cities provide many benefits to communities. Creative cities contribute significantly to meeting local and national policy goals such as economic innovation, social inclusion and environmental sustainability (Bradford 2004b). They play an important role in enhancing the dynamism, resilience and overall competitiveness of the national economies (Gertler 2004). They offer opportunities for cross-disciplinary learning, thus providing the stimulation necessary to promote innovation in a wide array of occupations and industries (Gertler 2004). With their power to engage different kinds of people and different kinds of knowledge, creative cities can develop innovative solutions to complex local issues (Bradford 2004a, Kalandides and Lange 2007, Landry 2000). Creative people play also a key role as 'dynamic agents of positive transformation' in communities. As mentioned above, under-utilized spaces and derelict neighbourhoods first become home to artists seeking for affordable working spaces, then the areas for the vibrant cultural life with the influx of people followed by improved public transport and municipal services, and finally an environment of tolerance characterized by differences (Crane 2007, Gertler 2004, Jones 2007). Creative people can also help to raise overall productivity in a regional economy by enhancing the entrepreneurial culture of the region because many of them are self-employed (Florida 2002, Gertler 2004).

Bradford (2004b) has classified the benefits of creativity into five categories: governance innovation; civic innovation; economic innovation; social innovation; and artistic and cultural innovation:

- *Governance innovation*: Governance innovation refers to breaking with tradition and harnessing diversity. Breaking from some elements of traditional municipal administration, creative places are becoming more inclusive and open to new collaborations and new ways of community involvement in the planning process.
- *Civic innovation*: This category of innovation refers to applying new problem-solving skills to contemporary urban challenges such as managing growth and diversity in the large cities, and shifting from the natural resource to the knowledge economy in smaller communities.
- *Economic innovation*: In the knowledge economy creativity that is based on ideas, design and networking is becoming more valued input and makes cities 'innovative milieux'.
- *Social innovation*: Social innovation refers to social transformation and social inclusion. The social context of cities can be transformed by citizen participation in arts and cultural activities, which is a route to inclusion of marginalized communities and to revitalized neighbourhoods. Social innovation makes cities more inclusive places.
- *Artistic and cultural innovation*: Creative cities support the arts and culture for their contribution to inclusion and different kinds of innovations. Creativity produces many forms of aesthetic expression that enable urban residents from different backgrounds to live more respectfully together.

The above-mentioned categories of innovation make cities successful and inclusive places while increasing their creative and innovative capacity. The synergies that can be generated by these different kinds of innovations also motivate cities and communities to innovate more.

What Are the Opportunities and Barriers to the Development of Creative Cities?

There are both opportunities and barriers in the development of creative cities and creative industries. The results of three research projects conducted in the UK, Canada and the Nordic Region of Europe (including Sweden, Norway, Finland, Denmark and Iceland) highlight the opportunities and barriers for creative city development as well as for investments in creative industries (Bradford 2004b, NESTA 2003, Fleming and NORDEN 2007).

According to the results of the research conducted in Canada, the opportunities for creative city development are found in the following fields (Bradford 2004b):

- multiculturalism (which facilitates cultural syntheses);
- growing knowledge that arts and culture contribute to positive outcomes across a range of urban fronts (including resident health, cross-cultural understanding, community safety and economic growth);
- community involvement (social inclusion challenges);
- the creative city process itself (focusing on building at a local level, where networks are the strongest and the possibility of aligning interests is the greatest);
- the educational system as an untapped resource in developing creative cities.

Although there are many opportunities to developing creative cities, there are equally many barriers, such as lack of awareness among policy and planning communities and the general public, poor collaboration within and between governments, and undervaluing of the contribution of the arts and culture. The Canadian experience shows the following barriers are of importance (Bradford 2004b):

- the lack of clarity on the meaning of creativity and its relevance in an urban setting;
- the lack of awareness in policy and planning circles about the creative city process;
- the absence of a practical toolkit for planning and implementing creativity in cities;
- the shortage of resources and skills at the municipal level to facilitate this process;
- the lack of creativity champions among a community's political, administrative, business and community leaders;

- research gaps in how artistic and cultural activities contribute to economic innovation and quality of life;
- the lack of clear and applicable indicators to capture the creativity of a city and the contribution of investments made to arts, culture and heritage;
- the exclusion or marginalization of some people and cultures.

Another set of barriers is related to structural and cultural obstacles to investment. Here the critical questions are whether investment and investor are ready. The results of two research reports on creative industries by NESTA (2003) in the UK and by Fleming and NORDEN (2007) in the Nordic Region of Europe show that there are some structural and cultural barriers to investment. The research conducted by NESTA sought to understand how the creative industries were perceived and valued by people, and involved interviews with 1004 people, representing a cross-section of the UK population. The results of this public survey showed that the public do not understand the creative industries' contribution to the economy, but value the products of the creative industries in terms of the contribution they make to the quality of their lives. While 75 per cent of respondents agreed that the UK's capacity to be innovative plays a crucial role in maintaining and enhancing the country's economic competitiveness, they do not recognize the creative industries as a major driver of the economy. The results of this survey show also that while the majority of investors believe that the sector has a large potential for growth, only one in five would invest in the creative industries. The same proportion of investors thought that creative industries business models were too risky to be worth in investing in. According to the results of this survey in the UK, while the public demands and appreciates creative industries' products, the investor community has little knowledge of them and is reluctant to invest in the sector, despite a huge proportion recognizing its recent growth and potential for the future.

The report by Fleming and NORDEN (2007) has similar results. Despite high growth in the creative industries in the Nordic Region, businesses in the sector are widely perceived to miss out on the opportunities to grow enjoyed by businesses in other 'knowledge sectors'. According to the results of this report there is a marked lack of awareness of investment opportunities and a lack of confidence and interest from institutional investors. Creative businesses too often lack the skills and know-how to realize the commercial value of their ideas. A mix of investment readiness and investor readiness factors constitute the structural and cultural barriers to investing in the creative industries (Figure 2.2). Networks, connectivity and collaboration are emerging critical issues in allowing for risk-sharing, gaining access to new ideas and markets, pooling resources, negotiating intellectual property rights (IPRs), and developing efficiencies.

However, the above-mentioned reports state also that there are ways to transform the barriers into opportunities. Mixing creative and business disciplines, developing new boundary-crossing collaborations (between technologists and content providers, and scientists and artists), and capitalizing on the uncommon ground of core general education and industry-specific skills can be mentioned

Investment Ready?
- Creative businesses (especially start-up and early stage businesses) too often lack essential investment readiness features, such as robust business plans, an acute awareness of markets, solid management structures/expertise, existing capital, clear realisation of Intellectual Property Rights (IPRs), and a track record of high growth.

- Creative businesses are innovators, developing products and services for which there is not always a proven market. Despite the success of the sector in developing new markets and thus growing year by year, investment is not forthcoming.

- Creative Industries businesses are often 'information poor', lacking basic routes to market knowledge, details of available specialist support, and business planning guidelines.

- Many creative businesses lack an appreciation that there may be finance opportunities for them and thus consider higher and faster levels of business development beyond them. An artificial ceiling is thus placed above Creative Industries businesses.

- Too many creative entrepreneurs consider investors (from both the private and public sectors) to be 'honour-bound' to invest in *their* idea or concept.

- Creative businesses often lack the specialist knowledge and expertise necessary to fulfill investment criteria: the sector is not presenting itself as the 'investment proposition' that, as a high growth sector, it surely is.

INFORMATIONAL **ASYMMETRIES**		**STRUCTURAL &** **CULTURAL BARRIERS** **TO INVESTMENT**

Investor Ready?
- Established 'investor communities' are reluctant to invest in small and early stage creative businesses due to a lack of knowledge of creative business growth potential and the inflexibility of current investment criteria.

- Long-held perceptions remain of creative businesses as too 'lifestyle-orientated', existing to support a way of life for the practitioner rather than as a commercially-driven concern. That Creative Industries businesses are often managed by highly skilled and entrepreneurial practitioners with commercial return a prime objective, is under-recognised.

- Some Creative businesses that have developed a relationship with investor communities have presented themselves as deserving of investment without attending to basic flaws in their business practice: investors are right to expect the Creative Industries sector to develop more 'investor-friendly' approaches for investment to take place.

- A lack of investor expertise reduces investor confidence, where a Creative Industries business will be dismissed as falling short of investment-readiness regardless of the viability of the business.

- The structural criteria of existing funds are too inflexible to engage many creative businesses – e.g. the amount of due diligence is disproportionately expensive for a business operating in a market for which little intelligence exists; and minimum investments might be too large for some creative firms.

Figure 2.2   Barriers to investing in the creative industries

Source: Quoted from Fleming and NORDEN 2007.

among the critical factors at the sector level (Fleming and NORDEN 2007). Nurturing and supporting the arts, access for all residents to creative opportunities in the city, and formation of strategic partnerships within the cultural sector, and between that sector and other creative actors across the city can be listed among the critical factors at the city level (Bradford 2004b).

How Can Public Policy Influence Creative Cities?

Public policy plays a critical role in nurturing a city's creative assets and infrastructure (Bradford 2004b, Gertler 2004, Fleming and NORDEN 2007). The policy context is comprised of a complex mix of initiatives at different levels from federal to provincial and local, and an effective coordination of policy initiatives between these levels (Gertler 2004). An effective coordination and collaboration among many policy actors – between government departments, across levels of government, and among governments, the private sector, community organizations, and engaged citizens including artists and cultural workers – enhance the creative capacity of cities (Bradford 2004b, Gertler 2004).

Gertler (2004) has classifed public policies that can influence creative cities at three levels – federal, provincial, and local – and in four policy categories – cultural, immigration and settlement, provincial, and municipal:

- *Cultural policy*: can help to ensure that arts and cultural endeavours are supported.
- *Immigration and settlement policy*: may have an impact on creative cities, especially since many immigrants settle in the same lower-income urban areas as artists.
- *Provincial policy*: can provide the 'connective tissue' between regions, in areas such as land use, green space protection and public transit.
- *Municipal policy*: has a significant role in city land use and development, in order to preserve the rich or mixed use nature of creative neighbourhoods.

Gertler (2004) has emphasized that public and private actions at the local level are the primary force for creating creative cities; however, the policies and regulatory decisions taken at the upper levels are equally critical. These policies not only provide the core funding and regulatory support for cultural activities and organizations, but also shape the broad background conditions and context that lay the foundations for a socially inclusive and cohesive path to the creative city. However, there are some critical issues in building a policy base (Fleming and NORDEN 2007):

- When building a policy base it is critical to take into consideration the existing assets rather than the desired set of assets. Addressing the desired asset base instead of reality would lead to building on the wrong base.
- A public policy approach should avoid being prescriptive. The most successful places have the least prescriptive public policy approach. The need is to work as an enabler rather than enforcer of cultural activity.
- Approaches to creative place-building should also avoid focusing on the value-adding role of the creative industries and use the sector as a 'tool' to deliver on a range of agendas, from social inclusion to cultural tourism. This kind of approach can create an instrumental burden that creative businesses may not wish to carry.

Another critical issue from a policy perspective is how to learn from innovations in other places and apply them locally while avoiding formulaic borrowing (Bradford 2004b). The experiences of successful cities show that the crucial step involves adapting 'models' developed elsewhere to the particular local context and applying an 'infiltration' within existing policies and programmes. Here, leadership emerges as a key factor for creative processes and successful implementations including both risk taking and securing resources.

What is Required to Build a Creative City?

'How to build creative cities?' has been one of the critical questions in the creative cities debate. The infrastructural conditions for competitiveness and creative growth, the critical factors to build creative cities, and the key principles in developing a city's vision and local planning processes towards creativity have been the hot topics of recent studies on creative cities from the planning perspective (Bradford 2004b, Duxbury 2004, Gertler 2004, Fleming and NORDEN 2007). A wide range of creative assets is required to build a creative city, from a creative vision to a creative strategy, from strategic resources to strong community networks, from creative city leaders to a creative governance and multi-level and cross-sectoral partnership and collaboration. To build a creative city also requires some infrastructural conditions for competitiveness and creative growth.

The infrastructural conditions for competitiveness and creative growth have recently been described by Fleming and NORDEN (2007). A cultural and creativity planning tool-box that includes '10 infrastructural conditions for creative growth' has been developed by Fleming and NORDEN for the Nordic Region of Europe (Table 2.3). Although these infrastructural conditions are recommended specifically for the Nordic Region, this planning tool-box provides a useful framework also for other cities.

While Fleming and NORDEN (2007) has described the necessary 10 infrastructural conditions for creative growth, Gertler (2004) has focused on the features of these infrastructures. According to Gertler (2004) two factors are critical to build creative cities: (i) investments must be made in the 'soft' and

Table 2.3   A cultural and creativity planning tool-box

The ten infrastructural conditions for competitiveness and growth	Evidence
1. A world class, high-profile cultural infrastructure	Such as galleries, museums, concert halls and events programmes. The wider the range of this infrastructure, the greater the competitive opportunities for creative businesses, provided that they are connected to these institutions through networks and collaborations.
2. A wide range of specialist creative industries support services	Some with a focus on growth such as business acceleration and investment programmes; others which are focused on continuous professional development of the individual.
3. A wide range of specialist and accesssible facilities for different parts of the creative industries	Such as through media centres, rehearsal space, studio space and workspace. Affordability and accessibility are crucial – across the creative industries value chain.

The ten infrastructural conditions for competitiveness and growth	Evidence
4. A strong and specialized Higher Education Sector	With outward-facing knowledge transfer, incubation and convergence programmes, strong links across creative and non-creative sub-sectors, and a commitment to inter-departmental approaches to creativity. Key is building management and entrepreneurial skills for undergraduates and supporting them effectively into business creation.
5. An innovative further and school education sector, plus a strong informal learning sector	The latter is vital to help entrants from non-traditional backgrounds. This needs to be married with services that help people identify career paths, offers brokering and mentoring programmes, offers in-work training and education opportunities and helps develop sustainable businesses/careers.
6. Spaces of convergence and connectivity	Where creative workers can meet, exchange and build relationships that can help with ideas generation and trade. It is vital that both creative workers and other users feel ownership of these places – allowing them to imprint their identity on the space so they can inhabit it on their own terms.
7. Global partnership and trade initiatives	With the most effective being based on the facilitation of business-to-business relationships and with the tone one of collaboration above competition.
8. Diversity advantage	Where complexly diverse communities are supported to project themselves as a major feature of the creative asset base of a place. Infrastructure that effectively provides spaces for exploring and promoting a place's diverse assets, are vital symbols and drivers of knowledge creation.
9. Strong spaces of cultural consumption connecting spaces of production	Consumers play a vital role in the development of creative industries, and most practitioners are consumers first and foremost. The aim is to develop highly networked, high energy creative clusters, where processes of cultural consumption are symbiotic with processes of cultural production.
10. A vibrant night-time economy	As well as cultural consumption of various kinds, a leisure infrastructure is an essential part of the urban creative milieu.

Source: Quoted from Fleming and NORDEN 2007.

'hard' infrastructures of urban creativity; and (ii) the pivotal public policy role in nurturing a city's creative assets and infrastructures. Soft infrastructure as a 'connective tissue' comprises the social networks and shared spaces facilitating interaction among creative people. Hard infrastructure refers to the physical environment of highways, public transit, sewer and water supply networks, and so forth. Governments establish the institutional and regulatory context for private sector and non-profit organizations to make their own unique contributions.

In another study, Duxbury (2004) has underlined the key principles in developing a city's vision and local planning processes. These key principles are as follows:

- Each city and community is unique in its identity, history, development and assets.
- Implementation of ideas is an art, based on knowledge and sensitivity to the community.
- City development must be rooted in authenticity, but cities should also be willing to learn from innovative ideas elsewhere (while avoiding formulaic borrowing).
- Durable planning and governance innovation depend on strong community involvement and shared ownership of the process and outcomes.
- Small projects sustained over time can make a difference.

However, Duxbury has mentioned that the presence of motivators to act, new ideas about desirable change, and sound principles of practice are not sufficient, for successful innovation the will and capacity to act in new and different ways are required. Creative city leaders must be willing to take risks and secure resources for their innovative solutions. Innovative change requires time, flexibility, strong community networks, a strong collective will and strategic resources.

Creative Urban Policies: Challenges for Governments

This comprehensive and multidimensional evaluation of creative cities show that the more attractive cities for new and creative activities have some common characteristics including an authentic and unique cultural heritage and identity, an environment that is tolerant to diverse people and lifestyles, a long tradition of creativity and innovation, and a highly developed 'hard' and 'soft' infrastructure. Successful cities seem to have also visionary individuals, creative organizations, a political culture, a strong leadership, a multi-level collaboration and an active community involvement in common. Successful cities seem to follow 'a determined, not a deterministic path' (Landry 2000). In these cities, planning is seen as a vision driven rather than narrowly technical process and involves turning a vision into practice by developing strategically focused and tactically flexible principles. These cities' visions focus on how to balance complex and

Table 2.4   Creative city strategies and challenges

Strategies for creative economy and entrepreneurship

- Grow the creative economy with emphasis on the design and business innovation and research clusters
- Build an integrated infrastructure to support economic development and foster an entrepreneurial climate
- Create business environments for investment

Strategies for social and cultural planning

- Build a creative community that attracts and retains creative people
- Promote diversity in the artistic life of the community
- Broaden public access to the local arts
- Invest and spark investment in the local arts
- Link programmes to other creative communities
- Set up and empower non-governmental organizations

Strategies for creative governance

- Build creative capacity
- Strengthen the organizational infrastructure
- Integrate diverse communities into 'mainstream' cultural institutions
- Develop the linkages with the business community and educational sector
- Strengthen the linkages between the culture and high-technology sectors
- Collaborate with non-profit cultural organizations to deliver diverse and excellent arts, and cultural services and activities throughout the city
- Support the non-profit facilities

Strategies for creative urban planning

- Integrate cultural planning interests in all planning processes of the city
- Build spaces of quality and variety
- Build public facilities, places for cultural and social exchange
- Adapt and re-use existing space and assets
- Revitalize public spaces and natural spaces through the arts
- Increase access for artist and community to a new and improved space and place
- Provide live and work areas in or connected to urban centres
- Provide highly flexible spaces that can accommodate a variety of uses
- Create creative and education environments for social investment

contradictory needs and to provide an active process of participation to develop consensus by minimizing or resolving conflict over implementation.

Today, many municipalities are planning and acting to re-vision and re-position their cities in order to make them more vibrant, more inclusive, and more supportive of cultural actors and new ideas. 'Creative urban planning' means 'synthesizing different traditions and seizing unexpected opportunities' (Bradford 2004b), and requires understanding the sophisticated relationship between creativity and urban space. The main challenge faced by many cities is how to translate this new understanding about creativity as a central driver of growth, change, and transformation into integrated and comprehensive urban strategies. Creative city strategies and challenges derived from the successful city experiences (see, among others, Duxbury 2004, Nieh 2005, Peck 2005) are summarized in Table 2.4. Creative urban strategies reflect the 'culture of creativity' of the city that is embedded in how urban stakeholders operate. As mentioned also by Landry (2000), creativity is not only about having ideas, but also the capacity to implement them, and to be a creative city requires dynamic thinkers, creators and implementers.

Being a creative city requires interconnected policies, plans, programmes and established practices, therefore, a collaboration among government departments, across levels of government, and among government, the private sector and community organizations. The second challenge for cities is building a 'creative urban governance' system. Being a creative city requires also taking some measured risks, widespread leadership, strategic principles and flexible tactics. Therefore, the third challenge for cities is building 'creative capacity'. Building creative capacity is a complex undertaking that often involves shifting mindsets, breaking down silos, re-balancing risk, visioning, building consensus, and creating the conditions for people to become agents of change rather than 'victims of change'. The fourth challenge for cities is also building a 'creative mental infrastructure'. A creative urban governance system and a new mental infrastructure call for another challenge: the fifth challenge for cities and governments is to develop a wealth of tools, strategies, policies and frameworks designed to build a culture of creativity and innovation.

Urban planning policy may play a more sophisticated role in the development of urban creativity. Urban planning may also play a strong role in building a vision of a dynamic, creative city with stakeholders and the community. The urban planning function, if imaginatively applied, may track the effects of creative change over the long-term, and when culture and creativity are used with a social purpose in urban planning policy, they contribute to sustainable development and societies.

Bibliography

Afolabi, M.O., Dionne, S. and Lewis III, H. 2006. *Are we there yet? A review of creativity methodologies*, paper presented to the St. Lawrence Section Conference, Cornell University.

Andersen, K.V. and Lorenzen, M. 2005. *The Geography of the Danish Creative Class: A Mapping and Analysis*. Copenhagen: Copenhagen Business School.

Andersson, Å.E. 1985. Creativity and regional development. *Papers of the Regional Science Association*, 56, 5–20.

Askerud, P. 2007. *Creative industries and urban development: Context and goals*, paper presented to the HKIP & UPSC Conference: When Creative Industries Crossover with Cities, Hong Kong, 2–3 April.

Baer, J. 1997. Gender Differences in the Effects of Anticipated Evaluation on Creativity. *Creativity Research Journal*, 10(1), 25–31.

Bagwell, S. 2008. Creative Clusters and City Growth. *Creative Industries Journal*, 1(1), 31–46.

Baycan-Levent, T. 2010. Diversity and Creativity as Seedbeds for Urban and Regional Dynamics. *European Planning Studies*, 18(4), 565–94.

Bayliss, D. 2007. The Rise of the Creative City: Culture and Creativity in Copenhagen. *European Planning Studies*, 15(7), 889–903.

Bechtoldt, M.N., De Dreu, C.K.W. and Nijstad, B.A. 2007. *Team Personality Diversity, Group Creativity, and Innovativeness in Organizational Teams*. SUS.DIV Working Paper No. 1.07.

Bianchini, F. and Parkinson, M. 1994. *Cultural Policy and Urban Regeneration: The West European Experience*. Manchester: Manchester University Press.

Boschma, R.A. and Fritsch, M. 2007. Creative Class and Regional Growth – Empirical Evidence from Eight European Countries. Jena Economic Research Papers 2007–066.

Bradford, N. 2004a. *Creative Cities: Structured Policy Dialogue Backgrounder*. Canadian Policy Research Networks: Background Paper F/46, August 2004.

—— 2004b. *Creative Cities: Structured Policy Dialogue Report*. Canadian Policy Research Networks: Research Report F/45, August 2004.

Chang, T.C. 2000. Renaissance Revisited: Singapore as a Global City for the Arts. *International Journal of Urban and Regional Research*, 24(4), 818–831.

City of Ottawa. 2003. *Ottawa 20/20 Arts Plan*. Available at: www.ottawa.ca/2020/arts/toc_en.shtml.

City of Toronto. 2003. *Culture Plan for the Creative City*. Available at: www.toronto.ca/culture/pdf/creativecity-2003.pdf.

Cooke, P. and Lazeretti, L. (eds) 2008. *Creative Cities, Cultural Clusters and Local Economic Development*. Cheltenham: Edward Elgar.

Crane, L. 2007. *Creative Industries Districts: An Innovative Concept for an Innovative Time*. Paper presented to the HKIP & UPSC Conference: When Creative Industries Crossover with Cities, Hong Kong, 2–3 April.

Cumbers, A. and MacKinnon, D. 2004. Introduction: Clusters in Urban and Regional Development. *Urban Studies*, 41 (5–6), 959–69.

Cunningham, S. 2002. From Cultural to Creative Industries: Theory, Industry, and Policy Implications. *Culturelink*, 19–32.

DCMS. 1998. *Cultural Industries Mapping Document*. London: Department of Culture, Media and Sport.

—— 2001. *Mapping Creative Industries Technical Document*. London: Department of Culture, Media and Sport.

Drake, G. 2003. This Place Gives Me Space: Place and Creativity in the Creative Industries. *Geoforum*, 34, 511–24.

Duxbury, N. 2004. *Creative Cities: Principles and Practices*. Canadian Policy Research Networks: Background Paper F/47, August.

EESC. 2003. European Economic and Social Committee. Available at: www.eesc. europa.eu.

Ellmeier, A. 2003. Cultural Entrepreneurialism: On the Changing Relationship Between the Arts, Culture and Employment. *The International Journal of Cultural Policy*, 9 (1), 3–16.

Evans, G. 2005. Measure for Measure: Evaluating the Evidence of Culture's Contribution to Regeneration. *Urban Studies*, 42, 959–83.

—— 2009. Creative Cities, Creative Spaces and Urban Policy. *Urban Studies*, 46(5–6), 1003–40.

Fasko, Jr, D. 2001. Education and Creativity. *Creativity Research Journal*, 13(3–4), 317–27.

Fleming, T. and NORDEN. 2007. *A Creative Economy Green Paper for the Nordic Region*. Nordic Innovation Center.

Florida, R. 2002. *The Rise of the Creative Class*. New York: Basic Books.

Florida, R. and Tinagli, I. 2004. *Europe in the Creative Age*. Final report to the Software Industry Center at Carnegie Mellon, the Alfred P.Sloan Foundation and DEMOS.

Foord, J. 2008. Strategies for Creative Industries: An International Review. *Creative Industries Journal*, 1 (2), 91–113.

Forte, F., Fusco Girard, L. and Nijkamp, P. 2005. Smart Policy, Creative Strategy and Urban Development. *Studies in Regional Science*, 35(4), 947–63.

Fusco Girard, L., Forte, B., Cerreta, M., De Toro, P. and Forte, F. (eds) 2003. *The Human Sustainable City: Challenges and Perspectives from the Habitat Agenda*. Aldershot: Ashgate.

Gertler, M.S. 2004. *Creative Cities: What Are They For, How Do They Work, and How Do We Build Them?* Canadian Policy Research Networks: Background Paper F/48, August 2004.

Glaeser, E.L. 2004. Review of Richard Florida's The Rise of the Creative Class. Available at: www.economics.harvard.edu/faculty/glaeser/papers/ ReviewFlorida.pdf.

Gornostaeva, G. 2008. The Film and Television Industry in London's Suburbs: Lifestyle of the Rich or Losers' Retreat? *Creative Industries Journal*, 1, 47–71.

Guilford, J.P. 1967. *The Nature of Human Intelligence.* New York: McGraw-Hill.

Hadamar, J. 1939. *The Psychology of Invention in the Mathematical Field.* Princeton, NJ: Princeton University Press.

Hall, P. 1998. *Cities in Civilization.* London: Weidenfeld and Nicolson.

—— 2000. Creative Cities and Economic Development. *Urban Studies,* 37(4), 639–49.

—— Creativity, Culture, Knowledge and the City. *Built Environment,* 30(3), 256–8.

Hartley, J. (ed.) 2005. *Creative Industries.* Malden, Mass.: Blackwell Publishing.

Helbrecht, I. 2004. Bare Geographies in Knowledge Societies – Creative Cities as Text and Piece of Art: Two Eyes, One Vision. *Built Environment,* 30(3), 194–203.

Herring, C. 2009. Does Diversity Pay? Race, Gender, and the Business Case for Diversity. *American Sociological Review,* 74(2), 208–24.

HKTDC (Hong Kong Trade Development Council). 2002. Creative Industries in Hong Kong. *Economic Forum,* 5 September. Available at: www.tdctrade.com/econforum/tdc.

Howkins, J. 2002. *The Creative Economy: How People Make Money from Ideas.* Harmondsworth: Penguin.

Hutton, T. 2004. The new economy of inner city. *Cities,* 21(2), 89–108.

ISoCaRP. 2005. *Making Spaces for the Creative Economy,* edited by Ng. Waikeen and J. Ryser. The Hague: International Society of City and Regional Planners (ISoCaRP).

Jing, Y. and Rong, Z. 2007. *Shanghai Expo, Creating the City Culture Map and Brand – Thoughts on Cultural Drive for City Development and Expo Planning.* Paper presented to the HKIP & UPSC Conference: When Creative Industries Crossover with Cities, Hong Kong, 2–3 April.

Jones, T. 2007. *Cultural Districts.* Paper presented to the HKIP & UPSC Conference: When Creative Industries Crossover with Cities, Hong Kong, 2–3 April.

Kalandides, A. and Lange, B. 2007. *Creativity as a Synecdoche of the City – Marketing the Creative Berlin.* Paper presented to the HKIP & UPSC Conference: When Creative Industries Crossover with Cities, Hong Kong, 2–3 April.

Kaufman, J.C. and Sternberg, R.J. 2006. *The International Handbook of Creativity.* New York: Cambridge University Press.

KEA. 2006. *The Economy of Culture in Europe.* Study prepared for the European Commission.

Landry, C. 1990. *Glasgow: The Creative City and its Cultural Economy.* Glasgow: Glasgow Development Agency.

—— 1996. *The Creative City in Britain and Germany.* London: Anglo-German Foundation.

—— 2000. *The Creative City: A Toolkit for Urban Innovators.* London: Earthscan Publications.

——— 2004. *Rethinking Creative City*. Available at: www.charleslandry.com.
——— 2006. *Lineages of the Creative City*. Available at: www.charleslandry.com.
——— 2007. *Creative Cities*. London: Earthscan Publications.
Lee, Y.S. and Yeoh, B.S.A. 2004. Introduction: Globalisation and the politics of forgetting. *Urban Studies,* 41, 2295–301.
Marcus, C. 2005. Future of Creative Industries: Implications for Research Policy. DG Research, European Commission, Brussels.
Markusen, A. 2006. Urban Development and the Politics of a Creative Class: Evidence from the Study of Artists. *Environment and Planning A*, 38, 1921–40.
McCann, J.E. 2002. The Cultural Politics of Local Economic Development: Meaning-Making, Place-Making and the Urban Policy Process. *Geoforum*, 33, 385–98.
McLaren, R.B. 1999. Dark Side of Creativity, in *Encyclopedia of Creativity,* edited by M.A. Runco and S.R. Pritzker. London: Academic Press.
MEA 2006. *Our Creative Potential: Paper on Culture and Economy*. Netherlands: Ministry of Economic Affairs and Ministry of Education, Culture and Science.
Merkel, J. 2008. Ethnic Diversity and the 'Creative City': The Case of Berlin's Creative Industries. Paper presented to Future Workshop, October.
Miles, S. and Paddison, R. 2005. Introduction: The Rise and Rise of Culture-led Urban Regeneration. *Urban Studies*, 42(5–6), 833–9.
Mommaas, H. 2004. Cultural Clusters and the Post-Industrial City: Towards the Remapping of Urban Cultural Policy. *Urban Studies,* 41, 507–532.
Montgomery, J. 2007. *The New Wealth of Cities: City Dynamics and the Fifth Wave*. Aldershot: Ashgate.
Morrison, Z. 2003. Recognising 'Recognition': Social Justice and the Place of the Cultural in Social Exclusion Policy and Practice. *Environment and Planning A*, 35, 1629–49.
Mumford, M.D. 2002. Social Innovation: Ten Cases From Benjamin Franklin. *Creativity Research Journal*, 14(2), 253–266.
Mumford, M.D. and Moertl, P. 2003. Cases of Social Innovation: Lessons From Two Innovations in the 20th Century. *Creativity Research Journal*, 15(2–3) 261–6.
Murray, K., Adams, D. and Champion, K. 2007. *Developing Creative Cities: A Perspective from the UK*. Paper presented to the HKIP & UPSC Conference: When Creative Industries Crossover with Cities, Hong Kong, 2–3 April.
Musterd, S. 2004. Amsterdam as a Creative Cultural Knowledge City: Some Conditions. *Built Environment*, 30(3), 225–34.
Musterd, S. and Deurloo, R. 2006. Amsterdam and the Preconditions for a Creative Knowledge City. *Tijdschrift voor Economische en Sociale Geographie*, 91(1), 80–93.
Musterd, S. and Ostendorf, W. 2004. Creative Cultural Knowledge Cities: Perspectives and Planning Strategies. *Built Environment,* 30(3), 189–193.

Musterd, S., Bontje, M., Chapain, C., Kovacs, Z. and Murie, A. 2007. *Accomodating Creative Knowledge: A Literature Review from a European Perspective*. ACRE report 1. Amsterdam: AMIDSt.

Nachum, L. and Keeble, D. 2003. Neo-Marshallian Clusters and Global Networks – The Linkages of Media Firms in Central London. *Long Range Planning*, 36, 459–80.

NESTA (National Endowment for Education, Science and the Arts). 2003. *Forward Thinking. New Solutions to Old Problems: Investing in the Creative Industries*. London: NESTA.

Newman, P. and Smith, I. 2000. Cultural Production, Place and Politics on the South Bank of the Thames. *International Journal of Urban and Regional Research*, 24, 9–24.

Nieh, D. 2005. Silicon Valley and Beyond: Towards an Architecture of Creative Communities, in *Making Spaces for the Creative Economy*, edited by Ng. Waikeen and J. Ryser. The Hague: International Society of City and Regional Planners (ISoCaRP), 10–25.

Peck, J. 2005. Struggling with the Creative Class. *International Journal of Urban and Regional Research*, 294, 740–70.

Porter, M.E. 1995. The Competitive Advantage of the Inner City. *Harvard Business Review*, 73, 55–71.

—— 1998. Clusters and the New Economics of Competition. *Harvard Business Review*, 76, 77–87.

Porter, M.E. and Stern, S. 2001. Innovation: Location Matters. *MIT Sloan Management Review*, (Summer), 28–36.

Pratt, A. 2008. Creative Cities. *Urban Design*, 106, 35.

Quinn, B. 2005. Arts Festivals and the City. *Urban Studies*, 42(5–6), 927–43.

Rantisi, N.M., Leslie, D. and Christopherson, S. 2006. Guest editorial. *Environment and Planning A*, 38, 1789–97.

Ray, C. 1998. Culture, Intellectual Property and Territorial Rural Development. *Sociologica Ruralis*, 38, 3–20.

Reis, S.M. 2002. Toward a Theory of Creativity in Diverse Creative Women. *Creativity Research Journal*, 14(3–4), 305–16.

Rhodes, M. 1961. Analysis of Creativity. *Phi Delta Kapan*, 42(7), 305–10.

Richards, G. 2001. *Cultural Attractions and European Tourism*. Wallingford: CAB International.

Rothenberg, A. 1990. *Creativity and Madness*. Baltimore, MD: Johns Hopkins University Press.

Sassen, S. 2001. *The Global City New York, London, Tokyo*. Princeton, NJ: Princeton University Press.

Santagata, W. 2002. Cultural Districts, Property Rights and Sustainable Economic Growth. *International Journal of Urban and Regional Research*, 26(1), 9–23.

Schumpeter, J.A. 1950. *Capitalism, Socialism and Democracy*. New York: Harper and Row.

Scott, A. 2000. The Cultural Economy of Paris. *International Journal of Urban and Regional Research*, 24, 567–82.

—— 2006. Creative Cities: Conceptual Issues and Policy Questions. *Journal of Urban Affairs,* 28(1) 1–17.

Sharp, J., Pollock, V. and Paddison, R. 2005. Just Art for a Just City: Public Art and Social Inclusion in Urban Regeneration. *Urban Studies*, 42(5–6), 1001–23.

Simmie, J. 2005. Innovation and Space: A Critical Review of the Literature. *Regional Studies*, 39(6) 789–804.

Simon, H.A. 2001. Creativity in the Arts and the Sciences. *The Canyon Review and Stand*, 23, 203–20.

Smith, G.J.W. 2005. How Should Creativity Be Defined? *Creativity Research Journal*, 17(2–3), 293–5.

Smith, R. and Warfield, K. 2008. The Creative City: A Matter of Values, in *Creative Cities, Cultural Clusters and Local Economic Development*, edited by P. Cooke and L. Lazeretti. Cheltenham: Edward Elgar, 287–312.

Snow, R.E. 1986. Individual Differences in the Design of Educational Programs. *American Psychologist*, 41, 1020–39.

Stenberg, R.J. (ed.) 1999. *Handbook of Creativity*. New York: Cambridge University Press.

Sternberg, R.J., Kaufman, J.C. and Pretz, J.E. 2002. *The Creativity Conundrum*. New York: Psychology Press.

Torrance, E.P. 1976. *Creativity in the Classroom*. Washington, DC: National Education Association Publication.

—— 1989. Scientific Views of Creativity and Factors Affecting its Growth, in *Creativity and Learning*, edited by J. Kagan. Boston: Houghton Mifflin, 73–91.

Törnqvist, G. 1983. Creativity and the Renewal of Regional Life, in *Creativity and Context: A Seminar Report*, edited by A. Buttimer. Lund Studies in Geography. B. Human Geography, No. 50. Lund: Gleerup, 91–112.

UNESCO. 2005. International Flows of Selected Cultural Goods and Services, 1994–2003: Defining and Capturing the Flows of Global Cultural Trade. Paris: UNESCO Institute for Statistics and UNESCO Sector for Culture.

UNCTAD. 2004. *Creative Industries and Development*. United Nations Conference on Trade and Development (UNCTAD). Washington, DC: UN.

—— Creative Economy Report. 2008: The Challenge of Assessing the Creative Economy Towards Informed Policy-making. United Nations Conference on Trade and Development (UNCTAD). Washington, DC: UN.

Wikipedia. 2009. *Wikipedia: The Free Encyclopedia*. Available at: www.wikipedia. org/.

World Bank. 2003. Urban Development Needs Creativity: How Creative Industries Affect Urban Areas. Washington, DC: Development Outreach, November.

Wu, W. 2005. *Dynamic Cities and Creative Clusters*. World Bank Policy Research Working Paper 3509.

Xu, Y. and Chen, X. 2007. *A Study on Creative Industries in Yangtze River*. Paper presented to the HKIP & UPSC Conference: When Creative Industries Crossover with Cities, Hong Kong, 2–3 April.

Yeoh, B.S.A. 2005. The Global Cultural City? Spatial Imagineering and Politics in the (Multi)cultural Marketplaces of South-east Asia. *Urban Studies*, 42(5–6), 945–58.

Yigitcanlar, T., Velibeyoglu, K. and Martinez-Fernandez, C. 2008. Rising Knowledge Cities: The Role of Knowledge Precincts. *Journal of Knowledge Management*, 12(5), 8–20.

Yip, S. 2007. *Creative industries strategy for cities: What planning can do (and cannot do)*. Paper presented at the HKIP & UPSC Conference: When Creative Industries Crossover with Cities, Hong Kong, 2–3 April.

Zukin, S. 1995. *The Cultures of Cities*. Oxford and Cambridge, MA: Blackwell.

—— 1998. Urban lifestyles: Diversity and standardization in Spaces of Consumption. *Urban Studies*, 35(5–6), 825–39.

Chapter 3

Creativity and the Human Sustainable City: Principles and Approaches for Nurturing City Resilience

Luigi Fusco Girard

The Creative City in a Multidimensional Perspective: Introduction

The Creative City

Cities are places of development, but also of poverty; of success, but also of failure; of hope, and also of desperation; of cooperation, and also of conflicts; of order, but also of disorder; and are generators of new risks (natural hazards, extreme weather events, and so on) (Fusco Girard 2006). More than ever, cities are places of contradictions, paradoxes and conflicts, where the economic wealth of a region/nation is produced, and, at the same time, ecological poverty and social poverty are increasing (UN-Habitat 2003). Worldwide, slums, informal settlements, and distressed urban areas are growing UN-Habitat 2003, 2006), burdened by unemployment, illegal economy and exploitation. Here social and ecological poverty are intertwined.

Therefore, cities are extremely complex places to manage/govern: they guarantee benefits to their residents/workers, but also produce many negative effects, such as pollution, environmental degradation, unemployment, social fragmentation and marginalization. Over time, these negative effects can increase, destabilizing the city's organization.

Nevertheless, the city represents the starting point for rebuilding a future which is more desirable on the whole for all people, creating hope for positive change. The city is also the starting point for: regenerating the economy (not linear but circular: a new ecological metabolism), providing social sustainability, and also regenerating democracy towards a *human* sustainable perspective. The future of itself humankind, and its environment, will be shaped in the city. The city is the most relevant threat to climate destabilization and, at the same time, it feeds the decay of social relationships. However, a better quality of life can be achieved in the city by improving the functioning of urban systems.

This hope finds its basis in the *creativity of the city*. The city itself is the product of human creativity (Soleri 1971). Cities have always been at the crossroads of creativity. In general, creativity has not developed in the countryside, but in the city mosaic, where different ethnic groups converge, as well as different cultures, ways of thinking and religions (Hall 1998). Strong creativity can to overcome all these contradictions, paradoxes and conflicts, in the cities of the North and South, in great metropolitan areas, and in medium and small towns.

Creativity as immaterial capital becomes the real 'strength' of a territory/city, more important than financial/infrastructural capital. Without that capital, the city becomes stagnant or declines, with it cities can face the enormous new challenges: economic competition, the environmental crisis and increasing inequalities (urban marginality and poverty). Crises force cities to be creative in building their future. Parts of their organizational structure are destroyed, while others are regenerated, thereby opening new evolutionary trajectories and opportunities (Schumpeter 1942). The creativity capital of the city is not only that of an *elite* of entrepreneurs and artists, but that of all its inhabitants, reflected in their lifestyles, the density of their relationships, and their self-organizational capacity and performances.

Creativity also concerns a *new way of living* in the city for its inhabitants: their creative capacity in combining/integrating old values into a modern vision (the ancient and the new; tradition and modernity). This creativity ensures the self-organizational capacity and resilience of the city, and therefore the possibility of the continuous recreation of new opportunities. A creative environment fosters people and helps them to become creative as 'entrepreneurs' in different fields.

The creative city is the one that is able to successfully face all the above-mentioned problems, improving the *choices* of governance/management/planning with the result of reproducing order also in conditions of turbulent (physical, economic, social) change, thus preserving and improving the quality of life for its inhabitants – the indicator of the success of creative actions.

The Aim of the Chapter

New solutions are absolutely necessary to face the dramatic problems of development and to allow development to really become 'liberation' from suffering, anxiety and dropping-out, all of which are growing in cities in the process of their urbanization. New creative initiatives are required to avoid development which might concern only a minority of persons, and to find a way to make it satisfy the ancient aspiration of *all* persons/people for 'happiness' (Layard 2005).

Really creative actions are interpreted here in a systemic and multidimensional perspective, as being able to integrate wealth production (business, profits) with ecological conservation and social promotion, in a win–win strategy. These initiatives should be taken particularly at a local level, continuously investing in innovations and in city resilience.

The aim of this chapter is to discuss city creativity in relation to four key interrelated problems: economic competition, ecological/climate stability, social cohesion and self-government promotion, by identifying a set of principles and tools to guide actions and innovative initiatives designed to better achieve from a systemic perspective the objectives of *human* sustainable city development, enhancing city resilience.

The focus will be on complex interdependences between creativity and resilience in implementing sustainability, and on principles, approaches and tools for nurturing city (economic, ecological, social and cultural) resilience. R*esilience* is proposed here as the notion that ties *creativity* to *sustainability*. So a creative city becomes a city that invests in economic, ecological and social resilience and – in particular – in *cultural resilience,* because cultural resilience allows for creative responses to changes and shocks, sustaining, from the bottom up, technological, economic and organizational innovations.

Investing in city resilience is the best defence against crisis. Creativity enhances resilience.

New relations among ideas and actions characterize creativity. The focus of this chapter is on the centrality of *relations*: relations among economic, ecological and social systems need to be improved or developed in order to maximize benefits that lead to *human* sustainable development. Creative initiatives improve, at the same time, environment (ecological resilience), employment (social resilience) and economic wealth (economic resilience). For example, action plans for mitigation and adaptation have to be linked to measures against poverty and for competitiveness. Resilience itself depends on the density of relations, correlations and connections: to be stressed and developed. Relations are the structural characteristic of systems.

New relations are proposed: in the economic system (between conventional and 'other' types of economy, between enterprise and the ecological/social environment, and between workers and enterprise); in the social system (between individuals and the community, and between public institutions and the third sector); in urban planning (between ancient centres, derelict areas and new peripheries); in design (between architecture and social-ecosystems); between top-down and bottom-up evaluations; and between aesthetics and citizenship.

These relations require innovative governance, based on (tests, simulations, and) real experiments. They also require a new diffused culture, less self-centred and less nourished by instrumental economic rationality, opened to a '*relational rationality*', in order to strengthen socio-cultural resilience. At the basis of the current processes there is not only a financial/economic crisis, but also its fundamental element – a culture which is: less and less careful about the public spirit; more and more indifferent towards a general vision of the future and to a long-term perspective; and more and more aware of particular interests to be achieved 'here and now' on the basis of an 'instrumental rationality'. This enlarges the competition and conflicts, the division between people, and between the rich and the poor, impoverishing every public space. Moreover, this culture also damages the economic and institutional system.

The evaluation process is proposed here as a general tool to implement and sustain new city governance in the direction of *human* self-sustainable city development: to promote cooperation, and the coordinative capacity of actions of multiple different subjects, and thus create a new culture, not self-centred, but able to stimulate new priorities in concrete choices. This process involves multi-criteria approaches by experts, decision-makers and decision-takers, and also by ordinary people: evaluation means, first of all, critical interpretation and comparison capacity. It gives to the city an 'intrinsic' organizational capacity.

The following sections will discuss the fundamental role of new governance in promoting resilience, creativity and sustainability through initiatives in local economic development, planning, architecture and culture.

Sustainability, Resilience and Creativity in the City

The Notion of the Creative City

The notion of the creative city and its various components has been extensively analysed in the literature (Acs 2002, Baycan-Levent and Nijkamp 2010, Carta 2007, Florida 2002, Florida 2005, Hall 1998, Jacobs 1969, Kunzmann 2004, Landry 2000, Landry 2006, Markusen 2006, Saxenian 1994, Scott 2000 and 2006, Simmie 2001).

The concept of the creative cities is fuzzy (Kunzmann 2004) and can be interpreted from many different perspectives. Peter Hall (1998) identifies in history some types of creative cities: technological-innovative; cultural-artistic; art and technology; and art, technology, and organization.

For example, creativity in the technological field has determined what are termed 'revolutions', which affect the city's organization. In the eighteenth century the Industrial Revolution deeply transformed the city. In the late twentieth century the information-technology revolution developed, globalizing the economy and determining a series of impacts, among which are the dematerialization of the economy itself, and the central role of culture/knowledge. Nowadays, the third revolution, that is, the energetic one, is taking place, and it will deeply change our cities (Droege 2006). Cities need a new comprehensive organization. What is necessary is an 'urban revolution' regarding many sectors of the city: the physical/ spatial (its form); the economic/financial (foundations, third sector, social enterprise, etc.); the ecological (its metabolism, including the social aspect); and the institutional (management/governance, public administration).

There is one kind of creativity of the city which proceeds through sharp breaks and discontinuities, which leads to the introduction of new strategies and to the rapid obsolescence of pre-existing ones. There is also a creativity of cities which feeds itself continuously, and which starts from the status quo, through successive incremental improvements (adaptive creativity).

But, in both these cases the empirical evidence shows that innovative elements are deep-rooted within *history*, within tradition. They do not fall from above or come from outside, but have a bond with the city's past history, with its soul and its spirit. They have been able to 'metabolize' the urban history.

Sustainable, Creative and Resilient City

In both cases it has been necessary to promote an 'innovative milieu', in order to valorize existing skills and talents. This 'innovative milieu' allows cities to be creative in the accelerated change: to be *resilient from the inside* (and not only because they receive exogenous resources or adapt the best practices developed elsewhere).

An essential element of this milieu is represented by better knowledge that allows different subjects to think in a new way, thus identifying new alternatives, new solutions, new choices.

The main condition for real success in implementing sustainable development is to invest in city creativity and resilience.

Sustainability, creativity and resilience are closely intertwined, as is clearly shown by best practices (that are concrete signs of the creativity of cities in implementing sustainable development) (Fusco Girard and You 2006). The image itself of a creative city reflects the interdependences among sustainability, resilience and creativity (Figure 3.1).

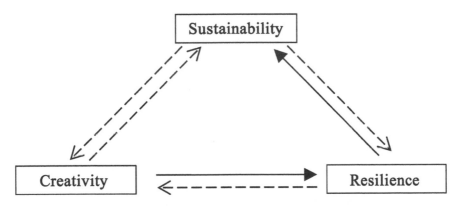

Figure 3.1   Interdependences among sustainability, resilience and creativity

Creativity and innovations enhance the capacity to face new risks and perturbations, that is, the resilience of the ecological, economic, and social systems. In other words, creativity enhances sustainability because it guarantees more resilience capacity to urban systems. Governance is required to manage city resilience (Lebel et al. 2006).

Multiple Perspectives of Resilience

Innovations can involve, for instance, a new circular metabolism (with reuse, recycling, and regeneration of materials) in urban ecosystems (ecological resilience), new economic competitiveness with the identification of original development trajectories in wealth production (economic resilience), and the opening up of new social bonds, community relationships (social resilience). The intensity of the resilience depends on specific innovations that are introduced into the urban system. They improve the comprehensive city self-organization and thus sustainability.

If resilience is the capacity of a system to maintain over time its original organizational structure (its identity and unity), absorbing shocks from outside, thanks to its self-organizational capacity it is possible to identify in ecological, economic and social resilience some common elements: the notion of memory, conservation, stability, correlations and feedbacks.

In particular, *social resilience* (Adger 2000 and 2007, Adger and Brown 2009) depends on the density of formal and informal social networks, which are able to preserve over time a certain organizational order: it depends on the existing community and on the sense of place (Relph 1976).

Economic resilience is the capacity to produce wealth, business and profits, by changing and experimenting with innovative production technologies, organizations and strategies. New connections among networks and new relationships are strongly improved through ICT.

Ecological resilience reflects the health and robustness of the system (Holling 1973, Costanza et al. 1991, Alberti and Marzluff 2004), i.e. the density of connections/relations between its different components that allow for circular processes (with a reduction of consumed materials, energy, water, etc.). More in general, to guarantee and regenerate city resilience, it is necessary to invest in the *cultural dimension*.

City cultural resilience is the internal energy, the inner force (or vitality) that allows the city to react to external forces, adapt to them, and conserve its specific identity in the long run, in spite of turbulent transformation processes, and to design new win–win solutions. Cultural resilience stresses the notion of the cultural memory of the community as a formative strength of collective consciousness, foundation of continuity, engine of the future and new actions, in orders to improve trust, cooperation and coordination of actions, and to promote a sense of community. A resilient city is also a creative city, able to reinvent a new equilibrium against destabilizing external pressures. It multiplies the potential of people to build new opportunities/alternatives. Cultural resilience depends on the capacity to think and choose in a systemic, multidimensional, open and relational way, linking short-, medium- and long-term perspectives, with attention to the 'memory' of the system in achieving common interests.

If these common elements/goals are recognized, cooperation and coordination of actions is possible. The approach becomes constructivist (not 'or ... or', but 'and ... and'), and characterized by gradualism. Cultural resilience is built on the basis of a way of thinking, founded upon a critical approach (i.e. on evaluation capacity), able to recognize and compare tangible and intangible elements – values, objectives, goals – by considering all existing interdependences and distinguishing a hierarchy or priority. Critical thinking is the capacity to learn from experiences and to appropriately select not only means but also objectives that have values and meanings, that are reference anchors in orienting innovations and creativity.

In other words, city cultural resilience depends on each inhabitant's capacity to transform data and information into critical knowledge and wisdom (Zeleny 2006), and to adopt a new way of life that rejects the current 'more, bigger, faster' culture.

The Four Challenges of Creative Governance

Creative and Effective Governance

Innovative governance reproduces order and compatibilities, avoiding the situation where comprehensive complexity may exceed the critical threshold of instability of the city system. It improves city resilience through the construction of a creative milieu that stimulates, in its turn, new creative actions of urban development, and thus sustainability.

Promoting urban creative actions is interpreted here in a systemic and multidimensional perspective: in particular, as the capability to go beyond entertainment, fashion, arts, theatres, and so on, in order to improve the living environment in different dimensions, producing new relations and thus spaces of cooperation in the economic, ecological and social fields. Ultimately, city resilience depends on the density of existing and innovative relationships.

Creative initiatives are all characterized by their capacity to unite, in a new synthesis, multiple elements that are generally considered as separated and distant, thus spreading benefits. To face the current crisis and unsustainable city growth, innovative governance is able to minimize human/social/ecological costs, producing new values/wealth through creative actions.

Sustainable strategy is characterized by the capacity to manage the growing complexity of the city as an evolving and dynamic system in constant flux, and solve conflicts with new capacity for synthesis, integrating multiple elements and components, generally considered separated and in conflict/contradiction, and identifying new connections, synergies and relational networks. It recognizes *best practice* as positive experience of change, the basis from which to learn through continuous and rigorous evaluation processes. Good and best practices are examples of concrete urban development actions which are authentically creative, because they combine opposite elements and actors in integrated win–win solutions, able to improve income, employment and environment, and then quality of life.

The approach of creative governance is experimental: it promotes experiments in local development, planning and architecture. They have to be assessed with sound evaluations. Evaluation and monitoring processes are assumed to be fundamental tools for new governance, planning and management, and in particular to promote 'educational coalitions' among public, private and third-sector institutions in order to increase city resilience.

Creative Governance and Experiments

The notion of *creative governance* has been widely examined in the literature (Balducci 2004, Healey 2004, Kunzmann and Sartorio 2005 and others).

Kunzmann (2004) underlines the need for a different kind of governance for metropolitan, medium-sized, and small towns. Hopkins (2008) proposes the perspective of city 'transition' governance, while Rotmans and Loorbach (2008) suggest a 'reflexive' governance to manage city changes. In any case, the governance of the city, as a dynamic complex and adaptive system, is oriented to manage this complexity by maximizing new economic and meta-economic values by means of stimuli to all value-creation processes. To be really efficient and effective it has to be characterized in a constructivist, participative, open to experiments, reflective and adaptive perspective. The listening capacity must be integrated with persuasive communication, argument, and coherence.

In particular, new governance is characterized by a process of searching, experimenting and learning (Rotmans and Loobach 2008). It orients, stimulates, influences choices in a flexible way on the basis of good reasons deduced from experience, pilot projects and best practice evaluations, and all are offered in a public debate (see below).

The more governance becomes based on an experimental approach, the more a great knowledge from experience is needed, that is, a well-founded evaluation capacity, from *the bottom up and the top down* (Figure 3.2). Power, political will and leadership allow breakthroughs to be made in routine management, assuming

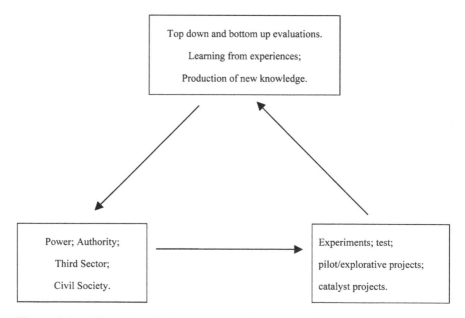

Figure 3.2   The triangle of power, experiments, and evaluation

choice risks that have to be supported by deepened and broadened knowledge (deduced from experiments). Three different levels of governance can be identified to manage the transition towards new organizational city assets: strategic, tactical, and operational (Fusco Girard and Nijkamp 1997, Rotmans and Loorbach 2008). They require different types of knowledge and evaluation processes.

Creative governance concerns four different fields, in order to face four different issues in a integrated and systemic approach:

- innovation in the institutional field to improve democracy at the local level: good institutions and good legislation are the basic conditions required to build the general milieu to make the city more resilient;
- the fight against fragmentation, inequalities and poverty, leading to city cohesion;
- the promotion of a new local economy and financial base, to restore eco-systems and reduce climate destabilization, through new technologies at micro- and macro-level, able to decouple wealth production from negative environmental impacts;
- urban planning (from reinventing the role of the city with a 'strategic vision' to eco-friendly planning and sustainable architecture; from the regeneration of neglected places, port areas and waterfronts to the regeneration of degraded peripheries and brownfield sites, together with historic city centre regeneration) and architectural design.

Innovation in the Institutional Field

The Subsidiarity Principle and Local Democracy

Besides the ecological and social crisis which combine with the economic one, there is a further emergency connected to the democratic organization of city life: namely, the complex relationships among citizen (civil society), private institutions and public bodies, and their different power. The management of power relations is an issue which concerns *governance*. Participation is a key attribute of governance oriented to redistribute power. While there is much talk about citizens' participation, indifference grows, and so does their 'distance' from the 'public thing', their resignation and mistrust. There is an increasing crisis of trust between citizens and institutions, which determines a fall in the cooperative capacity and democratic control of power.

Without people control, there is neither good government nor good governance.

We are used to discussing creativity applied to the dimension of art, to technological innovations (use of information and communication technologies (ICT) in greening the economy, in offering innovative e-services, etc.), or to economic competitiveness. But a more important application of creativity is in the institutional field, to promote community: trust and citizenship and self-organizational capacity, which all reflect and enhance systemic resilience. Innovative actions are required to implement the 'subsidiarity principle' (EU Maastricht Treaty, art.5), in particular in the 'horizontal' dimension, to reduce 'distance' between people and institutions, assigning resources and responsibilities to the lowest feasible level of government.

New institutions (North 1990) are needed, to guarantee to all the 'right to the city' (Brown and Kristiansen 2009) and to regenerate democracy (which is degenerating into oligarchies), by integrating representative with participative/deliberative processes, and promoting responsibility and people's self-government. All this is easier at a local/city level, through decentralization, founded on good and enabling legislation that promotes *human* sustainable city development. Decentralization makes city more resilient. Local community becomes the collective new subject to promote self-government of local common goods, through their self-management.

Evaluation processes by experts and by people should be promoted as a key tool for building decisions concerning *human* sustainable development in cities, enlarging the involvement of the different (public, private and social) subjects in new creative and cooperative networks, modifying choices/behaviours and producing social capital to make democracy work.

Tools for Participation

Another important attribute of local-level creative governance processes is the promotion of people's participation (real and not rhetorical) in the construction of public choices. This perspective is linked to the capacity to open up public choices to the third sector by involving associations, voluntary organizations, non-governmental organizations (NGOs), non-profit enterprises, social/cooperative enterprises and foundations. The third sector is able to produce relational values; it becomes a producer of sense, and meanings; it is not 'against', but it is 'for'; it is interested in long-term objectives and in intrinsic values, in common goods; it is connected to a particular geographical area; it is a bearer of creativity. It is a natural ally that sustains all initiatives, experiments/tools toward resilience and sustainability.

The Participatory Budget (introduced in Porto Alegre, and then in thousands of other cities) is an interesting example of creativity in the institutional field, aimed at democracy regeneration, starting from the neighbourhood/municipality's identification of priorities. It is an innovative tool able to stimulate coordination of actions, cooperation, trust, citizenship and, at the same time, the satisfaction of private needs, starting from poor. It can be applied with success in budget-constrained city construction as well as in urban management and planning (Simonter and Allegretti 2009), and in the integrated conservation of cultural heritage. It is open to poor people's involvement too, so as to encourage new relational networks.

Microcredit is a tool which is able to promote social inclusion, cooperative values and trust: it is another example of creativity applied in institutional fields, to reduce poverty and build relationships. It enables the achievement of both public interests and private ones (loans) at the same time.

All deliberative forums are examples of new democratic institutions that can integrate existing representative institutions so that long-term goals/objectives of development can be sustained in the most effective ways. They are founded on reasonable argumentation and convincing assessment processes, and not on merely voting or negotiating (Zamagni 2008, Zamagni and Bruni 2003).

In summary, the above-mentioned processes and tools, which are both effective and supported by the third sector, stimulate the capacity of self-organization, self-government and self-sustainability, thus increasing the intensity of social and civil networks. They determine the 'city resilience' and the capacity to identify new city trajectories of development in the global competition, stimulating participative democracy approaches founded on places.

Towards the Self-sustainable City

Many cities, from Barcelona to Malmo, from Seattle to Copenhagen or London are adopting plans to manage the impacts of climate change. New green areas, new green architecture, and green energies are being integrated into urban areas to improve the ecological city resilience capacity. This approach is much more effective if it is integrated to improve economic/social and *cultural* resilience too. Cultural resilience is the notion that tying sustainability to creativity stimulates an *autopoietic* attitude: that is, self-sustainability.

Self-sustainability requires decentralization, participation and specific organizational rules. The central role of knowledge and culture should be stressed (Duxbury and Gillette 2007), investing in cultural networks. School and University are strategic investments to enable each individual to transform data into information, information into knowledge, and critical knowledge into wisdom (Zeleny 2006): in other words, increasing the *cultural resilience capacity* of each subject means modifying the way of thinking towards a systemic, multidimensional and critical perspective, attentive to interdependences, deep connections (to wholeness and unity), the long term, general interests and intangible values.

All the above involves a way of thinking that is open to new priorities from among hard (economic) and soft (aesthetic, ethical) values; and able combine utility, beauty and fairness in current choices. Different actors of civil society can cooperate too, in transforming cultural values into civil ones.

ICT can also help to improve cultural resilience if it is used in creative way, to transfer not only information but also knowledge, meanings and sense.

Towards New Partnership Networks

Effective governance promotes virtuous market competition among different subjects, through specific incentives (audit certifications, brand names, awards, etc.) and ties, relations and interactions of a cooperative/coordinative nature. All these improve resilience and creativity. The first cooperative relations are among public institutions, private companies and civil society (third sector).

A significant task of public institutions is identifying *clear* goals, rules, and constraints in the perspective of the long term, so that enterprises can plan programmes, also in the long run, for new value creation chains. They should leave the market free to choose the more effective, innovative and competitive (technological) solutions, opening and sustaining new markets niches (through procurement rules, incentives, grants).

New platforms of systemic integration among companies, public bodies, research, financial institutions and professional organizations are required. New governance should build a *neural network* of connections, for the regeneration of the urban economy and for the reduction of unemployment, differences and poverty. Such public governance is oriented to promoting a suitable environment, to multiplying collaboration opportunities by linking different subjects. Networking is the most effective form of organization, to stimulate innovation and overcome existing constraints.

The Triple Helix model (Etzkowitz and Leydesdorff 2000) is now being proposed by local/regional institutions to link scientific incubators, parks, research laboratories, public and private financial institutions, and professional services in specific spaces.

A great effort should be made to invest in university research centres of excellence. These networks that interconnect private, public and social subjects can accelerate communication, stimulating the production of a new kind of knowledge, thus creating new added values and economic wealth.

They presuppose and, in turn, determine a capability to coordinate the actions of the various subjects involved in a specific project, thus multiplying the occasions for creative interchange. The denser the networks, the more possible it is to build a milieu that frees new energy and ideas. The creative city itself becomes an 'innovations incubator', in which different subjects are put together in space, and through dense networks gain a competitive advantage with respect to others, with positive impacts on employment (particularly of young people: the 'new poor').

The success of these cooperative networks and their continuous regeneration, in a strongly dynamic, interactive and chaotic context, depends on the density of social capital. Trust fosters the coordination of actions by different subjects and, therefore, the multiplication of the networks themselves. It is founded on concrete and repeated experience of actions, interactions and cooperation. On the other hand, networks, in their turn, reproduce trust.

Creativity in the Fight against Social Fragmentations, Inequalities and Poverty

Creativity and Ethics

There is another important perspective here: creativity as a tool in the struggle against poverty. Creative initiatives can use culture, in its different forms, to reduce poverty: see, for example, the role of regeneration of cultural heritage in attracting new activities and thus producing new jobs. Social 'disintegration', which is particularly evident in big cities (in their distressed and peripheral areas), is another aspect of poverty. It concerns the ever more fragile relational life, the loss of social linkages, the disintegration of common-life spaces, meaning the decay of the aspects that *link* people. In a fragmented city, both safety and quality of life are reduced. In the city people are free in their loneliness, living together without really meeting (Bauman 2003).

International migration increases difficulties in relationships between local and immigrant communities. Employment is the bridge between each human being and the others. Unemployment destroys this relationship. The above-mentioned phenomena reduce *city resilience*, i.e. the intrinsic capacity of the city to overcome crisis and to build hope (Newman et al. 2009).

Creative initiatives should rebuild the sense of community, which is the key element for improving resilience and promoting cooperation and thus sustainability. Without *community* there is no sustainable city development. Creativity is needed: to reduce marginalization, unemployment and illegal employment; to create new 'decent' jobs, in particular for young people; to transform young and poor people into entrepreneurs through access to credit (De Soto 2002, Yunus 1999 and 2003); and to build decent houses, infrastructures and services also for deprived people (see the goals of the United Nations (2000) *Millennium Declaration*), also through self-construction processes. Since the growth rate of distressed areas is increasing globally, the perspective is not the sustainable city, but 'slums without city' (Chatterjee 2002).

The Role of Civil Economy in the Fight against Inequalities and to Improve Social Cohesion

The civil economy sector (Zamagni 2008) consists of *non-profit* organizations (that coordinate economic factors in an efficient and competitive way to meet meta-economic civil-social objectives, without considering profit maximization), foundations, social and cooperative enterprises), ethical banks and ethical finance institutions, LETS (Local Exchange Trading Systems), time banks and fair commerce associations. They are all able to increase cultural resilience because they produce, in their exchanges, virtuous circular processes: social responsibility and public spirit. In a word, they replace the social capital that makes the economy and democracy work.

Civil economy provides relevant support in the fight against poverty and social fragmentation, because it contributes to the production of both social services and goods. It stimulates circularity processes and employment. Therefore, new governance should allow niches of economic organizations that are alternative to capitalist ones, enlarging and increasing them in the direction of 'economy civilization', that is in the direction of the 'economy of relationships' (Genovesi 1766). A new relation between profit and non-profit sectors can stress the production of meta-economic values in economic value creation processes: trust, care of common goods, social capital, social cohesion and resilience. All these can strengthen competitive capacity, as many best practices confirm (Silicon Valley, Baden Wurtemberg, etc.) (see Chapter 15 by Ravetz in the present volume). At the same time, a diffused civic culture improves urban cohesion and the sense of community, guaranteeing the resilience and adaptability of urban complex systems (that without trust might collapse under the strong pressures of change). In this way, the new proposed organizational assets in urban/regional systems, which are necessary, can be sustained by a bottom-up approach in order to be really implemented over time.

Human sustainable development requires a strong involvement of this sector, because it is the bearer of the *common good economy*, that is, a social exchange economy that is attentive to independent-of-use and existence values. It is a natural supporter of innovative and pilot projects, whereby the subsidiarity principle is implemented. This involvement (also in management plans) valorizes innovations in both the institutional and the economic field (see the section below), reducing poverty and inequalities.

Creative Initiatives to Face in Economic Challenge: From the Industrial Economy to the Knowledge Economy and to the Urban Ecological Economy

The Knowledge Economy Perspective

The New Economy is knowledge based (Gibson and Klocker 2005, Hospers 2003, Van Oort 2004, Yigitcanlar et al. 2008). The production of media, movies, videos, digital pictures, music, software, general and specific services, advertising, art performance, visual art, graphic art, fashion industry, interior design, and so on, is to acquire growing importance in urban economics and job creation as opposed to traditional industrial production (Cooke and Lazzeretti 2008, Franke and Verhagen 2005, Kong and O'Connor 2009). An increasing level of knowledge is being embedded into goods and services.

The above-mentioned process has always taken place, but it has never been characterized by the current weight of immaterial elements compared with material ones. Continuous creative capacities have become the engine of the new economy, and in particular to overcome the traditional trade-off approach. It is not (always) necessary to choose between, for example, the achievement of an objective of economic performance and the renunciation of social or ecological objectives but, with creative design, it is possible to achieve both simultaneously (Zeleny 2006 and 2009). Productivity, competitiveness, and attractiveness of cities and regions are improved through innovations (Florida 2005). They rely on local resources, that is,. on both human and social capital.

Human capital produces new ideas through knowledge, research and its regeneration.

Social capital multiplies communication and relationships among different subjects, and then it stimulates the production of new ideas. The integration of human and social capital increases an area's creativity potential, and it represents the starting point of self-sustainable endogenous development.

Cities all over the world are investing in cultural infrastructure as a catalyst to sustain local development and employment: science parks, cultural districts and hubs (Sacco and Ferilli 2006). Schools, universities, and research institutions are becoming the main investments for developing new knowledge and for transforming knowledge into actions: entrepreneurial capacity, self-employment, and so on, have become fundamental. They are cities' real wealth and replace the traditional urban economic industrial base (Hall and Pfeiffer 2000). Ancient heritage can also shelter incubators of innovations. Strategic local planning for culture (Barcelona Institute of Culture 2006, UCLG 2004) is a tool that can better coordinate all cultural initiatives in an effective network.

Cultural districts, as innovative initiatives, nurture creativity and move in a series of different steps towards innovations in products and services, regenerating the economic urban base.

Towards the Ecological Economic Base Integrated in the Socio-territorial Context

At the same time, cities have to face the unprecedented challenge of climate change in the context of the financial and economic crisis. All this requires new strategies and new solutions. Cities have to transform their traditional industrial economy into a new perspective, founded increasingly on urban ecology, so as to become more resilient and sustainable. The urban ecological economic base is the new perspective. It promotes eco-business through eco-profits, eco-efficiency ('factor 4' and 'factor 5': Von Weizsacker et al. 1998, 2009): clean technologies, solar industry (thin film solar, fuel cells, electric cars, etc.) with new green jobs (Costanza et al. 1991, Martinez Alier and Roca 2000).

In ecological economics the economic system is considered as a subsystem of the ecological system; the values of services produced by nature, in monetary and non-monetary terms, are recognized and assessed. Attention is paid to the distribution of wealth and not only to wealth production; and to resilience; and to load capacity.

Creative urban actions should promote industrial clusters (Porter 1998), in which wealth is produced to conserve environmental resources and quality, by specialized and interdependent activities in the energy field, transport, construction and the production of new and recycled materials. The *green economy*, characterized by clean energy, resources and water recovery systems, waste management and technological systems, applies ICT to energy and pollution management, to purification systems, and so on (UCLA 2006).

The green economy (which is rooted in development and use of products and services that promote environmental protection) stimulates the 'circularization' of the economy and thus innovations in the urban metabolism. Kitakyushu, Minamata and Kawasaki eco-towns are innovative experiences in the circular economy in Japan. The reduction of economic circuits at a local and regional level which reinforces the chain of value creation (in the construction and the food industry, etc.) is a well-known aspect of the new urban ecological economy. More in general, it is characterized by environmental high-technology industries, with networks of small and medium-sized enterprises which produce with low environmental loads new (and traditional) goods and services, by recycling and reusing materials, water, waste and energy. The circular economy cannot be implemented without innovations/creativity. The creative initiatives leading to the eco-industrial city not only produce new goods designed to be easily recycled: new added value and new jobs are created, while reducing pollution, in production/ maintenance, regeneration equipment for glass, iron, steel, plastic, aluminium, paper, rubber, wood and energy (Washington State 2007). City and industrial areas become really integrated because urban waste is transformed into inputs for the local industrial system, exchanging materials by-products, water and energy in a win–win perspective. The city (and in particular the port city) imports the industrial and urban waste for recycling/regenerating and exports products after biological, chemical and mechanical processing.

Eco-buildings and green industries, through industrial ecology approaches (Ayres and Ayres 2002, Ayres and Simonis 1994), can minimize impact in each step of the product life-cycle and the productive processes, reducing the total amount of waste, as well as reusing waste and products as raw materials for new production.

The new recycling and energy technologies are the entrance points to local economic development as they stimulate a distributed/small-scale, polycentric organization and decentralized models, with new networks. New technologies, required to improve the competitiveness of companies, cities and regions, have to be really compatible with nature, and then able to decouple economic growth from the production of negative environmental impacts (Hennicke 2004, Wuppertal Institute 1997). Hi-tech and science parks (focused on nano- and biotechnologies, digital media laboratories, etc.) are the core of new industrial clusters which should lead to sustainability, and which are to be promoted through public–private partnership. Green commerce, sustained by the third sector, integrates the city's ecological economic strategy.

A strong stimulus to urban creativity is represented by new energetic constraints: the increasing scarcity of fossil energy and the need to refer to new renewable energy sources (UNEP 2008). Reducing the impact that cities have on climate stability – the production of greenhouse gasses (GHG) – is the great challenge of the coming years (Droege 2006). The transition to a zero-carbon economy will be implemented by the adoption of technological innovations that will have a strong influence on the physical and spatial structure of cities, on their organizational architecture, and on their typology and morphology, considering that the particular kind of energy used has always had a strong interdependence with the physical-spatial assets (Droege 2006).

Continuous innovation is the core characteristic of the image of self-sustainable enterprise, which is able to redesign itself and improve its performance. As a complex dynamic system, green enterprise can adapt to changes through innovations, and learning from experience through evaluations, thus becoming more resilient.

Green co-evolutive enterprise minimizes waste, by reusing, recycling and regenerating materials through the new metabolism that stimulates, in its turn, new green activities that then become integrated in the territory. It has a recognized identity that produces an internal and an external bundle of relations and partnerships with workers and stakeholders, as in the models proposed by: Thomas Bata with the 'Bata management system' (Bata 1992); Adriano Olivetti, with the 'Community development approach' (1952); and Kyocera, with 'Amoeba management' (Ishida 1994). All these models are oriented to the organization and management of the *whole system*, including social infrastructures (housing, recreational, and educational facilities). Here, the economic budget is integrated with the social and the environmental budget.

The set of initiatives that leads to a new urban green economy provides more opportunities to combat unemployment, social differences, poverty, and marginalization.

Innovations in Financial Policy

Available funds should be employed to sustain more research and innovations on specific and concentrated initiatives. Foundations and, in particular, Bank Foundations have an important role in stimulating innovation processes, integrating resources from private donations.

In the initial stages, innovations in urban, financial and fiscal policies are required to sustain the transition towards the green economy: these concern tax incentives (allowances, abatements, etc.), grants and loans, generating revenue through cap and trade, and so on. Taxation on income should be reduced but increased on fossil-fuel consumption and on negative environmental impacts (carbon, gasoline, landfill taxes). Local institutions should benefit from additional fiscal resources coming from increased property values following creative initiatives.

Existing project-financing procedures are not effective, considering the high level of risks. More specific venture capital should be available. Many financial resources are necessary to implement creative initiatives and sustainable innovations: more venture capital is required, because often incentives are directed to sustaining existing activities, rather than stimulating innovations. Specific financial tools, absolutely necessary in order to involve the private sector (as in ESCo (Energy Service Company) proposals), can be tested in pilot projects that allow business, while simultaneously respecting the environment and combating poverty. Civil society should propose self-financing initiatives for specific projects.

A best-practice database is very useful for learning what has worked better in designing partnerships and voluntary agreements between the private and the public sector, allowing a reduction of financial resources and releasing funds to other destinations.

The third sector (involved in the process of the 'civilization' of the economy) can help in the above perspective, as it has an active role as collector of resources from foundations, ethical banks, local banks, and voluntary and charitable associations. New private–public and social partnerships are increasingly required.

Creative Initiatives in Urban Planning

Planning, Resilience, and Sustainability

An important field in which to implement innovative governance, that can tackle the four above mentioned issues, is urban planning.

The approach to the city as a complex dynamic and adaptive system (Bertuglia and Labella 1991, Portugali 1997) suggests considering urban planning as a process for organizing resistance to growing entropy – due to suburban sprawl, degradation, deindustrialization, and spatial segregation – nurturing resilience through rational and creative choices, so as to really move towards sustainability. It aims to reproduce order in a context of increasing disorder, by reducing vulnerability and risks; opening new opportunities for *all* people, adapting the city to new circumstances and forces; providing resources (land, space, energies, etc.) to respond to changing pressures; and reorganizing all its assets. Urban planning can contribute to a more effective, adaptable (resilient) city, reducing the ecological footprint and improving the capacity to better co-exist and co-evolve with other systems. A good comprehensive environment enhances human and ecological health conditions and also the capacity to attract investments and therefore to compete better. In particular, innovative solutions allow more socially and economically inclusive alternatives to be found, that is, strengthening the community and reducing inequalities, such as the increasing spatial segregation of immigrants.

Planning means not only embellishing the urban scene with new quality architecture, but also regenerating life in the streets, squares and neighbourhoods, valorizing local resources (land, heritage, landscape) by selecting new activities and use values for the city space, in order to create new economic/market values: new wealth.

Landscape is being incorporated more in urban planning and in economic strategy, as an engine of local development (UNESCO 2010). The regeneration of distressed, peripheral and slum areas is a priority in contemporary city planning, to combat the high growth of urban poverty. Creative urban initiatives start from the reconstruction of landscape in distressed peripheral areas, in abandoned industrialized sites, in port areas and in ancient city centres: a quality landscape attracts new services and investments and stimulates new partnerships and the transition to the new ecological economy.

Micro-creative activities, sustained through effective incentives, can reduce the role of the informal economy in distressed areas.

Permaculture principles (Holmgren 2004) about the importance of maximizing connectivity and relationships among systems, identifying holistic solutions, avoiding waste (considered as an indicator of poor design), and enhancing diversities are good guidelines to connect and reconnect peripheral, distressed and slum areas to the urban central tissue, not only with physical infrastructures but also with new economic activities and relations.

Urban planning aims to achieve a set of goals, which combine public and private objectives. Enterprise Zones, Empowerment Zones, Enterprise Communities, Zone Franches Urbaines, Contract of Quarters, Voluntary Agreement are just some examples of tools developed in the last few years to stimulate – in specific areas – new economic activities, through innovative relations between the private and the public sector. Such tools should involve the private sector to achieve the public interest in a contractual perspective. Evaluation methods are assuming a central role in comparing public interests with private ones, in search of the most satisfying balance. Through evaluations public bodies can transfer a percentage of land rent surplus or areas from private agents, to produce public infrastructures and services implementation. Simulations using global information systems (GIS) are useful for comparing new alternatives in the design and negotiation process, to choose the most satisfying solution.

Creativity in planning is also required to reduce urban rent in central areas: to stimulate the localization of innovative initiatives and then to attract new activities, employment, investment, population, tourists, avoiding having concentrations of people in single poles and multiplying polycentric assets. Specific areas for creative activities are going to be localized in the city through incentives and grants, particularly for young people.

More in general, city planning needs the cultural resource of 'creativity' (Ache 2000, Albrechts 2005, Hall 2002, Healey 2002, Kunzmann 2004, Sartorio 2005, Stein 2005) in these different steps and levels:

- to build a more attractive 'image' of the future city, a strategic '*vision*' founded on the 'spirit of the city', that reinvents the role of the city which, in its turn, fosters the inhabitants creativity; The 'vision' expresses the creativity of a city, as long as it combines in a synthetic way its historic identity with change, old and new values, rationality and emotions, conservation and development (strategic level);
- to identify the most inviting paths in order to implement this 'image' or '*vision*' over time (tactical level);
- to implement these paths concretely by means of experimental projects, with new rules and financial economic incentives (management level).

Spatial planning is becoming increasingly oriented by culture (Evans 2001 and 2005, Freestone and Gibson 2004, Hawkes 2001, Scott 2000): cultural districts, cultural industries and creativity incubators are proposed to catalyze economic development, thus sustaining urban cultural policies. Mega-events, such as International Expos, Forums, Exhibitions, Festivals, etc., localized in specific urban areas/spaces, are the stimulus for urban regeneration strategies, to enhance urban pride, trust, attractiveness.

Urban planning should also pay more attention to improving cultural resilience, thus guaranteeing sustainability.

Participation processes involving people in making urban spatial choices, through specific evaluation approaches and tools, helps to build citizenship. For example, the Participatory Budget can be implemented in designing new spaces, regenerating the existing ones, and formulating cultural heritage site management plans. Urban planning is becoming the entry point to improve democratic participation, and to stimulate urban democracy, through ITC tools. These tools can open new communication possibilities for city futures, offering information and knowledge. They become the most important infrastructure of the 'intelligent city' (as in the Songdo City Masterplan in Korea), and to spread services in the countryside.

Some Best Practice

Urban planning has to support the new city organization, founded on ecological principles that consider the city and the countryside together as a unique living system, subject to continuous fluxes of inputs and outputs. It can increase the capacity of the system to absorb negative impacts, enlarging natural capital and biomass, and sustain the circularization processes of the urban economy, leading to the co-evolution of city and nature (McHarg 1969, Register 2006, Soleri 1969, 1971 and 2006).

Creative planning reproduces '*places*' as spaces or new 'poles' where relations, bonds, and sense of belonging can be rebuilt. The polycentric model, the eco-village model, and the eco-town model are increasingly characteristic of the new urban planning with decentralized, small-scale, non-hierarchical organization/

assets. Green spaces/areas/corridors, green infrastructures and green transport are common elements of these models, which can be applied to cut net carbon dioxide emissions and stimulate a new urban metabolism.

Green architecture standards and approaches are now being incorporated into the planning rules of new neighbourhoods and in the rehabilitation of the existing heritage because urban planning is adapting to climate change: it is oriented to achieving climate stability through mitigation and adaptation action plans (see London or Malmo proposals), in line with climate plans (see Copenhagen initiatives).

Urban planning is paying increasing attention to new technologies to manage the transition from the fossil-fuel city to the zero-emission and zero-waste city. Strategic energy plans are becoming the basis of new urban planning: the experience of 'solar urbanism' in Freiburg, in Daegou (Korea), the 'solar city' in Linz, the green city in Hamburg, the plans of Masdar City (in the UAE) and Dontang City (in China) are inspiring new perspectives for urban morphology.

Stockholm (green capital), Hamburg (green economy city: see the Hafencity experience etc.), Barcelona (Forums city), Vienna (see for example Eurogate areas), Zurich (with new eco-quarters), Amsterdam (Zuidas eco-quarter, etc.) are all examples of innovative approaches in planning for a new urban ecological economy (Berrini and Colonnetti 2010). Urban planning, oriented to make the city more liveable and sustainable, supporting the restoration of ecosystems with economic wealth production and social goals achievement, produces new use values and strengthening independent-of-use values.

Innovative tools to finance new public spaces through private capital are currently being introduced in urban planning. Transfer Development Rights are a tool to realize urban parks, landscape conservation, and so on, in the transition to the post-fossil-fuel city, combining private and public interests. Compensation (Forte 2005, Forte and Fusco Girard 1997) and 'credits to build' (Rallo 2009) are other tools in the public/private partnership which might help to achieve, through agreements and contracts, a satisfying compromise between public and private interests. For example, social houses and social infrastructures are increasingly being financed by means of an 'exchange' between public and private values/ benefits. Reduced rents are offered to young people in exchange for services to elderly persons (as in the Alicante innovative experiences).

The urban transport system is to be reshaped to face new challenges. Transportation infrastructure and ICT are important sectors in which cities are investing to improve neural urban networks, to connect areas, activities and people in a polycentric dynamic model. Light rail, cleaner-air vehicles running on biofuel or hydrogen-powered fuel cells, rapid bus transit systems, optimizing traffic timing, etc. are being introduced in cities such as San Francisco, Portland, Stockholm and Jakarta to reduce travel time and emissions. Curritiba, Bogotà and Reykjavik are other examples of creative best practice in the transportation sector (UN-Habitat 2003).

The Role of Places

Urban planning is also faced with human/social ecology reconstruction: rebuilding social bonds, sense of community, social capital, and thus resilience. The new urban planning is focused on priorities such as production and the regeneration of *public spaces*, as specific areas of identity, social exchanges and life. They can also have a relevant role in managing relationships between the local community and immigrants.

'Places' are a particular kind of public multifunctional space. The city maintains its identity through its 'places' that are signs of its creativity. 'Places' are areas characterized by an extraordinary 'diversity' (different forms, typologies, morphologies, cultures, and traditions). A particular flow of relations between people and stones is maintained in 'places', and this determines their particular identity: the *spirit of places* (genius loci) (Norberg Schulz 2007). This is due to the specific combination of different material and immaterial elements, ancient and new architecture, stones and people. Places as spaces of social diversity are incubators of creativity, fundamental elements for urban community promotion and then for resilience capacity, with their independent-of-use values that sustain use values. These values are recognized by people through interpretations.

In this way, the communication flow becomes easier. Integrated conservation of the cultural heritage contributes to building a creative milieu, originally theorized by Jacobs (1969). The regeneration of the cultural heritage in ancient city centres is a 'creative' urban initiative, as long as it is implemented by interpreting the 'spirit of places', and transforms them into a new landscape, able to attract new activities and to improve economic competitiveness – by the production of new goods/services sold outside the area, so as to reduce unemployment and poverty, and foster better living 'together'.

A metabolized *spirit of places* becomes the engine of local development. Innovative relations/connections exist between ancient city centres and new degraded peripheries, which are the negation of a 'creative milieu'. The success depends on the capacity to link centre areas, peripheral quarters, dismissed areas and surrounding territory into a systemic perspective (through material and immaterial infrastructures) including their regeneration into comprehensive economic circularization processes.

Promoting Creative Architecture

Creativity and Architecture

Architecture is the last step in the implementation of urban planning. A strong relationship ties architecture to the city. Creative architecture should consider using innovations that have been introduced to conserve the environment by reducing waste and energy, and by improving the quality (beauty) of spaces, together with the accessibility of the worst-off people to a decent house and services. Creativity in upgrading existing assets consists of combining ecological objectives with environmental/aesthetic ones; with cost reduction, by using local materials, natural light, natural ventilation and recycled water; and with access to micro-credit and to available low-cost technologies.

Frequently architecture is trusted with the solution of urban problems (the 'architecture solution'), as if spectacular architecture embellishing the urban scene could really solve complex city problems. Architecture can contribute to regenerate urban economies in the globalized economic competition. If the brand is certified, there is development (Klingmann 2006). Architecture can play a strategic role as a catalyst for urban transformation: it is a great innovation yardstick. It can differentiate one place from another, it proposes icons; it can create a brand. The brand of the city is closely linked to its architecture, as experiences in cities such as Bilbao, Dubai and Shanghai can show.

Really creative architecture not only incorporates all new technologies to become connected with all (commercial, economic, social, health, energy, etc.) services, but is also able: to produce spaces that stimulate new systems of relations, and a sense of community (Soleri 1969); to link life and stones, people and man-made capital, and individuals and people. Creative architecture is increasingly being used as a strategic instrument to rebuild a local identity in a globalized context, to stress competitive capacity, and to stimulate trust in the future (see also the architecture proposed by, for example: Frank Gehry in Bilbao; Renzo Piano in San Francisco; Richard Rogers in Cardiff).

'Real' creative architecture achieves the best compromise between private and public interests. It preserves, valorizes and rebuilds landscapes, producing new places as spaces of life. This means that *creative architecture* produces 'places', not only scenarios for events, spaces for entertainment, divertissement, amazing shows, where users are involved with all their senses (as the economics of experiences suggests) (Pine and Gilmore 1999, Sundbo and Darmer 2008). It is a source of communication, connection, emotions and relations. In *'places'*, private and public interests, and hard and soft values are integrated. *'Places'*, as products of creativity, become in their turn the 'incubators' of community, creativity and resilience (Forte and Fusco Girard 2009).

Creative architecture is capable of integrating opposite elements: old and new elements, traditions and innovations, inanimate and living elements (green façades, roofs, etc.). It enhances the relationship between public and private spaces, encouraging multiple uses of the land, in order to respond to real local needs; it can foster a sense of liveability, vitality, the continuity of tradition, a sense of community: the perception of connections/relationships. Thus, it can contribute to urban resilience.

Architecture and the Urban Laboratory

Many cities have activated an innovative process starting with the regeneration of their port and industrial areas (often neglected), or with the historic centre. Many creative cities are port cities (e.g. Amsterdam, Rotterdam, Valencia, New York, Singapore, Malmo), which means urban spaces are 'open' to continuous economic, social, cultural exchange. They have stimulated innovations in architecture. Urban waterfronts and port areas are becoming the entry point to their sustainable development strategy, which shows concrete creativity in the production of new and sustainable urban landscapes. Well-known recent examples are port area projects by: Jacques Herzog for Hamburg; Steven Hall for Copenhagen; and Zaha Hadid for Antwerp. Each has a different interpretation of the relationship between ancient and new values.

Internationally, port areas are becoming the spaces where most of the actual creative actions aimed at promoting sustainability are being implemented. Well-known examples are investments in the waterfront regeneration – in Rotterdam, Barcelona, Liverpool, Valencia, Vancouver, Tokyo, Hamburg, Malmo, Amsterdam, Glasgow, Antwerp, Copenhagen, and so on. They can be interpreted as 'transition experiments' (Rotmans and Loorbach 2008) leading to a new ecological economic base. They express the *creativity* and also *resilience* of cities in response to the pressures of change, highlighting the capability of cities to transform themselves, while maintaining their identity. It should, however, be carefully assessed whether such experiments that integrate conservation and development can be repeated on a different scale and in a different context.

Architectural projects in neglected and port areas, considered as relevant practical examples of 'innovative experiment' in the perspective of the eco-city, should be able to stimulate, in their turn, creativity and resilience in the spatial and socio-economic context. They propose an innovative/positive image of the city, able to improve the entire urban system. Port areas are designed not only to produce business but also to achieve general goals, such as the conservation of the cultural and natural landscape: many port and coastal areas are listed as UNESCO sites (e.g. Malta, Liverpool, Istanbul, Bergen, Venice, Oporto).

The functioning of the natural ecosystem is the model for the new architecture so that it can contribute to a new city metabolism. This means that today architecture is also involved with the concept of creativity as the ability to reduce the use of materials, water and energy (avoiding energy consumption by: using natural cooling for air conditioning; taking advantage of natural lighting; taking heat from the earth ...); reusing and to recycling materials; and introducing renewable energies (Kibert 2005). It is cooled and heated by 'nature'.

'Zero-carbon' and 'zero-waste' architecture improves energy efficiency and the use of renewables, with lower operating costs and higher rent rates. In the face of the constraint of conventional energy, architecture often has a double aspect: high aesthetic value but a reduced environmental value, or, vice versa, a high energetic value (zero emissions, etc.) but a reduced aesthetic value (see, for example, the first passive house experiences).

A challenge for new architecture is to combine very innovative and well-designed green technologies with 'beauty'. In history 'beauty' has always characterized architecture, integrating good shapes with functions. Beauty is able to generate a 'force field' that reduces entropy and enhances city resilience, under certain conditions. Which ones?

An Aesthetic Strategy as a Creative Urban Initiative to Promote City Cultural Resilience

The Multiple Values of Beauty

Innovative governance, urban economy and planning/design strategies are effective if they are sustained by people's behaviour, that is, by a specific culture, which is less self-centred and more open to general interests. An 'aesthetic strategy' can be proposed as the entry point to this perspective.

The urban economy and planning, when aware of the environment and landscape, are able to produce order, quality, beauty. The beauty of a natural landscape reflects the harmony of the eco-system. Architecture that takes care of the organization of ecosystems, of the natural and cultural landscape, and that is really inspired by the wisdom of nature, is characterized by *beauty*.

A common characteristic of creative urban initiatives is the attention to the arts. The arts in the urban scene enhance the beauty of urban landscapes, and help in communication. The regeneration of 'places' (historical centres, waterfronts, port areas, cultural landscapes, etc.) is often included in a strategy for the '*beautiful city*', following a model already implemented in Paris in the last century, and then reinvented in Barcelona. A great number of experiences are now spread all over the world. 'Arts plans', 'heritage plans', 'culture plans' are some examples of tools to build actions that lead to creative and beautiful cities (see the experiences of, for example, Ottawa, Vancouver and Toronto), and that stimulate, in their turn, investments in the creative sectors as a priority in the new city economy. The beauty of cities has an economic value arising from its 'attractive power' (Botero 1588) that acts on tourist demand, on many other activities, on entrepreneurs and investments, and on specialized jobs, talents and people. But it has also a cultural, social and civil value.

Beauty is produced by creativity and fosters creativity/innovation. In fact, the quality of sites (beauty) has the 'power' to generate new ideas and innovation (creativity) (Forte and Fusco Girard 2009). Historic centres, which reflect the creativity of past generations, can become the sites of new creative processes, where promoted arts are linked to artistic industry. Here, specific incentives should encourage the localization of artistic production by also using abandoned spaces and stimulating the entrepreneurial skills of artists.

Beauty as a public good can inspire creative actions not only to reshape physical assets (new and old architecture/spaces) or improve the economic capacity to compete but also to enhance social cohesion and cooperation. Beauty is a particular private and public good that is becoming ever more appreciated. It has a subjective value but also a social value, because by increasing well-being and reducing stress, it helps to open up interpersonal relationships: it promotes attention, respect, care, setting in motion specific emotional exchanges.

Towards a 'Civil Aesthetic'

The strategy of city attractiveness through *beauty* stimulates the introduction of artistic man-made assets into the urban scene. Art is important. But it runs the risk of becoming mere 'urban make-up', producing only spectacular embellishment, a fragile decoration (Fusco Girard 1998) if it does not really reduce negative impacts on *everyday life*; if this strategy is unable to produce 'external effects' on the actual life of the city in terms of the capacity to live and work *together*; and if it does not open up less self-centred new perspectives.

In order to be effective, this strategy has to be integrated with actions which are not only connected with architecture, urban landscape, harbours, amenities, art exhibitions, big events, and so on, but which may also produce spillovers on *the art of life*, on the capability to transform each inhabitant into a citizen: into an artist (Rogers and Spokes 2003), and in particular into an '*artist of citizenship*'. Beauty as an intermediate good (between private and public) can stimulate a citizen's behaviour and be able to synthesize conflictual elements. Every action that is able to combine multiple and conflictual components in a new way, and to satisfy needs, is creative.

A specific example is the capacity to combine in choices beauty and utility, beauty and fairness, rights with duties, private interests with more general interests (the common good). This creativity belongs to the civic field and allows self-organization. It is founded on an interdependent multidimensional and critical way of thinking.

People's creativity (Landry 2000) can be built by stimulating each inhabitant to become creative as an 'artist' or an entrepreneur (able to produce new values in a way that others are not able to see and to do), in accordance with the 'civic aesthetic'. It supports, with its consequent different lifestyle, the new public governance development strategies in the short and medium-long term. It fights the causes of climate change at their roots, because it promotes a way of life more consistent with the ecosystem and its health, and, moreover, it stimulates a new demand for more eco-sustainable assets and services. This '*civic aesthetic*' helps to sustain the public spirit and urban participation in order to promote common goods and to choose the goals of common interest and therefore city democracy. It facilitates resistance to urban decline, and then contributes to city cultural resilience. Spaces characterized by beauty can really become spaces of social life and promote the sense of community, as long as this civic aesthetic approach is truly spread in the city. It is linked to the capacity to develop combined priorities in choices on the basis of economic, aesthetic and fairness criteria.

All schools and cultural networks should be involved in this process of 'educational alliance/pact', in particular with the local media and associations of the third sector. The challenge is to transform cultural/aesthetic and environmental values into civil values, and to build a sense of civic pride in beauty treasures (as examples of common goods): to pass from *care and respect* culture to r*esponsibility* culture; from emotional relationships stimulated by the beauty to interpersonal cooperative relationships, towards self-organization. The process, focused on some pilot experiences in 'creativity laboratories', starts by improving the 'sense of place' place' of the historic/cultural urban landscape (UNESCO 2010), in the perspective of specific agreements/pacts in its management. The 'aesthetic strategy' founded on a relational way of thinking and on a relational rationality open to beauty, utility and fairness and to the long term, is more effective if it is linked to critical interpretation and to evaluation culture, to relate multiple aspects in choices. Thus, it becomes able to reduce social fragmentation and to promote cohesion, because it stimulates circular relationships between common goods (intermediate between private and public ones) and people (community).

Creativity, Values and Evaluation: Creative City Initiatives and Evaluation Processes

Creativity and the Evaluation Process

Creativity and choices are closely connected: creativity multiplies alternatives. Different creative ideas generate different new values that need to be assessed. Evaluating means interpreting a general context, distinguishing different elements, foreseeing impacts of new ideas (before) using resources, land, spaces, and so on, and comparing alternatives with some anchor elements. By evaluating approaches, it is possible to deduce priorities among alternatives, considering multiple, multidimensional and conflicting criteria/objectives. Evaluation is necessary for decision-making processes, particularly in a time of crisis, with increasing scarcity of resources and energy, in order to improve governance, urban planning, design, and management.

Evaluation processes are fundamental tools for new governance leading to sustainability, and for checking creative and resilient initiatives. New governance is based on experiments, experiences, and best practice interpretation/comparison by experts, and also by the general public. Evaluation has a central role in the four above-mentioned perspectives, compensating for the lack of laboratory simulations of complex urban phenomena.

Innovative urban alternatives are characterized by high uncertainty, costs and risks. Lack of knowledge is the common element in all creative choices/ actions. Therefore, they require experimental, testing approaches and simulations

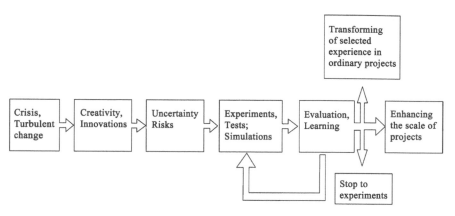

Figure 3.3   Evaluation as an engine of creativity

(see Figure 3.3) in order to learn from their successes or failures and about the specific characteristics of the dynamic urban system in supporting uncertain and/ or irreversible effects (critical capacity thresholds). Evaluations may suggest how to improve experiences: whether to transfer them into ordinary practices or totally change them.

Creative cities have to make greater investment in assessment processes as support for decision-making. The evaluation of the creative potential of a city is increasingly necessary for city/region development in order to properly identify the areas of strength and weakness.

Evaluation processes help to make decisions on 'what', 'where', and 'with whom' in order to implement creative initiatives, and know when to stop them. Evaluations are necessary tools in: different pacts, agreements, and city contracts; participation processes; finance and microcredit; taxation; sustainability focus groups; auditing; choices at a strategic, tactical and management level; and for investigations in general.

Evaluation is a fundamental tool for selecting (more innovative alternative) choices in urban planning and design which can combine many values, producing multiple benefits for many agents. They are to be evaluated in their quantitative and qualitative, direct, indirect and induced impacts, in the short, medium and long term, beyond any bureaucratic or strictly economic approach. An integrated assessment process not only helps to compare 'given' and defined alternatives but it also stimulates the identification and exploration of *new* alternative solutions. So, the evaluation process can become the *engine of city creativity*. An iterative decision-making process is activated through continuous feedback and improvements in the level of the achievement of objectives.

Creative Governance and Evaluations

Governance is based, first and foremost, on a 'good' evaluation capacity (Azzone and Dente 1999, Gibson et al. 2005). Creative initiatives are promoted through an evaluation process that selects among a number of different alternatives (Fusco Girard 2010.

Therefore, new evaluation approaches are required: a creative city promotes the culture of evaluation (Landry 2000, 2006) as a general rule to determine priorities in its actions/choices. Evaluation Offices and evaluation pools are introduced into its organizational structure (Landry 2000, 2006) to stimulate a creative approach, which is able to assess high risks and uncertainties, together with a less formal more in-depth approach.

At the same time, the creative city stimulates evaluations by all actors on the urban scene to understand the ex-ante and ex-post comprehensive impacts of actions, projects or plans (Plaza 2000): evaluation as the expression not only of expert knowledge but also of interpretation by the people. Democracy requires critical evaluations in order to participate in public debates, construct choices and control power. City cultural resilience is enhanced when all citizens have an evaluation capacity.

The creative city systematically collects data and information to improve knowledge for the critical assessment required in concrete urban programming and planning. Data, information and knowledge are to be structured in a systematic way so as to allow for comprehensive evaluations and comparisons of new ideas and their implementation and performance in satisfying spatial needs.

Thanks to 'good' evaluation processes the creative city is able to better interpret its actual context and the alternative reactions to it: it is able to make comprehensive comparisons about near-future scenarios, programmes, plans, and projects, in order to identify the best alternatives trajectories.

The functioning of the city as a complex dynamic system obliges us to assess the complex system of relations by means of a multidimensional evaluation approach that includes economic assessment. The integration between new dynamic multi-criteria analyses, multi-group analyses and GIS, which simulates alternatives, helps to cope with social conflicts, by inserting local parties (existing and new ones) into the common effort made to achieve the city's goals (Nijkamp and Perrels 1994, Nijkamp and van Delft 1977, Nijkamp et al. 1998, Fusco Girard and Nijkamp 1997).

Ex-post evaluations of experiments have to be compared with ex-post assessment of best practice. Ex-post evaluation of *best practice* (and of *worst practice)* gives the possibility of defining better approaches for ex-ante evaluation, able to support new plans, designs and management for the city. Best practice shows that the coordination of actions by different agents, the conservation of the environment and the pursuit of the general interest are also advantageous targets in economic terms (Fusco Girard and You 2006).

Table 3.1   Evaluation criteria and indicators

Economic resilience	Social resilience
• Funding from local foundations and local banks (for local activities)/year	• Increase of community's sense of cohesion (as a reflection of circular processes)
• Innovative public procurement supporting local industries	• Reduction in the percentage of unemployed people living in the area
• Regeneration capacity of economic activities (percentage of economic value created that is invested in innovative activities in the area/total number of activities)	• Experiences of self-organization capacity (percentage of people involved in forums, participative processes etc./year)
• Localization of new creative, flexible and adaptive activities that invest in resources productivity/total existing activities	• Implementation and upgrading of existing 'public spaces' and 'places'(number and surfaces/total)
• Density of networks among companies (number of voluntary agreements, partnerships/year)	• Conservation of elements expressing the area's cultural identity and memory
• Variation of the social sector economy (existing experiments and pilot projects in civil economy, social market economy, etc.)	• Number of events, festivities, ceremonies in the year, as an expression of collective/social memory
• Number of industrial production activities integrated in the spatial and social context (that invest in the circular economy, i.e. in reuse, recycling and regenerating resources)/total	• Average income and wealth distribution
• Investments in innovative research activities/total investments	• Involvement of the third sector in specific programmes/projects/activities (education, health, housing services, civil economy initiatives) and in public/private partnerships
• Number of University spin-offs/year	• Density of cooperative and partnership networks
• Incubators of new clean activities (recycling, regeneration of materials)	• Involvement of local people in urban planning
• Number of green, new energies, clean technologies design patents/year	• Capacity of learning from explorative experiences
• Number of cooperatives or social market enterprises or with innovative sustainable business model/total number of enterprises	• Openness of people to differences and diversities (degree of social diversity)
• Number of green businesses/total number of enterprises	• Level of interpersonal trust (surveys, pools, etc. relating to perception of belonging to a specific community)
• Density of networks among public authorities, enterprises and research centres (number of partnerships that are open to innovation flows)	• Number of donors/10,000 inhabitants.

Environmental resilience

- Reduction of vulnerability and risk levels
- Reduction of ecological footprint (conservation and increase of green areas, tree planting and maintenance, promotion of green roofs and green façades)
- Number of by-product exchanges among enterprises/total number of enterprises
- Conservation and improvement of landscape quality (as a reflection of circular processes)
- Reduced car travel demand and reduction of motor traffic
- Air and water pollution reduction (reduction of CO_2 emissions, etc.)
- Recovery/recycling/regeneration of waste material (percentage of plastic, metals, tyres, slag, cans, glass, paper, etc. reused, recycled, and regenerated)
- Water recycling (rain water percentage recovered, etc.)
- Self-organized city waste management/total waste management
- Percentage of local renewable sources (new electric power plants localization, based on energy innovation) used in productive processes
- Percentage of organic waste recycled (local composting production/year)
- Percentage of activities included in a smart energy grid (to use a variety of fluctuating energy sources)
- Localization of new industries with a low environmental load (that accept the 'zero emission' goal, that have adopted an Environmental management system, and that are ISO or EMAS certified/total number
- Number of modern eco-compatible buildings (LEED or DGNB certified)/total number of buildings

Cultural resilience

- Conservation/restoration of city 'places' (as an indicator of conservation of city identity and of sense of place promotion)
- Conservation/improvement of cultural landscape as a source of identity: public participation
- Symbolic values (rules, traditional conventions, people traditions, rituals, etc.) with which the whole community identifies itself
- Public spaces traditionally considered as meeting places (places where to hold celebrations, to meet people, etc.)
- Existence of specialized professional skills in the area
- Number of associations, NGOs, charities/100,000 inhabitants
- Number of young associations/total
- Number of cultural celebrations in the area/year
- Number of historical events linking specific areas with the city
- Number of persons/families involved in 'fair commerce' networks
- Art production for an extra local market
- Level of safety perception by people
- Evaluation capacity of inhabitants (participation in strategic environmental assessment, in environmental impact analysis, etc.)
- Percentage of people involved in specific urban laboratories
- Per capita investments in culture, research, education and formation/year

The task is to transform assessed best-practice knowledge (linked to specific experiences in space and time) into more general knowledge, to be communicated, transferred, and spread. Considering that creativity is interpreted here in relation to the promotion of economic, social and ecological resilience, specific indicators relating to the density of relationships (in different dimensions) are required. People's involvement in reaching the common interest, social inclusion, sense of community and collective identity become relevant to reflect the benefits of such relations (for example, credit availability from local banks stimulates local development with benefit for financial institutions, in a circular self-sustainable process, etc.).

The research on indicators that consider creativity and resilience (and are focused on circular relationships) is ongoing (Table 3.1, Fusco Girard 2010).

All these indicators in Table 3.1 allow for communication between public institutions and people in selecting alternatives (Bell and Morse 2008) and in sustaining certain choices (for example, in Strategic Environmental Assessment, in planning, etc.). They can be used in assessing pilot or demonstration projects/ experiments through which the subsidiarity principle is implemented.

Some Specific Evaluation Issues

The evaluation of creative/innovative projects is linked to the evaluations of values other than use values: of existence values, and of immaterial symbolic capital; and to intangibles (uniqueness in the organizational structure, cooperative activity, brand) that are becoming more and more important in value creation processes (Bianchi and Labory 2004, Eustace 2004, Zambon 2004). Other examples are the evaluation of networks, 'places', and natural services produced by natural ecosystems.

The evaluation of the innovative capacity of networks among public institutions, companies, research centres, banks, civil society and professionals requires the capacity to assess the complex and dynamic relationships/interdependencies among parties that can reinforce each other, thus multiplying benefits. They reflect different points of view and (shared) objectives. Both monetary and non-monetary approaches can be proposed. A complex set of indicators (able to express the vitality and identity of a place, the sense of community, etc. and, in particular, subjective perceptions) is needed to monitor continuously the effectiveness of coordinated actions in the urban/metropolitan context on these soft, intangible, immaterial values and their role in local development (Fusco Girard and Nijkamp 2009). These indicators enable the innovative management of cultural/environmental heritage site plans, and help in integrated urban design evaluations.

Thus multi-criteria, multi-groups, quantitative and qualitative evaluation tools are necessary to monitor and assess best practice and to propose certification for brand names in order to stimulate emulation and virtuous competition among cities.

At the operational design level, the implementation of specific evaluation methods can be suggested. For example, the LEED (Leadership, Energy and Environmental Design) approach could be extended to urban design and to ancient buildings restoration; and the BREAM (Building Research Establishment Environmental Assessment Method) (Anderson et al. 2009) and the DGNB (2009) (German Sustainable Building Certification) approach are tools able: to stimulate the reduction in the consumption of materials, energy, and water; to reuse, recycle and regenerate; and, at the same time, to promote a new way of thinking in design/planning, reacting creatively to constraints. In particular, the DGNB includes the economic and socio-cultural performances, in addition to the ecological ones.

These tools can be considered as a stimulus to check new creative solutions. They encourage the use of local materials, energies, recycled water (which is cleaned and used again and again), waste, etc. They allow wealth production to be decoupled from negative environmental impacts. An increasing use of the LEED approach to produce a set of global benefits, associated with the role of the construction sector in the economy, has been proposed by cities such as Austin, Seattle, Calgary, and so on.

Conclusions: Towards a New Governance for a Creative, Resilient and Sustainable City

The human dimension of development is interested in humankind's ancient aspiration to achieve happiness. This refers not only to income or a green environment but also (and first and foremost) to *relationships*: it is centred on human beings and their multidimensional relationships. The density of relations is a key element of urban resilience, together with the intensity of creative capital, to promote *human* sustainable development, in an ever more fragmented, uncertain and unstable transition phase. Resilience requires innovations to face new challenges and helps to achieve sustainability. Ecological, economic, social and cultural resilience modifies actions, interactions and evaluations at all (macro- and micro-) levels (Zeleny 2010).

The *human* self-sustainable city requires creativity in an integrated perspective, linking, for example, adaptation and mitigation action plans to face global change impacts (improving ecological resilience) to the reduction of unemployment, marginalization and spatial segregation, and to opening up new opportunities in urban economic attractiveness. The fight against climate change should be always intertwined with the reduction of poverty/segregation, and with wealth creation, in order to improve the comprehensive resilience of the city. A creative city, richer in the cultural sense, in its relational and intangible values, is able to face crisis from within, because it is vital and adaptable. It is capable of self-organization, because it is resilient. New governance is required to stimulate relations in order to better live *together* by reducing conflicts and increasing the quality of life for all people (rich, poor, local, immigrants).

Urban governance is really creative if it is able to manage the city as a complex dynamic and adaptive system and to achieve, in a win–win perspective, economic competitiveness, ecological conservation, and social/human costs reduction, because it produces/reproduces the city's resilience; if it is able to stimulate non monetary exchanges (such as social, cultural, symbolic ones) as parallel circular processes in creating added value, in promoting community and reducing poverty.

A set of general principles can be proposed for managing the transition toward a creative and resilient city. They can be summarized as follows:

1. *The co-evolution of city with nature and its cycles*: This 'relationality principle' is based on the use of technologies compatible with nature. It is characterized by a new relationship between nature and artefact (the 'harmony' with nature) (Register 2006) through low environmental loads technologies; reduction of materials, water and energy waste, and emissions; reuse of resources/materials; recycling/regeneration of materials and resources (plastics, steel, glass, aluminium, copper, etc.); use of renewable energies (see the good practice of Barcelona, Copenhagen, Reykjavik, Vaxjo, etc. in cutting carbon emissions) in order to sustain a *circular* urban economy that leads to a new urban '*ecological economy of relations*'. The contribution of the private sector is a necessity. Enterprise is the place where innovations in economic, social and environmental dimensions can produce the circular fluxes of the co-evolutionary approach, to be then extended to city and region. Co-evolution is also the perspective of the new architecture and planning.

2. *The centrality of culture as a catalyst of creativity* (Nadarajah and Yamamoto 2007):\Culture is the centre of all (institutional, social, economic, spatial) strategies. Culture is fundamental to free existing potential and nurture, in its turn, *city cultural resilience.*

   Cultural districts, creative and research hubs can become innovative investments to develop new knowledge for local development. The reuse, recovery and regeneration of historic cultural heritage plays an important role in sustaining a 'circular economy' of the city, stimulating new synergies. Innovative and cooperative neural networks are required to stimulate cultural resilience: among enterprises, the research sector, public and financial bodies, professionals, and civil society in order to multiply the outcomes and the impacts of investments. Where trust and social exchange exist, systemic resilience is self-produced.

   Culture is the true wealth of the city. It is reflected in the way to live, work, and interact. It is reflected in people-critical capacity. Evaluation, as an exercise in critical thinking, then becomes the essential element for change choices in all fields: production, consumption and public activity. Shared knowledge of good experiments is a source of empirical evidence for introducing innovative actions, employing new tools and assessing processes.

3. *The 'civilization of the economy'* with *involvement of the third sector* in promoting local development. This sustains the circular city economy from the bottom up, because it structurally promotes new virtuous circles of value exchanges. *Human* sustainable city development requires a strong platform that involves the third sector (and in particular young associations, and cooperative networks). The third sector promotes innovative projects that implement the subsidiarity principle and citizenship, which enhances systemic resilience, self-organization of people and thus implements the 'civilization of economic exchanges'.

    Creative governance needs new approaches and tools to be effective in promoting the implementation of the above principles. In the absence of laboratory simulation to analyse complex city phenomena, creative governance proposes experimental and explorative projects in the urban economy, planning and architecture, to be *attentively* assessed in terms of their successes and failures.

New governance recognizes the role of beauty (of the historic urban landscape (UNESCO 2010), of cultural and natural heritage) as generator of a 'force field', which enhances local competition capacity, achieved by improving connections, relationships, partnerships.

Creative governance that multiplies new opportunities might provide a firm basis for reducing poverty, injustice, and disparities investing in culture and cultural heritage. It should help in the building of the ability to reproduce participation, cooperation, trust: in building the 'citizenship art', which is the capacity to combine creatively the private and the public interest. Starting from the consciousness and reconstruction of 'sense of place' a *civil aesthetic laboratories* can be proposed to preserve and manage common resources, cultural heritage for future generations, and to improve the city's cultural resilience, leading to self-organization. Through self-management processes, circular relationships between local common goods and community can be stimulated or reinforced.

Examples of pilot projects concern the conservation of cultural heritage that guarantees the city's memory and identity, and the regeneration of run-down industrial or port areas, in the perspective of an ecological economic base. They integrate advanced culture and research services with the regeneration of materials, renewable energies, bio-architecture and green commerce. They should be linked to the development of distressed peripheral areas and their surrounding territory. These examples concern experimentation areas and pilot/demonstration projects that can contribute to the 'civilization' processes of the current economy. The creative city urgently needs laboratories and sound evaluations using all operative tools to achieve effective implementation.

Continuous evaluation and monitoring processes have become increasingly necessary tools in promoting innovative initiatives: to select really creative actions for regenerating the urban economy and the socio/ecological system; to concentrate scarce resources on innovative initiatives; and to identify their priorities. A list of criteria and indicators to assess resilience has been proposed as a first step for future studies. New governance is founded on 'Evaluation Offices', able to assess local resources (natural and cultural landscape, etc.) that are to be transformed, thus producing new values. The systematic identification of existing *best practice* (in the urban economic regeneration, planning, design, management, conservation of cultural heritage, and access to housing and services) helps the city to understand its position compared with experience of the other cities: its strengths and its weakness.

Sustainable development *best practices* are concrete examples of creative urban actions (Fusco Girard and You 2006). Their assessment helps us to deduce the process and tools (also legislative and institutional) that have allowed positive outcomes and opens up new perspectives and experiences in the economic, technological, civic, cultural, environmental, and institutional dimensions. They show that even in an ever more complex and fragmented context, creative urban actions can stimulate effective systemic integration and synergy. Generalized knowledge production derived from experiences improves choices for change directed towards sustainability, because it allows comparisons.

Evaluations as interpretation, comparison and awareness of the impact of alternatives concern governance, urban planning, design and management. They also concern all people in their capacity to construct wise and creative mediations among multiple and conflicting values/goals. They stimulate participation in public choices, and are tools to implement the responsibility principle. The creative urban initiatives for the building of the 'good city', that is, the *human* sustainable city, are those that stimulate all people to be inspired by a creative spirit.

Bibliography

Ache, P. 2000. Vision and Creativity: Challenge for City Regions. *Futures*, 32, 435–49.

Acs, Z. 2002. *Innovation and the Growth of Cities*. Cheltenham: Edward Elgar.

Adger, W.N. 2000. Social and Ecological Resilience: Are They Related? *Progress in Human Geography*, 24, 347–64.

—— 2007. Ecological and social resilience, in *Handbook of Sustainable Development*, edited by G. Atkinson et al. Cheltenham: Edward Elgar, 78–90.

Adger, W.N. and Brown, K. 2009. Adaptation, Vulnerability and Resilience: Ecological and Social Perspectives, in *Companion to Environmental Geography*, edited by N. Castree et al. Oxford: Wiley-Blackwell, 109–122.

Alberti, M. and Marzluff, J.M. 2004. Ecological Resilience in Urban Ecosystems: Linking Urban Patterns to Human and Ecological Functions. *Urban Ecosystems*, 7(3), 241–65.

Albrechts, L. 2005. Creativity in and for Planning. *Disp*, 162(3), 62–9.

Anderson, J. et al. 2009. *Developing an LCA-based tool for infrastructure projects*. Paper presented at the 25th Annual ARCOM Conference: Nottingham.

Available at: dspace.lboro.ac.uk/dspace-jspui/bitstream/2134/5989/1/ARCOM_ Conference_September_2009_SG_v9.pdf.

Ayres, R.U. and Ayres, L.W. 2002. *The Handbook of Industrial Ecology*. Cheltenham: Edward Elgar.

Ayres, R. and Simonis, U. 1994. *Industrial Metabolism*. Tokyo: United Nations University.

Azzone, G. and Dente, B. 1999. *Valutare per governare. il nuovo sistema dei controlli nelle pubbliche amministrazioni*. Milano: ETAS.

Balducci, A. 2004. Creative Governance in Dynamic City Regions. *Disp*, 158, 21–6.

Barcelona Institute of Culture. 2006. *Barcelona Strategic Plan for Culture: New accents 2006*. Barcelona. Available at: www.bcn.es/plaestrategicdecultura/pdf/ StrategicPlanBCN.pdf.

Bata, T. 1992. *Knowledge in Action: The Bata System of Management*. Amsterdam: IOS Press.

Bauman, Z. 2003. *Voglia di Comunità*. Bari: Laterza.

Baycan-Levent, T. and Nijkamp, P. 2010. Diversity and Creativity as a Research and Policy Challenge. *European Planning Studies*, 18(4), 501–04.

Bell, B. and Morse, S. 2008. *Sustainability Indicators*. London: Earthscan.

Berrini, M. and Colonnetti, A. 2010. *Green Life*. Bologna: Editrice Compositori.

Bertuglia, C. and Labella, A. 1991. *Sistemi Urbani*. Milano: Franco Angeli.

Bianchi, P. and Labory, S. 2004. *The Economic Importance of Intangible Assets*. Aldershot: Ashgate.

Botero, G. 1588. *Delle Cause della grandezza e magnificenza delle città*. Venezia.

Brown, A. and Kristiansen, A. 2009. *Urban Policies and the Right to the City*. UNESCO UN-HABITAT. Available at: unesdoc.unesco.org/images/0017/001780/178090e. pdf.

Carta, M. 2007. *Creative City: Dynamics, Innovations, Actions*. Barcelona: List.

Chatterjee, G. 2002. *Consensus versus Confrontations*. Nairobi : UN–Habitat.

Cooke, P. and Lazzeretti, L. 2008. *Creative Cities, Cultural Clusters and Local Economic Development*. Cheltenham: Edward Elgar.

Costanza, R., Cumberland, J., Daly, H., Goodland, R. and Norgaard, R. 1991. *An Introduction to Ecological Economics*. Boca Reton: S.Lucie Press.

De Soto, H. 2002. *The Mistery of Capital*. New York: Basic Books.

DGNB. 2009. *German Sustainable Building Certificate*. Stuttgart.

Droege, P. 2006. *The Renewable City*. Chichester: Wiley.

Duxbury, N. and Gillette, E. 2007. *Culture as a Key Dimension of Sustainability*. Working Paper, 1, Creative City Network of Canada.

Etzkowitz, H. and Leyderdorff, L. 2000. The Dynamics of Innovation. *Research Policy*, 29, 109–23.

Eustace, C. 2004. The Intangible Economy, in *The Economic Importance of Intangible Assets*, edited by P. Bianchi and S. Labory. Aldershot: Ashgate.

Evans, G. 2001. *Cultural Planning, an Urban Renaissance?* London: Routledge.

—— 2005. Measure for Measure: Evaluating the Evidence of Culture's Contribution to Regeneration. *Urban Studies*, 42(5–6), 955–83.

Florida, R. 2002. *The Rise of the Creative Class: And How it's transforming Work, Leisure, Community and Everyday Life*. New York: Perseus Book Group.

—— 2005. *City and Creative Class*. New York: Routledge.

Forte, F. 2005. *Struttura e forma del piano urbanistico comunale perequativo*. Napoli: Edizioni Scientifiche Italiane.

Forte, F. and Fusco Girard, L. 1997. *Principi teorici e prassi operativa nella pianificazione urbanistica*. Rimini: Maggioli.

—— 2009. Creativity and New Architectural Assets: The Complex Value of Beauty. *International Journal of Sustainable Development* 12(2/3/4), 160–91.

Franke, S. and Verhagen, E. 2005. *Creativity and the City*. Rotterdam: NAi Publishers.

Freestone, R. and Gibson, C. 2004. *City Planning and the Cultural Economy*. Paper presented at the City Future Conference, Chicago, 8–10 July 2004.

Fusco Girard, L. 1998. *Conservazione e sviluppo*. Milano: FrancoAngeli.

—— 2006. La città, luogo di ricostruzione della speranza, in *Ripartire dalla città*, edited by F. Mazzocchio. Roma: Ave, 24–36.

—— 2010. Sustainability, Creativity, Resilience: Toward New Development Strategies of Port Areas Through Evaluation Processes. *International Journal of Sustainable Development*, 13(1/2), 161–84.

Fusco Girard, L. and Nijkamp, P. 1997. *Valutazioni per lo sviluppo sostenibile della città e del territorio*. Milano: Franco Angeli.

—— 2009. *Cultural Tourism and Sustainable Local Development*. Aldershot: Ashgate.

Fusco Girard, L. and You, N. 2006. *Città attrattori di speranza*. Milano: Franco Angeli.

Genovesi, A. 1766. *Lezioni di commercio o sia d'economia civile*. Napoli: Regio Cattedratico di Napoli.

Gibson, C. and Klocker, N. 2005. The Cultural Turn in Australian Region Economic Development Discourse: Neoliberalizing Creativity? *Geographical Research*, 43(1), 93–102.

Gibson, R., Hassan, S., Holz, S., Tamsej, J. and Whitelaw, G. 2005. *Sustainability Assessment*. London: Earthscan.

Hall, P. 1998. *Cities in Civilization*. London: Orion Books.

—— 2002. Planning: Millennium Retrospect and Prospect. *Progress in Planning*, 57, 263–84.

Hall, P. and Pfeiffer, U. 2000. *Urban Future 21*. London: Spon.

Hawkes, J. 2001. *The Fourth Pillar of Sustainability: Culture's Essential Role in Public Planning*. Melbourne: Common Ground Publisher.

Healey, P. 2002. On Creating the City as Collective Resource. *Urban Studies*, 39(10), 1777–92.

—— 2004. Creativity and Urban Governance. *Disp*, 158, 11–20.

Hennicke, P. 2004. *Towards Sustainable Energy Systems: Integrating Renewable Energy and Energy Efficiency is the Key*. Paper presented at the International Conference 'Renewables 2004', Wuppertal/Eschborn.

Holling, C.S., 1973. Resilience and Stability of Ecological Systems. *The Annual Review of Ecology, Evolution, and Systematics*, 4, 1–23.

Holmgren, D. 2004. *Permaculture: Principles and Pathways Beyond Sustainability*. Hepburn Victoria, Australia: Holmgren Design Press.

Hopkins, R. 2008. *The Transition Handbook: From Oil Dependency to Local Resilience*. Totnes: Green Books.

Hospers, G.J. 2003. Creative Cities: Breeding Places in the Knowledge Economy. *Knowledge, Technology and Policy*, 16(3), 143–62.

Ishida, H. 1994. Amoeba Management at Kyocera Corporation. *Hulman Systems Management*, 13(3), 183–95.

Jacobs, J. 1969. *The Economy of Cities*. New York: Random House.

Kibert, C. J. 2005. *Sustainable Construction: Green Building Design and Delivery*. Hoboken, NJ: Wiley.

Klingmann, A. 2006. *Brandscapes: Architecture in the Experience Economy*. Cambridge, MA: The MIT Press.

Kong, L. and O'Connor, J. 2009. Creative Economies, Creative Cities, Asian European Perspectives. *Journal of Cultural Economics,* 34(2), 147–50.

Kunzmann, K. 2004. An Agenda for Creative Governance in City Region. *Disp*, 158, 5–10.

Kunzmann, K.R. and Sartorio, F. 2005. More Creative Governance. *Disp*, 162(3), 3–4.

Landry, C. 2000. *The Creative City*. London: Earthscan.

—— 2006. *The Art of City Making*. London: Earthscan.

Layard, R. 2005. *Happiness: Lessons from a New Science*. London: Penguin.

Lebel, L. et al. 2006. Governance and the Capacity to Manage Resilience in Regional Social-Ecological Systems. *Ecology and Society*, 11(1), 19. Available at: www.ecologyandsociety.org/vol11/iss1/art19/.

Markusen, A. 2006. *Cultural Planning and the Creative City*. Paper presented at the Annual American Collegiate Schools of Planning Meetings, Fort Worth, TX, 12 November 2006.

Martinez Alier, J. and Roca, J. 2000. *Economia Ecologica, Politica Ambiental*. Ciudad de México: Fondo de la Cultura Economica.

McHarg, I.L. 1969. *Design with Nature*. New York: The Natural History Press.

Nadarajah, M. and Yamamoto, A.T. 2007. *Urban Crisis, Culture and Sustainability of Cities*. Tokyo: United Nations Press.

Newman, P., Beatley, T. and Boyer, H. 2009. *Resilient Cities*. Dordrecht: Springer.

Nijkamp, P. and Perrels, A. 1994. *Sustainable Cities in Europe*. London: Earthscan.

Nijkamp, P. and van Delft, A. 1977. *Multicriteria Analysis and Regional Decision Making*. Dordrecht: Springer-Verlag.

Nijkamp, P., Oirshot, G. and Oosterman, A. 1998. *Regional Development and Engineering Creativity*. Amsterdam: Free University.

Norberg Shulz, C. 2007. *Genius Loci*. Milano: Electa.

North, D. 1990. *Institutions, Institutional Change and Economic Performances*. Cambridge: Cambridge University Press.

Olivetti, A. 1952. *Società, Stato, Comunità. Per una Economia e Politica Comunitaria*. Milano: Edizioni Comunità.

Pine, J. and Gilmore, J. 1999. *The Experience Economy*. Boston, MA: Harvard Business School Press.

Plaza, B. 2000. Evaluating the Influence of a Large Cultural Artifact in the Attraction of Tourism, The Guggeheim Museum Bilbao Case. *Urban Affair Review,* 36(2), 264–74.

Porter, M. 1998. Clusters of New Economics of Competition. *Harvard Business Review,* 76(5), 77–90.

Portugali, M. 1997. Self Organizing Cities. *Futures,* 29(4), 77–90.

Rallo, D. 2009. Perequazione: realizzare la città pubblica con i 'soldi' dei privati, in *Perequazione e Qualità Urbana. Transfer of Dewlopment Right and Urban Form,* edited by C. Trillo. Firenze: Alinea, 237–40.

Relph, E. 1976. *Place and Placeness*. London: Pion.

Register, P. 2006. *Eco-cities*. Gabriola Island, WA: New Society Publisher.

Rogers, M. and Spokes, J. 2003. Does Cultural Activity Make a Difference to Community Capacity? *Community Quarterly,* 1(4), 1–8.

Rotmans, J. and Loorbach, D., 2008. Transition management: Reflexive governance of societal complexity through searching, learning and experimenting, in *Managing the Transition to Renewable Energy*, edited by J.C.J.M. Van der Bergh and F.R. Bruinsm. Cheltenham: Edward Elgar, 15–46.

Sacco, P. and Ferilli, G. 2006. Il distretto culturale evoluto nelle economie post industriali. *DADI, IUAV, WP,* 4, 2–28.

Sartorio, F. 2005. Strategic Spatial Planning. *Disp,* 162(3), 26–40.

Saxenian, A. 1994. *Regional Advantage, Culture and Competition in Silicon Valley and Route 128*. Cambridge, MA: Harvard University Press.

Schumpeter, J. 1942. *Capitalism, Socialism and Democracy*. New York: Harper.

Scott, A. 2000. *The Cultural Economy of Cities*. London: Sage Publications.

—— 2006. Creative Cities: Conceptual Issues and Policy Questions. *Journal of Urban Affairs,* 28(1), 1–17.

Simmie, I. 2001. *Innovative Cities*. London: Spon.

Simonter, Y. and Allegretti, G. 2009. *I bilanci partecipativi in Europa*. Roma: Ediesse.

Soleri, P. 1969. *Arcology, the City in the Image of Man*. Cambridge, MA: MIT Press.

—— 1971. *The Sketchbook*. Cambridge, MA: MIT Press.

—— 2006. *The City in the Image of Man*. AZ.: The Cosanti Press.

Stein, U. 2005. Planning with all your Senses – Learning to Cooperate on a Regional Scale. *Disp,* 162(3), 62–69. Available at: www.nsl.ethz.ch/index.php/en/content/view/full/1102.

Sundbo, J. and Darmer, P. 2008. *Creating Experiences in the Experience Economy*. Cheltenham: Edward Elgar.

UCLA. 2006. *The Economic Development Potential of Green Sector*. Los Angeles: The Ralph and Goldy Lewis Center for Regional Policy.

UCLG (United Cities and Local Government). 2004. *Local Agenda 21 for Culture*. Barcelona: UCLG.

UNEP. 2008. *Global Trends in Sustainable Energy Investments*. New York: UNEP.

UNESCO. 2010. *Recommendation on the Historic Urban Landscape. Preliminary Report*. Paris: UNESCO.

UN-Habitat. 2003. *The Challenge of Slums: Global Report on Human Settlements*. London: UN-Habitat.

UN-Habitat. 2006. State *of the World's Cities Report 2006/7*. London: UN-Habitat.

United Nations. 2000. *Millennium Declaration*. New York: United Nations.

Van Oort, F.G. 2004. *Urban Growth and Innovation*. Aldershot: Ashgate.

Von Weizsacker, E.U., Lovins, A.B. and Lovins, L.H. 1998. *Factor Four*. London: Earthscan.

Von Weizsacker, E. et al. 2009. *Factor %*. London: Earthscan.

Washington State. 2007. *The Green Economy Jobs Initiative*. Seattle: CTED.

Wuppertal Institute. 1997. *Futuro sostenibile*. Bologna: EMI.

Yigitcanlar, T., Velibeyoglu, K. and Baum, J. 2008. *Creative Urban Regions*. London: Information Science Reference.

Yunus, M. 1999. The Grameen Bank. *Scientific American*, 281(5), 90–95.

—— 2003. *Banker to the Poor*. New York: Public Affair.

Zamagni, S. 2008. Prefazione, in *Dove lo stato non arriva*, edited by C. Cittadino et al. Firenze: Pasigli, 5–15.

Zamagni, S. and Bruni, L. 2003. Una economia civile per città felici, in *L'uomo e la città*, edited by L. Fusco Girard et al. Milano: FrancoAngeli, 595–609.

Zambon, S. 2004. Intangible and Intellectual Capital, in *The Economic Importance of Intangible Assets*, edited by P. Bianchi and S. Labory. Aldershot: Ashgate, 153–83.

Zeleny, M. 2006. *Human System Management*. Amsterdam: IOS.

—— 2009. On the Essential Multidimensionality of an Economic Problem: Toward Trade off Free Economics. *AUCO Czech Economic Review*, 3, 154–75.

—— 2010. Bata Management System: A Built in Resilience against Crisis at Micro Level. *AUCO Czech Economic Review*, 4, 102–17.

Chapter 4

A Systems View of Urban Sustainability and Creativity: Vulnerability, Resilience and XXQ of Cities

Waldemar Ratajczak

Holistic View of the City

The literature offers a great many works devoted to the theory of the city considered from a variety of perspectives: social, economic, and environmental.

In recent decades, the concept employed to explain the complexity of the city is that of a dynamic complex system (see for example Albeverio et al. 2008, Batty 2005, Kidokoro et al. 2008, Portugali 2000).

Figure 4.1 presents the complexity of a city from a planning point of view. A city is a complex system and its subsystems are interrelated, with the nature and strength of the interrelations varying over time. Their intensity in cities of developing countries differs from the intensity of those in advanced economies. They also differ depending on whether the city has a regional, national or global significance.

A city as a complex system has to maintain the state of equilibrium between factors favourable to its growth and those posing some threat to it. In the light of the latest conceptions, cities should seek to fulfil the principles of sustainable development, which also apply to spatial aspects (Benton-Short 2008, Cagmani 2007, Ooi 2005, Pelling 2003).

A (mega)city is a complex system embracing the following elements:

- a composition;
- a structure; and
- surroundings.

As a complex system, a (mega)city has got two essential properties which set it apart from one that is merely complicated. These are:

- emergence; and
- self-organization.

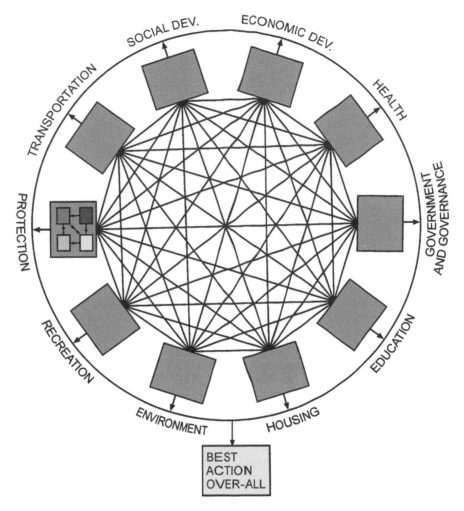

**Figure 4.1   Conceptual model of the planning process – the multiple systems
             of a city**

Source: Hamburg et al. (1968).

Emergence is the appearance of a type of behaviour that cannot be anticipated
from the knowledge of the parts of the system alone. The other property, self-
organization, means that there is no external controller or planner engineering the
appearance of those emergent features; they appear spontaneously.

     Those two general properties are responsible for the fact that cities usually
exhibit different development trajectories and specific characteristics of their own,
such as presence, potential and pulse.

From Vulnerability to the XXQ Concept

Cities as complex systems undergo a variety of processes that on the one hand can present a threat to their existence, and on the other hand, robust growth.

Cities which have adopted the principles of sustainable development are now subject to forces which Peter Nijkamp (2008) has termed XXQ forces: forces that shape cities of exceptionally high quality. As a result, such urban places can be called an XXQ type of cities.

The dependence between a city's susceptibility to threats (vulnerability), robustness when hit by a crisis (resilience), sustainable development, quality of urban life (liveability), and the effect of XXQ forces can be expressed in the form of the following implication relation:

Vulnerability<< Resilience ⇒ Sustainability ⇒ High Level of Liveability (HLL) ⇔ XXQ   (1)

This formula can be read as follows: If the resilience of a city greatly exceeds its vulnerability, it develops as a sustainable city; as such, it displays a high level of liveability, and thus may be considered a city which fulfils extra high-quality conditions.

Vulnerability

Vulnerability has many faces (see Figure 4.2). The literature offers several ways of understanding this notion:

1. as a particular condition or state of a system before an event triggers a disaster, described in terms of criteria such as susceptibility, limitations, incapacities or deficiencies, for example, the incapacity to resist the impact of the event (resistance) and the incapacity to cope with the event (coping capacities);
2. as a direct consequence of exposure to a given hazard;
3. as the probability or possibility of an outcome of the system when exposed to an external event associated with a hazard, expressed in terms of potential losses, such as fatalities or economic losses, or as the probability of a person or a community reaching or surpassing a certain benchmark, such as the poverty gap.

Recently the concept of vulnerability (see Figure 4.3) has been used in urban policy analyses to describe a specific dimension of socio-economic and environmental disadvantage. Unlike the notions of economic deprivation or environment deterioration, which focus on the present condition, *vulnerability is a forward-looking concept*. In its general meaning, the idea of vulnerability relates to the way in which events can affect a certain system (a city), and specifically the likelihood of the system experiencing a loss or negative outcomes in the future because of particular events or actions (Alasia et al. 2008).

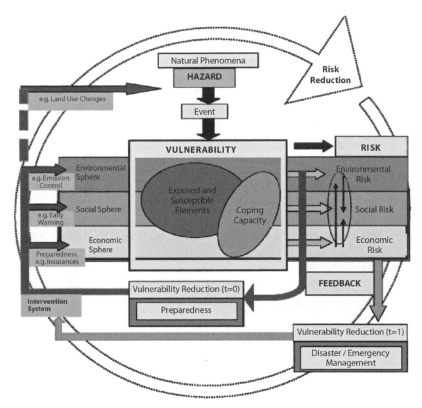

Figure 4.2   BBC model incorporating exposure and coping capacity into vulnerability

Source: Villagrán (2006).

Measuring the vulnerability of cities is no easy task. Answers to the following key questions may help to clarify the choice of a proper method:

- Who and what is vulnerable?
- Vulnerable to what?
- Who wants to know and why?
- What circumstances and context shape the daily life of the affected?

Generally, there are four basic methods for computing vulnerability indices:

1. using variables (components of the index) which describe a city as vulnerable. The variables are measured in different ways, therefore their summing up requires standardization (Briguglio 2000);
2. mapping on a categorical scale and calculating an average (SOPAC);
3. regression analysis (Alasia et al. 2008);
4. principal components analysis (Cutter and Finch 2008).

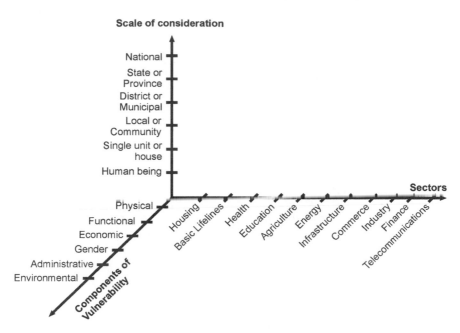

Figure 4.3   **Framework for the vulnerability concept, introducing the independent dimensions of scale of consideration, components of vulnerability, and city sectors**

Source: Villagrán (2006).

The conceptual model of the dependence between the vulnerability and resilience of a megacity has been lucidly characterized in the Munich Re Foundation 2007 Report (p. 9). It is presented in Figure 4.4. A (mega)city as a complex socio-economic system is an area of interaction of environmental, social and economic factors (cf. Wisner et al. 2003).

There are two groups of decision-makers in a (mega)city who play a key role in moulding the relations between vulnerability and resilience, viz. its residents and institutions (or authorities). Their response to a threat depends on the extent of their preparedness and potential for adaptation to critical situations. Because a (mega)city is a dynamic system, the relations between vulnerability and resilience vary over time and are not always harmonized. The future of any (mega)city displaying a certain level of vulnerability depends on the dynamics of its resilience.

Resilience

Like vulnerability, also resilience is understood and defined in a variety of ways (see e.g. Adger 2000, Alberti et al. 2003, Holling 1973, Norris et al. 2008). For the purposes of this chapter, the notion of resilience given by the Resilience Alliance (RA) is particularly useful. RA defines it as 'the amount of change the

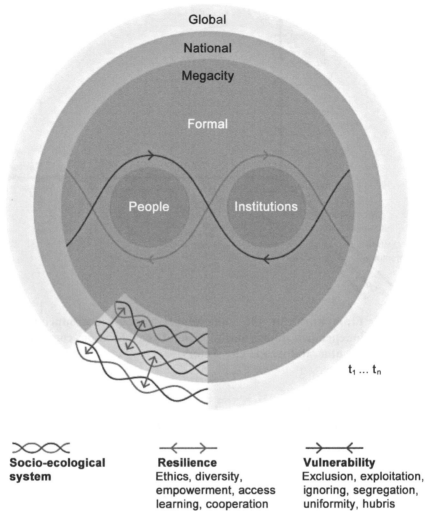

Figure 4.4   Megacity dynamic stress model

Source: Munich Re Foundation 2007 report.

system can undergo and still retain the same controls on function and structure; the degree to which the system is capable of self-organization; and the ability to build and increase the capacity for learning and adoption' (RA 2007). In other words, recovery, flexibility and learning are key concepts in resilience.

The dependence between the vulnerability and resilience of a city is clearly presented in Figure 4.5. It shows the city to possess a measure of vulnerability to the kind and force of stressors, but also to be capable of a response in the form of a measure of resilience.

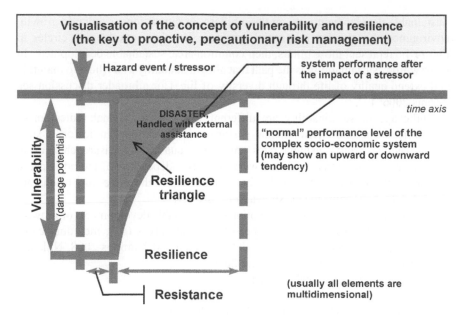

Figure 4.5   Bogardi's notion of vulnerability spanning damage potential, resistance, and resilience

Source: Villagrán (2006).

It is obvious that the city should strive to minimize the resilience triangle. The triangle should at best have an area of zero. This, however, is impossible owing to the inertia of the city as a complex socio-economic system. Differing in their locations, political systems, socio-economic advancement, wealth, and so on, cities are also characterized by differences in the area of their resilience triangles.

Sustainability

Considered from a general point of view, sustainable development (SD) can be treated as a pattern of resource use that aims to meet human needs while preserving the environment, so that these needs can be met not only in the present, but also in the indefinite future. Therefore, SD encompasses two basic aspects:

1. ecological, or a reduction of human impact through the ecologization of economic processes and the implementation of integrated systems of environmental protection;
2. civilizational, or a search for and implementation of new forms of economic development, new technologies, new forms of energy and social communication, and new forms of man's non-economic activity.

Sustainable development rests on three pillars, or dimensions: economic growth, environmental protection and social progress. The three overlapping circles in Figure 4.6 represent those pillars; their common area, sustainable development. In other words, 'sustainability is the path that allows humanity as a whole to maintain and extend quality of life through diversity of life' (for a broader discussion see Adams 2006: 13).

Indicators of sustainability are different from the traditional indicators of economic, social and environmental performance. Traditional indicators – such as stockholder profits, asthma rates and water quality – measure changes in one part of a community as if they were entirely independent of the other parts. Sustainability indicators reflect the fact that in reality the three dimensions are very closely interconnected, as shown in Figure 4.7.

During the last two decades the notion of sustainable development (or growth) has been employed in a great number of scientific areas (43), including cities (Erickson 2006, Kahn 2006, Lerch 2007, Newman and Jannings 2008, Nijkamp and Perrels 1994, Nijkamp and Vreeker 2000).

For a city, being sustainable means:

> improving the quality of life in the city, including ecological, cultural, political, institutional, social and economic components, without leaving a burden on the future generations. A burden which is the result of a reduced natural capital and an excessive local debt. Our aim is that the flow principle, that is, based on an equilibrium of material and energy and also financial input/output, plays a crucial role in all future decisions upon the development of urban areas. (Urban21 Conference, Berlin, July 2000)

To describe a sustainable city in formal terms, a sustainable city index can be constructed. Its general form, moulded on Figure 4.7, can look as follows:

$$\text{sustainable city index} = \beta_1 \text{ (environment)} + \beta_2 \text{ (economy)} + \beta_3 \text{ (society) (2)}$$
where β_1, β_2, and β_3 stand for index weights

Naturally, this symbolic equation needs empirical concretization through the identification of concrete sustainability indicators from the fields of the environment, economy and society. Then appropriate weights can be designated β_{11}, β_{12}, ..., β_{22}, ..., β_{33}, and they can be preceded by either a plus or a minus sign.

> Working with sustainability indicators therefore provides the potential of translating the ideals and values of sustainable development into measures for assessing the progress that cities are making. Such assessment in turn provides the basis upon which to seek new directions for policies and programmes that are important to shifting cities towards more sustainable paths to development and growth. (Ooi 2005: 79)

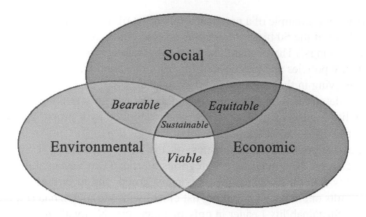

Figure 4.6 **Scheme of sustainable development at the confluence of the three dimensions of sustainability**

Source: Adams (2006), Wikipedia (n.d.).

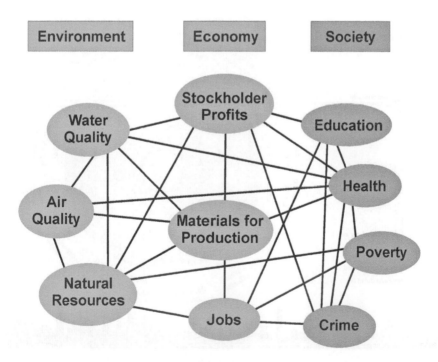

Figure 4.7 **Communities are a web of interactions among the environment, the economy, and society**

Source: Sustainable Measures (1998–2010).

The best-known example of a practical application of the sustainable city concept is the ranking of the 50 biggest US cities carried out in 2005 by the SustainLane US City Rankings. 'The SustainLane US City Rankings focus on the many ways in which city policies and practices differ from one another, and how this affects the people living in those places' (Karlenzig et al. 2007: 5). Overall rankings for 2008 were determined by averaging 16 individual category rankings (based on a sustainability indicator), each of which was multiplied by a proper weight. (See Sustain Lane 2004–08 for the final results).

Figure 4.8 presents characteristics of the most sustainable US city, viz. Portland, Oregon. In as many as 10 categories, Portland was Sustainability Leader, in three, it was in the Sustainability Advances group, and in further three, in the Mixed Results one. The 2008 ranking list closes with Mesa, Arizona (Figure 4.9), which was Sustainability Leader in only one category, National Disaster Risk. In five categories (out of 12) it was classed in the Sustainability in Danger group, in three, in the Mixed Results group, and in further three, in the Sustainability Advances one.

The rankings published by SustainLane are an impulse for city authorities to undertake measures to improve the sustainability of their cities, both those opening

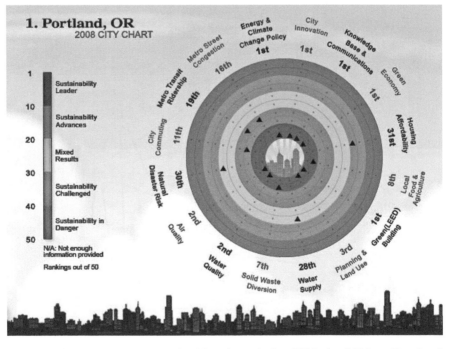

Figure 4.8   The most sustainable city of the USA in 2008 – Portland (Oregon)

Source: Sustain Lane (2004–08), information for Portland.

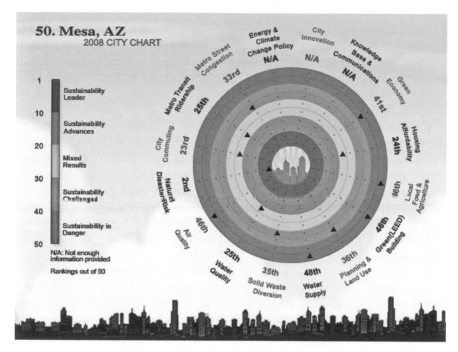

Figure 4.9   The least sustainable city of the USA in 2008 – Mesa (Arizona)

Source: Sustain Lane (2004–08), information for Mesa.

the ranking and those closing it. For instance, Portland was also the leader of the 2007 ranking (which means it kept its position in 2008 as well), while Columbus, Ohio, was the last; in 2008, however, it moved to 30th place – as many as 20 places up the list.

Liveability

The conception of liveability is not an unequivocal term in the literature on the subject; neither is it understood and felt in the same way by societies representing different cultures. That is why the predominant type of research reported in the literature is empirical studies embracing a wide spectrum of intuitive understandings of liveability.

Standing out among the many definitions of the liveability of cities is the one given by Jones et al. (2003): 'The concept of liveability can be loosely defined as "all those aspects of a place which make people happy to live and work here", but it embraces a complex and inter-related set of issues which includes poverty and inequality, housing conditions, safe working environments, environmental pollution, policing, personal security and safety issues, transport facilities, cultural and recreational amenities, and the quality of the public realm.'

The notion needs operationalization through the construction of suitable liveability measures. There are many such measures (cf. VCEC 2007) which embrace costs of living in a city and the quality of life in it. They seek to establish what living conditions a concrete city offers its potential residents. That is why liveability measures are more significant for public authorities rather than for an individual or the city population as a whole.

Liveability measures are also used to prepare liveability rankings of cities. The best known is that drawn up by the Economist Intelligence Unit (EIU), which rests on 40 indicators grouped into five weighted categories: stability, health care, culture, education and infrastructure. According to the EIU 2008 list, the world's five most liveable cities were Vancouver, Melbourne, Vienna, Perth, and Toronto (cf. Economist 2008). Another ranking, known as Mercer's Quality of Living Survey, employs 39 factors divided into ten categories (cf. Mercer 2008). In 2008 it identified the five most liveable cities as Zurich, Vienna, Geneva, Vancouver, and Auckland.

An exceptionally broad and deep approach to the assessment of the liveability of cities, albeit only European ones, has been adopted by the European Union's 'Urban Audit: Towards the Benchmarking of Quality of Life in 58 European Cities'. Its set of statistics contains more than 200 statistical indicators covering 21 principal categories (cf. European Union 2000).

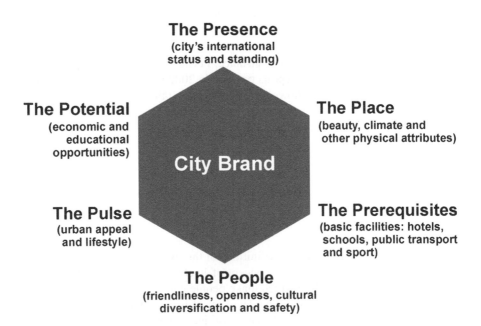

Figure 4.10   The city brand hexagon

Source: Anholt City Brands Index (2007).

Closely associated with the liveability of cities is the city brand, which emphasizes the most essential and attractive attributes of cities and also includes different aspects of their liveability. The brands they have acquired rest on both rationality and emotion, which are both very difficult to quantify. The internationally known City Brands Index (CBI) constructed by Simon Anholt is based on six dimensions forming a city brand hexagon presented in Figure 4.10 above. It offers an opportunity for establishing, and gaining an insight into, a city's position in the world's urban system.

Figure 4.11, in turn, presents the brand hexagons of the first and second cities in the Anholt 2007 ranking, viz. Sydney and London. It is readily apparent that Sydney outranked London in the categories of 'place' and 'prerequisites'. Similar observations can be made about the whole set of 40 cities taken into consideration.

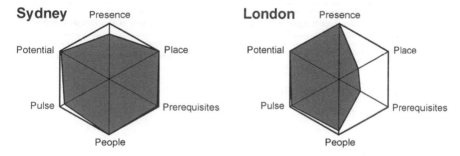

Figure 4.11   The brand hexagons of Sydney and London
Source: Anholt City Brands Index (2007).

There is no doubt that the aim of the authorities of each city is, or should be, to raise its level of liveability. The higher the liveability, the greater the city's competitiveness. At present there are at least two movements (or policies) seeking to improve the liveability of cities, namely Slow Cities and Smart Growth.

The Slow Cities movement was established in 1999 in Italy. The adoption of Slow City thinking is generally perceived to improve the liveability of specified areas within towns and cities, and to offer an alternative conception of urban life.

The Smart Growth movement has developed in the United States. It recognizes a relationship between the development of a city (community) and the quality of life. It seeks to minimize damage to the environment and build liveable towns and cities. However, this movement also has its critics (O'Toole 2001). The most important principles of the two movements can be presented as in Figure 4.12.

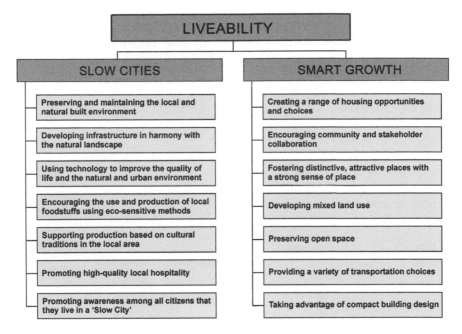

Figure 4.12   View of liveability

Source: Jones et al. (2003).

The XXQ Concept

XXQ is understood as a symbol of a (mega)city with an extra high quality of life. It is an effect of a peculiar feature of this complex system – its self-organization. Self-organization is gradable, and the high level of self-organization of a (mega) city in its various functional aspects ensures it a low level of social vulnerability, a small surface of the resilience triangle (cf. Figure 4.5 above), a growth pattern resting on sustainability principles, and finally a higher-than-average level of liveability.

The XXQ concept has been introduced into the literature as a formal notion by Peter Nijkamp in several of his publications, and most recently in his *JRRSA* paper (Nijkamp 2008). Its chief underlying idea is that (mega)cities can be viewed as self-organizing innovative complexes (SICs) exhibiting a few important features:

> (1) a reliance on creativity, innovativeness and leadership, (2) competitive advantages to be created by R&D, (3) productivity and competitiveness as critical success factors, (4) a market orientation determined by product heterogeneity and monopolistic competition, (5) a development path marked by evolutionary complexity and behavioural learning principles. (Nijkamp 2008: 25; Nijkamp and Ubbels 1999)

A SIC does not emerge by chance; rather, it is an effect of the cooperation of forces which Nijkamp presented in the form of a pentagon model (Figure 4.13).

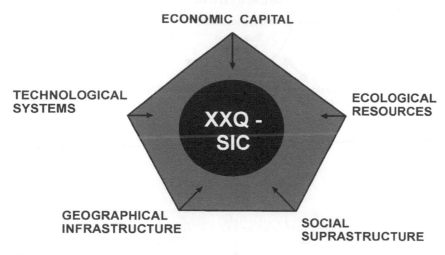

Figure 4.13   A pentagon model of XXQ forces for SICs
Source: Nijkamp (2008).

In this model:

- economic capital is a prerequisite for a strong performance of a sustainable urban area;
- ecological resources represent the environmental basis of the sustainable development of a (mega)city;
- technological systems is a term embracing not only advances in technology, but also the creation of an innovative culture involving both the producers and the consumers, as well as the promotion of a sustainable image of the (mega)city through public involvement;
- geographical infrastructure covers such network properties of cities as their accessibility and connectivity;
- social suprastructure is a factor underlying a socially sustainable society, especially in such of its aspects as creativity and diversity.

Conceptual in nature, the pentagon model can be treated as a strong tool for explaining and even understanding the emergence of SIC-XXQ. The process of SIC-XXQ formation can be strengthened or weakened by the endogenous potential of a (mega)city, its learning ability, and finally is capability of creating innovation understood as knowledge used to generate new knowledge (Drucker 1993, Komninos 2002, Landabaso and Reid 2003). The chief features of this process are presented in Figure 4.14.

Figure 4.14　The 'intelligent city': Cooperates locally to compete globally
Source: Landabaso and Reid (2003) (modified).

As an effect of the *co-opetition* among the (mega)city components, there develops a 'virtuous growth circle' driven by three of the factors in Nijkamp's model, viz. technological systems, geographical infrastructure, and social suprastructure. The other two factors – economic capital and ecological resources – are located outside the virtuous circle. The operation of the virtuous circle and the interaction between it and external factors located outside it enhance the (mega) city's innovation capacity. As new innovations are generated and developed, the (mega)city gains a competitive advantage which makes it a destination place for new economic activities, new residents, and so on. In other words, the (mega)city develops according to the SIC condition with XXQ. This means that the (mega)city turns into a self-organizing innovative complex following sustainable development principles, with stable jobs in the long term and higher standards of living.

Conclusions

Contemporary (mega)cities are complex dynamic systems and their specific properties make them take an evolutionary path of development with all the consequences of this process. The most important among them seem to be: self-organization, innovativeness, ability to learn, as well as bio- and cultural diversity.

(Mega)cities can be found in a variety of places in geographical space (physical and socio-economic), hence they have to cope with different conditions of development, both favourable and adverse.

The chapter shows there to be a logical relation between a (mega)city's vulnerability and its status defined as XXQ. All elements of the relation contribute significantly to establishing the city's position, but resilience is especially vital among them, as shown by many real-life examples.

Bibliography

Adams, W.M. 2006. *The Future of Sustainable: Re-thinking Environment and Development in the Twenty-first Century*. Report of the IUCN Thinkers Meeting: 29–31 January.

Adger, W. 2000. Social and Ecological Resilience: Are They Related? *Progress in Human Geography*, 24(3), 347–364.

Alasia, A., Bollman, R., Parkins, J. and Reimer, B. 2008. An index of community vulnerability: Conceptual framework and application to population and employment changes. Research Paper 88.

Alberti, M., Marzluff, J., Shulenberger, E., Bradley, G., Ryan, C. and Zumbrunnen, C. 2003. Integrating Humans into Ecology: Opportunities and Challenges for Studying Urban Ecosystems. *Bioscience*, 53 (12), 1169–79.

Albeverio, S., Andrei, D., Giordano, P. and Cancheri, A. 2008. *The Dynamics of Complex Urban Systems*. Heidelberg, New York: Springer.

Anholt City Brands Index. 2007. *How the World Views its Cities*. Third Edition. Anholt.

Batty M. 2005. *Cities and Complexity*. Cambridge, MA, London: MIT Press.

Benton-Short, L. and Short, J.R. 2008. *Cities and Nature*. London: Routledge.

Briguglio, L. 2000. An economic vulnerability index and small island developing states. Working Paper, Seminar on Island Studies.

Cagmani, R. 2007. City Network as Tools for Competitiveness and Sustainability, in *Cities in Globalization: Practices, Policies and Theories*, edited by P.J. Taylor et al. London: Routledge, 105–29.

Cutter, S.L. and Finch, C. 2008. Temporal and Spatial Changes in Social Vulnerability to Natural Hazards. *PNAS* 105(7), 2301–06.

Drucker, P.F. 1993. *The Post-capitalist Society*. New York: HarperBusiness, A Division of HarperCollins Publishers.

Economist. 2008. *Economist Intelligence Unit Global Liveability Rankings*, January. Available at: www.economist.com/markets/rankings/displaystory. cfm?story_id=11116839.

Erickson, D. 2006. *MetroGreen: Connecting Open Space in North America Cities*. Washington, London, Covelo: Island Press.

European Union. 2000. Urban Audit: Towards the Benchmarking of Quality of Life in 58 European Cities. Available at: ec.europa.eu/regional_policy/urban2/urban/audit/ftp/vol3.pdf.

Hamburg, J.R., Creighton, R.L. and Scott, R.S. 1968. *Evaluation of Land Use Patterns: Urban Development Models*. Conference procedings. Washington, DC: Highway Research Board.

Holling, C.S. 1973. Resilience and Stability of Ecological Systems. *Annual Review of Ecology and Systematics*, 4, 1–23.

Jones, P., Hillier, D. and Comfort, D. 2003. Liveability – Slow Cities and Smart Growth: Peter Jones, David Hillier, and Daphne Comfort look at the concepts of 'Slow Cities' and 'Smart Growth' and, drawing on some illustrative examples of embryonic thinking and practice, and ask how applicable they are to development and regeneration in the UK. *Town and Country Planning*, 72(11), 355(4).

Kahn, M.E. 2006. *Green Cities: Urban Growth and the Environment Brookings*. Washington, DC: Institution Press.

Karlenzig, W., Marquardt, F., White, P., Yaseen, R. and Young, R. 2007. *How Green is your City? The SustainLane US City Rankings*. Gabriola Island, BC: New Society Publishers.

Kidokoro, T., Harata, N., Subanu, L.P., Jessen, J., Motte, A. and Seltzer, E.P. 2008. *Sustainable City Regions: Space, Place and Governance*. Hicom, Japan: Springer.

Komninos, N. 2002. *Intelligent Cities*. London and New York: Spon Press.

Landabaso, M. and Reid, A. 2003. Developing Regional Innovation Strategies: The European Commission as Animateur, in *Regional Innovation Strategies: The Challenge for Less-favoured Regions*, edited by K. Morgan and C. Nauwelaers. London and New York: Routledge. 19–39.

Lerch, D. 2007. *Post Carbon Cities: Planning for Energy and Climate Uncertainty*. Sebastopol: Post Carbon Press.

Mercer. 2008. *Quality of Living Global City Rankings 2008 – Mercer Survey*. Available at: www.mercer.com/referencecontent.htm?idContent=1307990#Top50_qol.

Munich Re Foundation. 2007. *From Knowledge to Action*. Report.

Newman, P. and Jannings, I. 2008. *Cities as Sustainable Ecosystems: Principles and Practices*. Washington, Covelo, London: Island Press.

Nijkamp, P. 2008. XXQ Factors for Sustainable Urban Development: A Systems View. *The JRRSA*, 2(1), 1–34.

Nijkamp, P. and Perrels, A. 1994. *Sustainable Cities in Europe*. London: Earthscan.

Nijkamp, P. and Ubbels, B. 1999. *Infrastructure, Suprastructure and Ecostructure: A Portfolio of Sustainable Growth Potentials*. Serie Research Memoranda, 51. Amsterdam: Vrije Universiteit.

Nijkamp, P. and Vreeker, R. 2000. Sustainability Assessment of Development Scenarios: Methodology and Application to Thailand. *Ecological Economics*, 33(1), 7–27.

Norris, F., Stevens, S., Pfefferbaum, B., Wyche, K. and Rose, L. 2008. Community Resilience as a Metaphor, Theory, Set of Capacities and Strategies for Disaster Readiness. *American Journal of Community Psychology*, 41(1–2), 127–50.

Ooi, G.L. 2005. *Sustainability and Cities: Concept and Assessment*. Singapore: IPS and World Scientific Publishing Co.

O'Toole 2001. *The Vanishing Automobile and Other Urban Myths: How Smart Growth will Harm American Cities*. Bellevill, IL: Demographia.

Pelling, M. 2003. *The Vulnerability of Cities*. London: Earthscan Publications.

Portugali, J. 2000. *Self-Organization and the City*. Berlin: Springer-Verlag.

RA. 2007. *Resliance Alliance*. Available at: www.resalliance.org/576.php.

Sustain Lane. 2004–08. 2008 US City Sustainability Rankings. Available at: www.sustainlane.com/us-city-rankings/overall-rankings.

Sustainable Measures. 1998–2010. *What is an Indicator of Sustainability?* www.sustainablemeasures.com/Indicators/WhatIs.html.

UCN. 2006. *The Future of Sustainability: Re-thinking Environment and Development in the Twenty-first Century*. Report of the IUCN Renowned Thinkers Meeting, 29–31 January 2006, The World Conservation Union.

Villagrán, J.C. 2006. *Vulnerability: A Conceptual and Methodological Review*. Publication Series of UNU-EHS 4. Bonn.

Wikipedia (n.d.) *Sustainable Development*. Available at: en.wikipedia.org/wiki/Sustainable_development.

Wisner, B., Blaikie, P., Cannon, T. and Davis, I. 2003. *At Risk: Natural Hazards, People's Vulnerability and Disasters*. UK: Routledge.

PART II
Creative and Entrepreneurial Urban Economy

Chapter 5

Entrepreneurial Creative Clusters in the Global Economy

T.R. Lakshmanan and Lata Chatterjee

Introduction and Overview

In the last three decades, innovations in *material* (e.g. transport, communications and production) and *non-material* technologies and infrastructures (trade liberalization, financial innovations, etc.) have led to a two dimensional *transformation – economic and urban –* in the highly industrialized countries of North America, Europe and East Asia.

The economic transformation is the familiar *globalization –* an explosive expansion of cross-country economic interactions, division of labour, complex webs of production chains, and a globally distributed production system.

However, the 'place' or urban location has not been eroded away in the global economy as a geographical and socio-economic entity, and seems to matter in this globally distributed production system. Indeed, the emerging increasingly knowledge-intensive global economy is *centred on cities,* which serve as spatial *economic and urban* platforms for private, public and social sector change agents who create and commercialize new knowledge and operate the global economy and its worldwide business and non-business networks. In these cities or urban clusters, creative environments emerge, characterized by (i) the occurrence of knowledge spillovers from one group to another, promoting innovation and growth,[1] and (ii) flexible interactive networks – interpersonal, inter-enterprise, and inter-organizational – stimulating innovation.[2]

In this process, some cities undergo transformation, acquiring new (and shedding old) economic and political capacities, functions and identities, and become *Entrepreneurial Cities*. In the context of the above structural change, the economic, social and political agents in these cities engage in *entrepreneurial*

---

1 This idea of cities as locales of creativity is not new, as evident from the writings of historians from Toynbee to Braudel (1992). More recently, Jane Jacobs (1969) described these creative spatial spillovers in the context of New York and Toronto.

2 Camagni (2004) emphasizes the 'physical' and 'social' proximity of persons working in the enterprises within the urban clusters. He suggests that the physical and social proximity leads to what he calls 'relational capital', by which people are able to connect with one another, to develop notions of trust, exchange knowledge and innovate.

behaviour (such as foresight, risk taking, discovery and innovation). Consequently, this chapter labels such dynamic cities as 'Entrepreneurial Clusters or Cities' in the Global Economy.

This chapter offers a survey of the recent rise, evolution and the defining characteristics of entrepreneurial cities, viewed as complex entities, which *both direct and adapt to structural change*. The next section opens with a brief review of the factors underlying the 'resurgence' of cities as sites of innovation-led growth in an era of globalization and an increasingly knowledge-intensive economy. A variety of concepts drawn from several disciplines have surfaced to characterize the acquisition and maintenance of new economic, political and social capacities by these cities in order to be dynamically competitive in the global economy. These concepts, often used interchangeably, include: agglomerations/growth clusters, 'innovation milieu', entrepreneurial cities and creative cities. Three models of such innovation-led urban/regional development, namely, the 'New Economic Geography', New Growth Theory, and a Spatial–Relational Interaction model of Innovation-led Agglomeration, are briefly presented. The current conceptual state-of-the art in this extensive literature is summarized.

The third section highlights the new and evolving dimensions of *decision and management context* in the dynamic entrepreneurial clusters, as they are buffeted by the technical, economic and financial globalization and the broader social evolution in recent decades in the form of new urban lifestyles and the rise of new relationships and patterns of engagement between urban citizens and urban government. The argument made here is that if structural change accompanying the rise of creative cities is to take hold, a number of prior conditions must be met:

- such cities must not only acquire *new* (but also shed *older*) economic and political capacities;
- the acquisition of such new capacities is possible only when urban public and social sector change agents complement the work of private sector agents and all jointly produce *urban public value and structural competitiveness*;
- more inclusive rules of engagement and decision-making among urban public, private and social sector actors must emerge in the form of *new models of urban governance*.

The fourth section profiles the entrepreneurial functions of urban public and social sector agents in these entrepreneurial creative cities. It also compares and contrasts the motivations, characteristics, activities and composition of these urban private, public and social sector actors, as they *co produce urban value*.

The fifth section describes the new economic and political capacities needed in these entrepreneurial clusters and the evolving context of urban governance. It also addresses the problem of coordination of the economic/political functions by urban private/public/social actors in a non-hierarchical environment. Such coordination can be in the form of a partnership under the *principle of flexible rationality*. There is thus a shift from the urban government to urban governance.

The resulting institution or governance mechanism is a network model that exploits the complementarities of the three sectors in the creation of public value.

Part VI profiles selected American experience in the rise of entrepreneurial creative clusters and their evolution in terms of capacities, functions, new forms of urban governance and the creation of dynamic competitiveness.

Models of Innovation-led Urban and Regional Development

Over the last two to three decades – contemporaneous with the resurgence of cities in the global economy – there has been a flurry of interest in both urban academic and policy communities in concepts such as agglomerations/clusters, 'new industrial areas', 'innovation milieux', 'creative cities'. These terms have been widely used (often interchangeably) both by researchers (business analysts, economists, management theorists, economic geographers, urban and regional development theorists) and by policy types (mayors, development executives, etc.). Given the multiple discipline origins of these theoretical and policy speculations, it is helpful to organize them into *three* conceptual models of innovation-led urban and regional development.

The central idea behind these models of innovation-driven urban evolution is that, in an increasingly knowledge-intensive economy, some agglomerations enjoy *increasing returns* in the form of dynamic location advantages, enabling innovation and dynamic competitiveness of these cities in the global economy. There are three broad modelling streams which try to capture these advantages:

1. the 'New Economic Geography' Model, which accounts for agglomeration effects (within and across industries);
2. the New Growth Theory, which throws light on the endogenous self-reinforcing mechanisms of knowledge creation;
3. a Spatial–Relational Interaction Model of Innovation-led Agglomeration, which suggests that spatial and relational proximity promote knowledge creation and commercialization in urban agglomerations.

Prior to the review of these models advanced in the last two decades, it is appropriate to highlight the set of theoretical ideas on agglomerations prevalent before the arrival of these models. Enquiries into the nature and sources of increasing returns which produce urban agglomeration come from different traditions in economic geography and location economics from the late nineteenth century. In particular, the insights of Marshall (1890), Hoover (1948), Isard (1956), Vernon (1960), Chinitz (1961) and Jacobs (1969) are relevant.

Marshall (1890) suggested that firms are able to take advantage of *agglomeration economies* by locating in large clusters. The micro foundations of the Marshall's agglomeration economies (in terms of reasons why firms in an industry locate in proximity) suggest (i) *input sharing* among firms, enabling a larger input supplier market with lower average costs and a larger range of inputs for all firms, (ii) *matching* in the sense in such large agglomerations, workers with a wider range of skills can be matched with the diverse requirements of employers. Further, workers find it less risky in such locations with many employers, and (iii) *knowledge spillovers or learning*, in the sense that in these dense locations, there are knowledge spillovers, with workers being the primary vehicles of these transfers.[3] As contrasted with the Marshallian external economies for a particular industry, Jane Jacobs (1969) emphasized the power of industrial diversity in a region on subsequent economic performance. Like historians such as Pirenne, and Braudel (1992), Jacobs argues that the multiplicity and cross-fertilization of 'ideas' in variegated environments of a large population and economic cluster stimulate creativity and innovation.[4]

The 'New Economic Geography' Model

It is in this context that Krugman (1991) applied a general equilibrium modelling framework to the *geography of the economy* under conditions of increasing returns to scale and labour mobility and reinterpreted the findings of Marshall on agglomerations. In the resulting 'New Economic Geography' (NEG) model, spatial concentration and dispersion emerge as a consequence of market interactions among individual firms under conditions of scale economies. Over time with lower and lower transport costs, there is a circularity or cumulative causation (especially under conditions of larger local demand in the agglomerations) leading to further strengthening of existing large agglomerations, higher nominal wages, and greater variety of goods.[5]

---

3   Agglomeration economies have been further characterized in the work of Hoover and Isard as (i) *localization economies* (being proximate to other producers of the same good or service), which embrace factors that reduce the average cost of producing outputs in that location, and i(i) *urbanization economies* (being proximate to producers of a broad range of goods and services, and (iii) which are external economies accruing to enterprises because of savings from large-scale operation in a large agglomeration (and thus independent of industry) in the form of knowledge gains from diversity (Jacobs, 1969). Research by Vernon (1960) and Chinitz (1961) on the causes and consequences of spatial clustering of economic activities in a metropolis such as New York focused on issues linking growth and agglomeration.

4   Further, such diversity fosters specialization in inputs and outputs, yielding higher returns (Quigley, 1998). Glaeser and Gottlieb (2009) suggest that industrial diversity in US cities is important to subsequent economic performance. Rosenthal and Strange (2004) suggest that that the extent of diversity in manufacturing industries at the start of the period was not important in determining employment and subsequent performance of mature industries, but did matter in attracting new industries such as scientific instruments and electronic components.

5   This leads eventually to higher land rents and congestion etc., and thereby eventually to decreasing returns to scale in these cities.

While the NEG model provides insights to agglomeration phenomena that were unavailable before, it exhibits serious weaknesses in explaining same key contemporary aspects of agglomeration development. This model accounts for only pecuniary economies, and makes no mention of either human capital or technological spillovers. While it offers an explanation of the rise of agglomerations, it has few, if any, answers for the contemporary knowledge-creation process and innovation-led growth in urban agglomerations.

The New Growth Theory

In the same two decades, there has been a development of theories of 'endogenous growth' over time and its determinants. There can be spillovers from investments of capital (physical or human) by firms or individuals to the capital held by others (Romer, 1986) and when these spillovers are very pronounced the private marginal product of such capital can remain high. This rapidly growing body of research shows that knowledge is the motor of development and identifies the endogenous self-reinforcing processes of knowledge creation. When this endogenous growth notion is applied to regions, the resulting analyses assume that the idea of increasing returns is spatially embodied in agglomeration economies (McCann, 2005).[6]

A Spatial–Relational Interaction Model of Innovation-led Agglomerations

The key idea here is that innovation or the creation and commercialization of new knowledge in a dynamic region is based on interactions among autonomous but interdependent economic agents. In such economic regions there are very many linkages and interconnections among (large and small) private economic agents and between them and public entities, and civil society actors (who jointly make up the contemporary knowledge-intensive global economic system).

From a microeconomic view, two new competitive factors appear in such an innovative region and alter the incentive structures of dynamic small and medium enterprises (SMEs) (Alter and Hage 1997). First, innovation becomes a *more pivotal competitive* factor than reduction of *transaction cost*s and productivity enhancement typically emphasized by the New Institutional Economics (NIE) theorists such as Williamson (1985) and Chandler (1977). Second, the firms in dynamic regions become concerned with the reduction of a *new class of costs* they confront in this knowledge economy. These are the *adaptive costs* incurred by the firm, as it monitors the environment for changes in technology and products,

---

6   Some economic geographers and heterodox economists argue that while the NEG and the New Growth Theory are mathematically complex they are conceptually simplistic in the sense that they ignore historically and spatially contingent factors, which may be important in the regional evolution. Thus institutional arrangements and cognitive and cultural proximity may be as important as the spatial proximity emphasized in these models (Boschma and Lambooy 1999, Pred 1977).

identifies competitive strategies, and implements such strategies quickly enough to retain or improve market share (Lakshmanan and Button 2009).

These linkages and interactions allow firms and other economic agents to complement their core competencies with requisite knowledge and capacities creatively, speedily and flexibly. These linkages are really 'embedded in the social network.[7] Firms (often small and medium-sized) in such regions develop flexible and interdependent relationships with suppliers and competitors and increasingly depend on intangibles, like know-how, synergies and untraded knowledge (Von Hippel 1988). Such productive relationships among firms are promoted by *spatial and relational proximity in the agglomerations* (Camagni 2004).

Dynamic agglomerations exhibit the attributes of 'learning systems', such as entrepreneurial ability and relational skills.[8] These attributes of learning systems reflect the cultures of local entrepreneurial economic agents, (firms, public or civil society enterprises) stimulating them to continually create and innovate (Acs et al. 2002, Lakshmanan and Chatterjee 2006, 2003, Storper 1995).

Conceptual State of the Art on Innovation-led Regional Development?

While the three modelling approaches differ in their areas of emphases in terms of assumptions, focus, and formal formulation, two areas of agreement appear: (i) an agglomeration or an urban cluster facilitates regional performance, and (ii) notions of regional knowledge generation, spillovers and accumulation are present in these models.

From the extensive literature generated in connection with the three modelling frameworks above, one can summarize that the increasing returns associated with what we describe here as (entrepreneurial) creative urban clusters are dynamic location advantages which derive from:

- *physical proximity* among economic actors, facilitating interactions and enabling access to, appropriation and sharing of *tacit knowledge* thus promoting innovation;
- *relational Proximity* of economic agents facilitating cooperative behaviour, collective learning, and socialization of innovation risk;

---

7   Indeed, even the market relations of the economists' world are socially embedded in the sense that they depend upon assumptions, norms and institutions shared by the actors and do not themselves derive from economic decisions (Granovetter, 1985, Polyanyi, 1957). The recent interest in embeddedness and social capital as key supporting assets of productivity has been inspired by the spatial clustering and dynamism in places such as Italy's Emilia-Romagna and California's Santa Clara Region.

8   This theoretical framework discusses the concept of *tacit knowledge* (non-explicit or non-prepositional). Key issues here are: How is tacit knowledge produced? How is it found, appropriated and shared?

- *lowering of Adaptive Costs* among firms competing in an environment of rapid pace of change of knowledge;
- *institutional Proximity* among the firms in the agglomeration in terms of shared rules, codes and norms of behaviour which will promote cooperation in interactive learning processes (Amin and Cohendet, 2004).

The New Decision and Management Context in Entrepreneurial Clusters

Figure 5.1 highlights two key drivers of urban change, namely, (i) the technical, economic, and financial globalization processes and, (ii) the broader social evolution in recent decades in the form of new urban lifestyles and the rise of new relationships and patterns of engagement between urban citizens and urban government.

These change drivers are creating *new and evolving decision and management contexts* in these dynamic urban regions or entrepreneurial cities (Figure 5.1). The emerging urban decision and management context is characterized by:

1. *Worldwide interurban economic competition*, with economic actors in each urban area needing to develop and implement economic growth goals in an entrepreneurial fashion. In such a decision and management context, the entrepreneurial city has to acquire new functions and identities – with a transformation of traditional roles, identities, tasks of the urban government in order to be more competitive and engage in wealth creation (Table 5.1). Further, these cities need to develop new physical infrastructures, policy regimes, and human and organizational capital in order to promote international competitiveness.
2. *Rise of new urban economic actors in the public and social realm*: The creation of new physical and non-physical urban infrastructures, and restructured urban economic and physical space appropriate to the requirements of the emerging knowledge economy is possible only when the urban public and social sector economic agents complement the work of private sector agents (traditionally associated with economic growth). This process of joint action by urban private/public/social actors (with their complex interdependencies) produces urban *public value* (Moore 1996).
3. *The failure of the traditional urban hierarchical model and the drive for more inclusive urban decision-making*: As various urban stakeholders drawn from public, private and social sectors push for and seek balance among their conflicting claims, the old hierarchical urban government model fails. New rules of engagement and negotiable relationships among the autonomous but interdependent private, public and social stakeholders have to develop. Mechanisms of coordination or governance of such diverse and complex types of interactions among these different urban actors need to be designed.

4. *New Models of Urban Governance*: In this emerging inclusive urban decision context comprising diverse economic actors adjusting to ongoing global structural change, new modes of governing economic interactions among such actors – or governance institutions – develop. Our argument is that the political, social and private sector actors participate in different kinds of networks – within each sector and between sectors. The purpose of engagement in these networks is twofold: (i) the reduction of uncertainty in a volatile economic environment, and (ii) the promotion of ex-ante coordination of actors from various urban sectors, which collectively produce new capacities and the consequent structural competitiveness for that city.

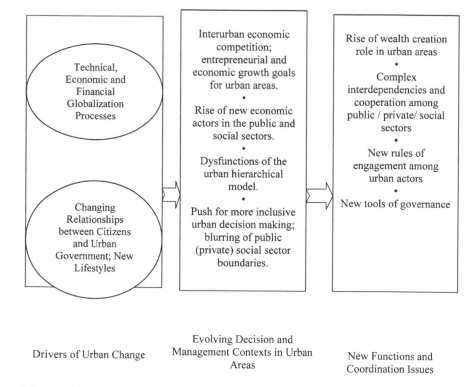

Drivers of Urban Change   Evolving Decision and
                                 Management Contexts in Urban   New Functions and
                                 Areas   Coordination Issues

Figure 5.1   Change processes, and new urban decision and management contexts

Table 5.1   New functions and coordination issues in the entrepreneurial city

- Transformation of traditional roles, tasks and activities and identity of urban government in the new competitive city
- 'Guided Capitalism' Phase II (Urban Government response to market failure for spatially produced goods)
- Role of creation of wealth and urban public value through interdependence and cooperation in an entrepreneurial role in a complex economic web (including market, government and civil society)
- New Rules of Engagement among urban public/private/social sector economic actors
- Need for horizontal, cooperative relations
- Dysfunction of top-down decision model and rise of new types of relationships
- Reduction of innovation costs

Social and Political Entrepreneurs in Urban Development and Regeneration

All economic, political and social activity occurs in a place; space and time are critical elements in the entrepreneurial decision process. Urban entrepreneurs take three interrelated but distinct decisions – *what* to produce, *where* to produce and *when* to produce. The difference between success and failure in the innovative efforts of entrepreneurs of *what* to produce depends on their correct assessment of the locational and timing decisions. History is replete with examples of entrepreneurs who failed because they were ahead of their time. Failure also results from the choice of inappropriate locations and successful entrepreneurs have good intuition of *where* to apply their innovations. The literature on locational issues in entrepreneurial decision-making has focused on the behaviour of market actors, and on the impacts of such activity on the regional economy (e.g. Acs et al. 2002, Develaar 1991, Malecki 1994, Stohr 1989). In this chapter we extend the discussion of urban entrepreneurship to the roles of non-market – civil society and public sector – entrepreneurs and the manner in which all three types of actors influence each other in their entrepreneurial decisions in urban space.

Urban social entrepreneurs (SEs) respond to existing urban conditions and use innovative strategies to change social realities, and the urban environmental context broadly defined, for the collective good. First, social entrepreneurs take a strategic view – visualizing and judging the potential of urban localities and aiming to bring about urban transformation. SEs *alter perceptions* that members of civil society, the public sector and the business community have of development potentials of urban localities. Second, they focus on improved service delivery in specific localities and of the role that improved service delivery plays in promoting development potentials of those localities. Thus, SEs recognize an urban social need and relevant innovative solutions, promote and market their ideas changing the perceptions and attitudes among public, private and social sector actors, and marshal personal and community resources through institutional innovation, risk taking, and performance-oriented implementation.

Social entrepreneurs have, for well over a century, raised social and political consciousness about urban 'market failures' and 'government failures', and demanded solutions for welfare improvements in specific urban locations. They have also provided a variety of innovative solutions (that have improved the life chances and mobility of urban residents and fostered social change in urban areas) and have been primarily viewed as humanitarians. Such late nineteenth- and early twentieth-century social innovators as Jane Addams (Settlement House, Housing, etc., in Chicago), Horace Mann (promotion of public education), Chamberlin (first public provision urban water, sewer, and power provision in Birmingham, UK), Olmstead (urban parks and environment), and Patrick Geddes (English New Towns) are viewed as visionaries. Given the *laissez faire* ambience of those times (and the recent resurgence of neo-liberal views), it is not surprising that the economic, political and societal impacts of the reforms of SEs, and their role as entrepreneurs of altered urban realities, were less emphasized. Consequently, most of the literature on SEs, then and now, discusses what individual social entrepreneurs do or their motivations, and their achievements (Bornstein 2004).

The translation of innovative ideas of a few SEs or visionaries into socio-economic change in urban space is, however, contingent. Not all visionaries with creative ideas are social entrepreneurs. Their ideas about new urban development orientations, and supportive institutions need to be promoted and marketed in an ambience where agents of change confront defenders of the status quo.[9] When we say the 'time for an idea has come' what we really imply is that there is some individual or a collective which has successfully implemented a radical idea – promoting an idea, overcoming social resistance, taking risk, investing money and non-monetary resources and other elements of entrepreneurial behaviour. Thus, SEs change the behaviour of other decision-makers in society, by building on the ideas and research of those before them, and by their own truly original ideas. They make societal change possible through their innovative solutions and persistent strategies. Their efforts have an input into the perceptions and behaviour of other individuals, and collectives at large, and spread through the body politic as the *desirable and the doable.* A major impact that SEs have is in creative institution building for consciousness-raising, marshalling support and for service provision.

The success of SEs depends on their possessing two critical, and related, attributes – namely their capacity to network and to understand the social milieu in which their innovative actions have to be embedded. Networks are 'Interconnected dyadic relationships where the nodes may be roles, individuals or

---

9   We can draw a parallel here between the difficulties in the transition from an invention to an innovation in the economic domain, and the ideas of such social visionaries and the implementation of their ideas in the social domain. A variety of strategic and organizational resources need to be deployed in order to overcome the resistance of vested interests (Lakshmanan 1993).

organizations'. Networking remains crucial since entrepreneurship is a continuous process of innovative activity requiring information on new opportunities and constraints. While entrepreneurship is a continuous process, entrepreneurs are not continuously entrepreneurial. They act in the entrepreneurial role when new opportunities for intervention are identified. Networking provides information about risks, uncertainties, peer evaluation for successful venture creation and growth (see Table 5.1). It allows successful entrepreneurs to mobilize social resources and increase their stock of social capital.

Networks are socially embedded relationships and network ties can be of three types – information networks, exchange networks and networks of influence. All three types of network are common to entrepreneurs in general, though the relative weighting of these network ties varies by the type of entrepreneurial activity. Information networks are critical for SEs since they lack monetary resources (relative to business and public entrepreneurs) to implement their ideas. They need to convince a larger citizenry of the benefit and feasibility of their innovative solutions to alter existing urban realities. Their networks primarily work through informal channels of mentoring and social contact and their ability to effect change is based on *people* power rather than on *monetary* power even though access to monetary resources are facilitative. One of the many constraints faced by SEs is the lack of funding for innovative projects. Networks of influence and exchange with public and business entrepreneurs are also critical for the success of SEs and their networks extend to entrepreneurs of the other two sectors.

SEs focus on institutional development, primarily through establishing innovative start-ups or radical modification of existing 'not for profit' institutions. The activities of SEs are implemented through non-governmental organizations (NGOs) that start small and are community based. However, these start ups have the potential to grow in their original location and, through processes of diffusion, spread to other cities in the nation and internationally. Thus there are both innovative and imitative entrepreneurs in urban transformation through SEs. The success of innovative SEs is interlinked through knowledge networks with imitative entrepreneurs often brokered through the public sector or other resource-rich SEs such as private foundations. Institution building is the major instrument in their efforts to bring about change in perceptions of the *what and how of social innovations*.

Endogenous growth theory posits the perpetuation of positive trends through dynamic knowledge accumulation. For urban welfare, this implies endogenous growth in a neighbourhood will be sustained if the community continues to innovate and to make it more attractive for propulsive firms, diverse service providers and optimistic people to locate there either through immigration or retention of local residents. SEs perform a catalytic role in propelling growth *through stemming a downward spiral* – resulting from processes of de-skilling, widening social pathologies, loss of confidence in the area, and disinvestments by public and private sectors – by propelling quality of life improvements they are instrumental in causing an upward spiral through positive feedbacks.

Table 5.2 Urban, political, social and economic entrepreneurs:
A comparative profile

	Political entrepreneurs (PEs)	Social entrepreneurs (SEs)	Economic entrepreneurs (EEs)
Attributes and motivations	Seek political pay-offs, risk takers with strategic vision, detecting potentials in localities; flexible and patient in order to change attitudes of different types of people; PEs can influence generative allocations of urban entrepreneurial resources; (NY COMPSTAT), PEs can cause rent-seeking activities.	Aim is social value creation to improve quality of life; combine innovations, resourcefulness and opportunity to transform attributes of urban space to improve social efficiency and equality; focus on accountability to urban constituencies.	Profit bottom-line; bring to market new products and processes; reconfigure spatial competitiveness.
Activities	Change reward structures (e.g. rules for entrepreneurial activities); leverage social, public funds from private and civil society sectors; allocate revenues for innovative solutions; knowledge transfers; various partnerships with SEs and EEs; co-production of urban development with economic and social sectors. Foster opportunities and removes barriers for SEs and EEs.	Problem identification, consciousness raising; large input of vision, determination, community support but limited money; demanders of new policies; focus on risk reduction and on (change of urban location values); providers of targeted services; consensus building; communicate to and gain support from clients.	Spatial arbitrage; risk taking and new locations and ventures; production of negative externalities.
Composition	Political and elected leaders, administrators, special commissions, etc.	Not-for-profit, non-profit, private voluntary sector.	Global corporations, SMEs, real-estate interests, financial iNstitutions.

	Political entrepreneurs (PEs)	Social entrepreneurs (SEs)	Economic entrepreneurs (EEs)
Networks	Node in the flow of knowledge linking SEs and EEs; node for resource transfers through grants, loans, loan guarantees, fostering SE and EE networks and their cooperation.	Networks and connectivity within and between communities. Networks based on 'trust; with PEs and EEs; leverage community and market power for locality improvement.	Inter-enterprise networks. University/firm networks. Town/ government networks. Untraded interdependencies among sectors.

Successful urban transformation occurs only when there are economies of scope in that a variety of social entrepreneurs focus on complementary areas of activities such as housing, employment generation, crime reduction, environmental improvement through parks and recreation facilities, establishment of health clinics, drug rehab centres, and so on, in a specific location. While entrepreneurial communities do not simultaneously start all these innovative activities, activities confined to one or two types of service modifications will not be able to bring about urban regeneration in a neighbourhood or city. Communities need to be viewed through an evolutionary prism. The combination of economies of scale and scope allows communities to move into a self sustaining path of a community or a self sustaining path of a virtuous cycle over time. The more rapid the spread of complementary innovative activities, the more dynamic the neighbourhood becomes;[10] virtuous cycle over time as noted earlier. The more rapid the spread of complementary innovative activities, the more dynamic the neighbourhood becomes.

Concepts of market failure, negative externalities, and the theory of public goods provide theoretical grounding for the involvement of the public sector in urban service provision. Thus the urban public sector agents became involved during the industrial era as producers and deliverers of distributive and redistributive urban services – e.g. water, sewer, transport, housing, education (Lakshmanan and Chatterjee 1977, 2006).

Schumpeter (1984) claimed self-interest motivated actors in both the public and the market spheres. Just as market agents are primarily motivated by profit when producing specific goods, public officials favouring public projects are

---

10   When social sector agents make mistakes in judgment about resources available from governmental and non-governmental sources, overestimate the forces of change, underestimate local urban dynamics, and so on, they fail to stimulate change through new institutional development. Institutions die in their nascent stage and their efforts are lost or are adopted and modified by SEs at a later time.

primarily motivated by the desire to get elected, for enjoying political power and prestige and the other perks of office. The public sector actor's production of socially useful services are means for attaining self-interested goals. Society benefits from service innovations (as elaborated in Downs in the *Economic Theory of Democracy*, 1987). We can argue that public entrepreneurs engage in strategic action in the sense described by Habermas (1984) in which actions are taken for self-interested goals. Increase in societal resources and social welfare is a by-product of the ambitious, self-interested actions of innovative individuals in both sectors.[11]

Table 5.2 compares and contrasts the characteristics, objectives, composition, activities and the networks used by contemporary urban political, social and economic entrepreneurial agents.[12] Urban social entrepreneurs are motivated by several objectives: social value creation; the improvement of urban quality of life; the transformation of the attributes of urban space in order to enhance urban efficiency and equality; and enhancement of the accountability of different urban constituencies. By contrast, political entrepreneurs (PEs) in urban areas seek political pay-offs and direct their actions accordingly. They are risk takers with strategic views of the urban area and development potentials of localities; they are flexible and patient in order to market their ideas about these potentials and change relevant attitudes of other social, political, and economic actors. PEs can either, (in the Baumol (1990) sense), cause rent-seeking activities or influence generative allocations of urban entrepreneurial resources.

Economic entrepreneurs (EEs) in an urban area comprise of the endogenous SMEs who partner with PEs and SEs as they incubate and grow. The large Global Network Corporations (GNCs), who have economic clout and are being wooed for inward investment in the urban area, make up the other component of the urban EEs. The specific patterns of engagements and partnership of GNCs with the PEs (and with SEs in some cases) vary with the institutional context and nature of the

---

11   In contrast to this generalization, there are examples of socially-well meaning entrepreneurs in both sectors who have taken innovative decisions guided primarily by their social conscience, rather than self-interest; they are by and large exceptions to prove the rule.

12   Public policies can influence the allocation of entrepreneurship more effectively in urban areas than it can influence its supply. There are examples of productive and unproductive entrepreneurship in cities. For example, the selective use of arson for land clearance in urban sites and the resulting acquisition of capital from insurance claims may be entrepreneurial in nature but private gains are made at the cost of social loss. Gains from drug trafficking, pimping and other forms of social pathologies can be entrepreneurial but it is of the unproductive type. It is not uncommon to find corrupt officials and police aiding such non-beneficial entrepreneurial activities in the market with a partnership between creative corrupt officials in the public sector and illegal entrepreneurs in the private sector. In depressed, underserved urban areas entrepreneurial talents can be devoted to unproductive uses because of economic and socio-political rewards for such activities.

private/public sector interfaces in the countries where the urban area is located (Savitch 1988).

In terms of activities, urban SEs identify urban problems and opportunities, engage in consciousness raising, and demand new urban policies; they offer large input of vision, determination, and community support, but limited money; SEs engage in consensus building, communicating and gaining support from clients; SEs provide services targeted to residents and localities and reduce risks in specific localities (thereby enhancing development arbitrage – thus linking all three types of entrepreneurs in urban regeneration. PEs change reward structures (e.g. rules for entrepreneurial activities), and foster opportunities and removes barriers for SEs and EEs; they leverage social and public funds from public, private and civil society sectors; allocate revenues for innovative solutions; engage in knowledge transfers, and various partnerships with SEs and EEs; PEs engage in co production of urban development with economic and social sector entrepreneurial actors.

SEs, PEs and EEs build, maintain and use respective networks in their activities to advance their objectives. The networks that EEs use in dynamic regions, and those used by urban SEs in interacting with PEs are based on 'trust'. They are often deployed to leverage community power for locality improvement. Political entrepreneurial actors serve as nodes in the flow of knowledge linking PEs and SEs, for resource transfers through grants, loans and loan guarantees, and for fostering SE and EE networks and their cooperation

Entrepreneurial activities of both the public and civil society sectors are complementary even though the SEs and PEs have different bottom lines. Indeed, the convergence of the bottom lines of SEs PEs and EEs in a specific urban location confers on that location dynamic competitiveness and helps promote endogenous growth in the entrepreneurial city.

While all types of entrepreneurs act in environments of uncertainty and ambiguity, PEs are interested in reducing development uncertainties through place-specific investments in infrastructure, housing, transport and the like. PEs often partner with SEs in provision of services, because SEs are commonly early movers in sectors such as initiating health clinics, crime watch programmes, drug rehabilitation, and housing for the homeless. Creating transitional housing for the homeless, and drug rehab programmes are equity motivated, but also generate place-specific development pay-offs and efficiency gains. By reducing uncertainty and risk in certain urban localities, the activities of SEs can attract economic entrepreneurs to locate their ventures in (prior) risky areas. Dynamic efficiencies can be realized when creative ways of achieving equity are realized with the ability to use existing information on activities of SEs and PEs, and perform arbitrage – thus linking all three types of entrepreneurs in urban regeneration.

New Capabilities and Governance Models in Entrepreneurial Clusters

Two conditions, as noted earlier, must occur, if urban public, private, and social economic actors are to successfully collaborate in order to create urban value First, these actors need to acquire new economic and political capacities in order to function strategically and efficiently in the increasingly knowledge-oriented global economy. Second, these urban actors in the entrepreneurial city need to develop a framework or an institution for governing their interactions, their mutual engagement, and for enabling horizontal cooperative relationships among them.

New Economic and Political Capacities Needed to Support the Entrepreneurial City's Wealth Creation Role

To operate effectively in the globally competitive urban economy, the entrepreneurial cluster must acquire new economic and political capacities to create and maintain its competitiveness and thus support its wealth creation role. In terms of *economic capacities*, the entrepreneurial city must be able (a) to generate and apply knowledge across many urban policy areas, to enable local producers of goods and services to secure international markets and attract mobile international capital, (b) create new urban 'places' or spaces within a town, (c) coordinate different types of supportive infrastructure investment, (d) engage in restructuring urban spaces and assembling of land to be leased, and (e) negotiate with international capital in order to serve urban objectives (Table 5.3).

Table 5.3   New capacities needed in the entrepreneurial city

Economic Capacities Needed
- Ability to generate and apply knowledge across many policy areas
- Acquisition and maintenance of structural competitiveness
- Supportive economic environment (an enterprise culture, permanent innovation, labour market flexibility)

New political capacities needed
- From government to governance
- Inclusive stakeholder governance
- New mechanisms for developing strategic vision, economic decision-making and policy formation
- Methods of performance assessment and accountability (organizational intelligence/ performance accountability)
- Institutional innovation

The new *political* capacities relate to the management of the complex forms of reciprocal interdependence among the three urban (private, public and social) autonomous sectors. These interdependencies may occur in the case of specific projects jointly pursued or programmes pursued across different temporal and spatial horizons. It is under these complex circumstances that there is a movement from *government to governance*. The notion of government is tied up with who governs and who has the power to act. The more relevant issue in urban governance in contemporary urban areas, is: how is the capacity to make and implement decisions developed?

Governance is the ability to coordinate the various urban stakeholders' activities and make and implement decisions. In other words, materially interdependent, but formally autonomous agents in the private, public and social sectors must coordinate their actions in order to secure a joint beneficial urban outcome. Effective economic performance requires combining a top-down strategic vision of development with the bottom-up (market-driven) performance-oriented actions. Thus performance assessment becomes an important part of the political capacity in the city. The entrepreneurial city will require tools of ongoing intelligence, performance indicators and performance evaluation. These objectives are to be translated into specific indicators of performance which are evaluated after a period of time. Such capacities – strategic planning and performance evaluation – are part of the necessary political capacity at the urban level in the entrepreneurial city. Further, a variety of *institutional innovations* – in the form of community development corporations, business incubators, local development groups and public–private partnerships – are being developed to support the urban governance in the entrepreneurial cluster city.

The New and Evolving Context of Urban Governance

The evolution in the purposes, styles and institutions of governance in the American cities over recent decades reflects the consequences of two developments: an initial period of urban government 'failure' in the provision of (distributive and redistributive) services, and a subsequent period of increasing urban private service provision and competition in the provision of urban services. The consequence of these two developments has been a premium on innovation, new product development and consumer responsiveness in the area of urban services provision. Over the same period, there have been changes in the relationships between the urban public sector, the urban citizens and the various social sector agents. As the different urban stakeholders sharing urban space press and try to balance their conflicting claims, new rules of engagement and more negotiated relationships among the urban private, public and social sector stakeholders develop. In this context, the older urban hierarchical governmental model of vertical coordination is unable to function.

There is thus a transition from *urban government to urban governance*. Governance in this context refers to emerging novel forms of collective decision-making at the urban level. These emerging forms involve new types of relationships between urban government and its citizens, and between the urban public, social, and private sector agents. Further, issues about the purpose of government, transparency and accountability, and the construction of legitimacy in the new forms of urban governance arise and need to be addressed. (Goss 2001, Lakshmanan and Chatterjee 2006). Finally, new organizational forms such as partnerships, strategic alliances, inter-organizational webs, and virtual organizations appear. Local governance is thus carried out – through 'self-organizing' organizational networks (Rhodes 1997). Such networks link a variety of urban public sector agents, developmentally oriented private sectors and urban social sector agents.

If the purpose of public sector activities is to create a measurable improvement in urban social outcomes, such activities must add *public value* (Moore 1996). Resources utilized in the production of public sector activities are justified only when such value is added. Moore's (1996) contribution is that public sector entrepreneurial agents' activities can create innovative ways to improve urban social outcomes (in a manner similar to the way private entrepreneurs secure private value). In this view, public and social sector agents add public value by drawing on social resources such as *the resources* of *consent, compliance and public actions* (Goss 2001). By winning the consent of different stakeholders to an innovative course of action in the urban realm, these public and social sector agents engage in entrepreneurial dynamic processes. Such processes lead to innovation and increases in *public value* in the urban sphere.

As the scope of urban governance in the globally competitive urban economies widens beyond service provision to strategic and tactical activities which expand global markets for local goods and services and the creation of urban wealth, public, private and social sector economic agents cooperate in this process of endogenous urban economic development. These agents from the three sectors act not only in their respective hierarchical, market and network cultures, they also manage and work across the boundaries, in the spaces between hierarchies, markets and networks (Goss 2001).

Table 5.4   The shift from urban government to urban governance

Old	New
Command and control	Enabling and empowering
Top-down	Shared, participative
Technical/bureaucratic	Complex adaptive system
Direction	Jazz
Routine activities	Knowledge creation and diffusion

Table 5.5   The contrasts between the old and the emerging governance model

The traditional urban government	The new model of urban governance
1. Coordination of economic activities: Market and Hierarchy	1. Inclusive approach: Networks: Market, hierarchy and heterarchy
2. Intersectoral relationships: Command and control	2. Negotiations and persuasions.
3. Problem formulation: Public vs. private	3. Public and private and social sector actors.
4. Organizational strategies; Focus on use of physical and human assets to achieve professional tasks with uniformity, but without discretion and flexibility (which are necessary for responding to change)	4. Focus on urban society outcomes and ongoing resources of all sectors to create public value.
5. Transactional efficiency processing information and adapting to external signs such as prices or political pressures	5. Transformational efficiency continuous process of renewal and innovation, creating new resources and mobilizing under-utilized urban resources; helping citizens prevent and solve complex problems
6. Programmes/agencies	6. Governance tools

The key shift in the new urban governance model (Tables 5.4 and 5.5) comprises of less emphasis on the traditional *transactional efficiency* and a greater emphasis on *transformational efficiency*. The latter is a continuous process of renewal and innovation, creating new resources and mobilizing under-utilized resources, and helping citizens prevent and solve problems. In today's knowledge economy, the capacity to learn and innovate in a changing environment is important. In a rapidly evolving economic environment, the ability to make wise strategic decisions and effectively implement them depends on reducing uncertainty, engaging in ex-ante coordination of the different major economic agents who are involved in the production of that urban good or service. When different types of urban economic agents are involved in this process, they engage in collective learning. The capacity to innovate and learn is thus crucial for these three groups of urban economic agents, who have a lot of complex interdependencies, yet are autonomous with different resources and who have different time horizons and different space horizons (Table 5.2 above). In such contexts the familiar methods of coordination among the various urban economic agents – the anarchy of the exchange of markets or the hierarchy of government – become inadequate.

As a consequence, there is an expansion of networks at the expense of markets and hierarchies in such situations. The earlier discussion (in the third section) about the SMEs in dynamic regions characterized by uncertainties, interdependencies and collective learning noted that even among economic actors networks are expanding as coordinating the innovative activities (Camagni 2004, Capello 1996, Lakshmanan and Button 2009). As economic political and social sector agents engage jointly in securing urban competitiveness, the expansion of networks in the urban governance at the expense of market and hierarchies is logical.

The institutional mechanism which coordinates such complex cooperative relationships among economic actors is the network. Networks represent a new institutional form, developed in response to the changing context of desired economic interactions among economic agents in innovative regions. Networks differ from the hierarchical coordination of the firm since each participant has autonomy, and from the market since networks have 'visible hands' in the form of complex decision-making groups at multiple levels. Networks comprise sometimes only firms, at other times of firms, public sector actors and social sector actors.

In this situation of a multisectoral model of governance embracing different political, social and economic sets of actors, the inter-organizational relationships extend beyond the markets and hierarchies to a more negotiated system intended to advance innovation, flexibility and 'entrepreneurial culture'. Intersectoral relationships appear in the form of inter-firm networks, public–private partnerships, and a variety of multilateral 'negotiated economy'. Such relationships, self-organized over a period of time, have been called *heterarchy* by Bob Jessop (1997). By heterarchy is meant: rule through diversity, or how diverse agents are put together. Our preference is a term such as *hyperarchy* – after the hyper links of the internet. Basically the idea here is the rule though networks in an urban-centred knowledge economy.

Evolution of the US Multisectoral Model of Urban Governance

It is important to point out that the development of this multisectoral model of governance and the mechanisms of economic coordination have arrived in American urban areas over more than two decades of experimentation and collective learning. The US Federal government, from President Carter's time in the late 1970s to the recent times of President Clinton, supported urban government experiments in the new forms of entrepreneurial city governance. In other words, many urban regions gained experience in interagency and intersectoral coordination. What is experience, except another name for past mistakes and subsequent learning? US Government provided these urban areas with the resources to make the mistakes and learn from this process. The Federal government provided economic incentives over this 20-year period. for experimentation with economic development, intersectoral and interagency cooperation and coordination. Take for example, the Economic Development Administration (EDA) Title Two – business incubation; Housing and Urban Development (HUD) Title Nine – special economic adjustment assistance; Community Development Corporations; the Community Development Block Grants (CDBG), which is urban development action grants.

These programmes provided much needed resources, much needed funds awarded on a competitive basis not only to the urban government, but also to urban private entrepreneurs, and to urban social entrepreneurs for targeted service delivery. All this created an ambience in which social and political entrepreneurial agents could get experience in urban development and regeneration, and all three sectors could function *entrepreneurially, competitively and cooperatively*. An illustrative innovation of this era is the adaptive reuse of redundant structures. for example, the Piano Factory in Boston, which created a mixed income housing for artists, elderly, and so on. And in the process, they created this ability to coordinate a variety of economic agents – which resulted in an ensemble of socially embedded institutions and practices to realize community competitiveness.

Evolution of Entrepreneurial Clusters: Illustrative American Experience

Various American examples of entrepreneurial clusters have clearly differential performance capabilities. There are cities such as New York, which are key nodes in the global economy and serve as strategic poles of global economic governance (Sassen 2000). A dynamic social entity such as New York not only acquires and deploys the knowledge necessary for its endogenous growth but also the social capacity to monitor, analyse and guide the efficient performance of various dimensions of the city's and global economies (Lakshmanan et al. 2000). These activities constitute social learning and require the creation of institutions, policy-making, and incentives for continual learning and adaptation to changing circumstances

More generally, how well a city performs depends upon factors such as its market power in terms of its attractiveness to international capital. Big cities such as London and New York may pose such an attraction to international capital, which allows the public and social sector actors in those cities more elbow room in intersectoral negotiations (with international capital) and deciding their own strategies. Such ability to design their own context instead of adjusting to it characterizes the recent experience of large cities such as New York, Paris and London (Savitch 1988). The extensive inter-governmental support for the regeneration of the city can strengthen the performance of the entrepreneurial city (e.g. Paris), enhancing the participation of the various interest groups, extending their voice, and strengthening the city. In contrast, the low-ranking global cities may be trapped in the bottom, with limited strategies for attracting mobile capital.

Table 5.6 selectively provides a flavour of the variety of institutional forms found in one entrepreneurial city, namely, Boston. Moreover, the success or failure of these institutions depends on how consistent they are with the individual city's socio-political climate and economic context – all of which are highly dynamic in the entrepreneurial cities. The discussion here provides only a framework for elucidating the characteristics and variety of institutions. For each institutional form, the table identifies *objectives sought, types of policy instruments (financial*

Table 5.6   Entrepreneurial city's objectives, institutions, instruments and infrastructure

Institutional forms	Types of Instruments*			Infrastructure		Objectives
	Financial	Fiscal	Physical	Administrative	
Community Development Corporation (CDC).	Community bloc development grants, other project grants, venture capital pools.	Tax abatement for city.	Capital improvement in streets and neighbourhoods.	Streamlining licensing and permitting.	Downtown revitalization.
Local development councils and business incubators.	Use of various debt instruments such as revenue bonds, venture capital pools.	Tax increment financing.	Industrial parks on greenfield and brownfield sities.	Skilling networking.	Start-up companies.
Public–private partnership for employment creation.	Loan guarantees.	Special assessment districts.	Historic and landmark preservation.	Regional economic development groups.	Advertisement and tourism development.
Empowerment zones and enterprise communities. EZ/EC (Partnership but Fed/State) city gov't and private.	Federal grants for financing capita.	Property improvement tax abatement; income tax credit; sales tax exemptions.	Transportation assistance for commuting; worker training.	Creation of business technology centre; market associations.	Retain existing industry; attract new investment; foster local entrepreneurship.
Reibuild Boston Energy Initiative (partnership of city government, power companies, community organizations and energy consultation).	Funding for energy companies.	–	Retrofitting buildings; consulting new construction.	Linkage to business owners via Boston Office of Business Development.	Upgrade energy efficiency of new and retrofitted facilities; reduce factor costs, attract new investment.

Note: * A selected and illustrative list. Not all instruments were used by all cities. The package varied between cities and in time within a city.

and the underlying fiscal), and physical and administrative infrastructures used to promote objectives. To take a specific example, start-up companies were co-produced, in areas adjacent to universities (e.g. MIT, Harvard, Boston University) in Cambridge and Boston Massachusetts combining University research and development (R&D) activities with those of local development councils, adopting a business incubator model, involving the participation of entrepreneurial firms tapping into public and venture capital seed funds. Towards this end, infrastructure investments were made in industrial parks on greenfield and brownfield sites by both the public, not-for-profit and for-profit private sector institutions to sponsor flexible and sophisticated human capital creation. The financial instruments for the development of this infrastructure capital could be drawn from the issuance of debt instruments such as revenue bonds and tax increment financing by the public sector and capital development loans by the universities. These are complicated processes not only in the variety of institutions involved, but also in the diversity of the vested interests of each institution. The public sector is increasingly adopting a catalytic and enabling role in protecting public interest while reducing its own service delivery responsibilities.

Table 5.6 does not provide an exhaustive list because the variations in time and space are numerous and specific institutional forms derive from the geographical and historical context of the city. Since globalization affects cities differentially, responses vary according to local considerations of ameliorization, local resource capacities, and local abilities to mobilize social and political capital and institutional history. The objective here has been to provide a flavour of the great diversity of institutions involved in the American entrepreneurial city and to show the complexity involved in co-production strategies growing out of the continuing experiment of fostering partnerships between the numerous public and private sector stakeholders. Policies are most effective when the public sector can catalyze private and social institutions with diverse instruments to work together for the common urban welfare.

Bibliography

Acs, Z.J., de Groot, H.L.F. and Nijkamp, P. 2002. *The Emergence of the Knowledge Economy*. New York: Springer-Verlag.

Alter, C. and Hage, J. 1997. *Organizations Working Together: Coordination in Interorganizal Networks*. Beverly Hills: Sage.

Amin, A. and Cohendet, P. 2004. *Architectures of Knowledge*. New York: Oxford University Press.

Baumol, W. 1990. Entrepreneurship: Productive, Unproductive and Destructive. *Journal of Political Economy*, 98, 893–921.

Bornstein, D. 2004. *How to Change the World: Social Entrepreneurs and the Power of New Ideas*. New York: Oxford University Press.

Boschma, R.A. and Lambooy, J.G. 1999. Evolutionary Economics and Economic Geography. *Journal of Evolutionary Economics*, 9, 411–29.

Braudel, F. 1992. *Perspective of the World Civilization: Civilization and Capitalism in the 15–18th Century*. Berkeley, CA: University of California Press.

Camagni, R. 2004. Uncertainty, Social Capital and Community Governance: The City as Milieu, in *Urban Dynamics and Growth*, edited by R. Capello and Peter Nijkamp. Amsterdam: Elsevier, 121–50.

Capello, R. 1996. Industrial Enterprises and Economic Space: The Network Paradigm. *European Planning Studies*, 4(4), 485–98.

Chandler, A.D. 1977. *The Visible Hand: The Managerial Revolution in American Business*. Cambridge, MA: Harvard University Press.

Chinitz, B.J. 1961. Contrasts in Agglomeration: New York and Pittsburgh. *American Economic Review*, 51, 279–89.

Develaar, E.J. 1991. *Regional Economic Analysis of Innovation and Incubation*. Vermont: Avebury.

Downs, A. 1987. *Economic Theory of Democracy*. Reading, MA: Addison Wesley.

Glaeser, E.L. and Gottlieb, J.D. 2009. The Wealth of Cities: Agglomeration Economies and Spatial Equilibrium. *Journal of Economic Literature*, 983–1028.

Goss, S. 2001. *Making Local Governance Work*. New York: Palgrave.

Granovetter, M. 1985. Economic Action and Social Structure: The Problem of Embeddedness. *American Journal of Sociology*, 91, 481–510.

Habermas, J. 1984. *The Theory of Communicative Action, Volume 1: Reason and the Rationalization of Society*. Boston, MA: Beacon Press.

Hoover, E. 1948. *The Location of Economic Activity*. New York: McGraw-Hill.

Isard, W. 1956. *Location and Space-Economy: A General Theory Relating to Industrial Location, Market Areas, Land Use, Trade and Urban Structure*. Cambridge, MA: MIT Press.

Jacobs, J. 1969. *The Economy of Cities*. New York: Random House.

Jessop, B. 1997. The Entrepreneurial City: Re-imaging Localities, Redesigning Economic Governance, in *Realizing Cities: New Spatial Divisions and Social Transformation*, edited by N. Jewson and S. MacGregor. London: Routledge, 28–41.

Krugman, P.R. 1991. Increasing Returns and Economic Geography. *Journal of Political Economy*, 99, 483–99.

Lakshmanan, T.R. 1993. Social Change Induced by Technology: Promotion and Resistance, in *The Necessity of Friction*, edited by Nordal Ackerman. Berlin: Springer Verlaag, 135–59 (reissued as paperback, Westview Press, 1999).

Lakshmanan, T.R. and Button, K.J. 2009. Institutions and Regional Development, in *Handbook of Theories of Regional Growth and Development*, edited by R. Cappello and P. Nijkamp, New York: Elsevier, Chapter 22.

Lakshmanan, T.R. and Chatterjee, L. 1977. *Urbanization and Environmental Quality*. Resource Paper (RP 77–1). Washington, DC: Association of American Geographers.

—— 2003. *The Entrepreneurial City and the Global Economy*. Paper presented to the International Workshop on Modern Entrepreneurship, Regional Development and Policy, The Tinbergen Institute, Amsterdam, 23–24 May.

—— 2006. The Entrepreneurial City in the Global Marketplace. *International Journal of Entrepreneurship and Innovation Management*, 6(3), 155–72.

Lakshmanan, T.R., Andersson, D.E., Chatterjee, L. and Sasaki, K. 2000. Three Global Cities: New York, London and Tokyo, in *Gateways to the Global Economy*, edited by Å. and D.E. Andersson. Cheltenham, UK: Edward Elgar, 49–82.

Malecki, E. 1994. Entrepreneurship in Regional and Local Development. *International Regional Science Review*, 16(1–2), 119–54.

Marshall, A. 1890. *Principles of Economics*. New York: Prometheus Books.

McCann, P. 2005. Transport Costs and New Economic Geography. *Journal of Economic Geography*, 5(3), 305–18.

Moore, H. 1996. *Creating Public Value, Strategic Management in Government*. Harvard, MA: Harvard University Press.

Polyani, K. 1957. *The Great Transformation: The Political and Economic Origins of our Time*. Boston, MA: Beacon Press (Original published in 1944).

Pred, A. 1977. *City-Systems in Advanced Economies: Past Growth, Present Processes and Future Development Options*. London: Hutchinson.

Quigly, J.M. 1998. Urban Diversity and Economic Growth. *The Journal of Economic Perspectives*, 12(2), 127–38.

Rhodes, R. 1997. *Understanding Governance, Policy Networks, Reflexivity and Accountability*. Buckingham: Oxford University Press.

Romer, P. 1986. Increasing Returns and Long-run Growth. *Journal of Political Economy*, 94, 1002–37.

Rosenthal, S.S. and Strange, W.C. 2004. Evidence on the Nature and Sources of Agglomeration Economics, in *Handbook of Regional and Urban Economics: Cities and Geography*, edited by J.V. Henderson and J.F. Thisse. Amersterdam: North Holland, 2119–72.

Sassen, S. 2000. *Cities in a World Economy*. Second edition. California: Pine Forge.

Schumpeter, J.A. 1984. *Capitalism, Socialism and Democracy*. New York: Harper Collins.

Savitch, H.V. 1988. *Post-Industrial Cities: Politics and Planning in New York, Paris, and London*. Princeton, NJ: Princeton University Press.

Stohr, W. 1989. Local Development Strategies to Meet Local Crisis. *Entrepreneurship and Regional Development*, 1(3), 293–300.

Storper, M. 1995. The Resurgence of Regional Economies, Ten Years Later: The Region as a Nexus of Untraded Interdependencies. *European Urban and Regional Studies*, 2, 191–221.

Vernon, R. 1960. *Metropolis 1985*. Cambridge, MA: Harvard University Press.

Von Hippel, E. 1988. *The Sources of Innovation*. Oxford: Oxford University Press.

Williamson, O.E. 1985. *The Economic Institutions of Capitalism*. New York, Free Press.

Chapter 6
Sustaining the Creative City: The Role of Business and Professional Services

Peter Daniels

Introduction

The modern (creative) economy thrives on ideas rather than physical capital so that the principal producers of wealth are information and knowledge (Drucker 1993). Thus, the economic value of goods and services is increasingly determined by their knowledge content; the facts or ideas acquired by study, investigation, observation, or experience. For at least a decade science, technology and industry policies have been promoted under the banner of 'knowledge-based economies' or 'creative-knowledge cities' where growth is driven by high-technology investments, high-technology industries, more highly-skilled labour, and associated productivity gains (Brinkley 2008, OECD 1996). Industries that are producing and distributing knowledge and information rather than goods have become the benchmarks for national, regional or urban economic well-being. The importance, for example, of knowledge-intensive business services as sources of knowledge creation and exchange with local actors in urban regional economies is now very evident (Koch and Stahlecker 2006, Smedlund and Toivonen 2007, Wood 2002).

Creativity[1] is central to successful economic development in the twenty-first century as demonstrated through the efforts of cities to expand their knowledge-based industries and creative economy and to foster arts and culture (see for example Brown et al. 2007). While notions of the creative city or creative activities are far from consensual or unambiguous (see for example, Costa et al. 2008), the degree to which policy-makers can assemble a 'creative city', i.e. an urban environment capable of generating creativity, innovation and thus economic growth, has been the subject of much thought and analysis in recent years (Currid 2006, Oakley 2004, Pratt 2008, Scott 2006). While it is unlikely that such environments can be constructed from scratch, influential advocates such as Florida (2002, 2005, 2008) suggest that it is possible to foster the creative city. For Florida cities need to become magnets for the creative class as exemplified by the three 'Ts', namely Talent (have a highly talented/educated/skilled population), Tolerance (have a

---

1 Personal ideas turned into public ideas, products and services, see www.creativeeconomy.com/#.

diverse community, which has a 'live and let live' ethos), and Technology (have the technological infrastructure necessary to fuel an entrepreneurial culture).

This triumvirate is central to the Florida philosophy and it can at least be inferred from his diagnosis that the *in situ* attributes of a city's human resources are either already up to the required level of creative capacity or can be supplemented by attracting the appropriate talent. The idea that the human resources of a city or region can be retrained and upskilled as a substitute for importing – or perhaps complementing the import of – the creative class is not really explored. The reality for many cities, especially those struggling with significant industrial restructuring, is that they will likely be below even the minimal expectations for a 'magnet' that is attractive to the creative class. This being the case, there is no alternative but to contemplate endogenous enhancement of their human resources. Saxenian (2006) for example acknowledges, using detailed time-series evidence for Silicon Valley, that sustaining regional advantage requires the recombination of both local and distant know-how via local knowledge-producing and circulating mechanisms. Indeed, as Florida (2005, 2008) also shows, cities that are relatively weak with respect to the three 'Ts', such as Louisville, Buffalo or New Orleans, have struggled to clamber on to the creative cities bandwagon while those with superior cultural, creative and technological milieu, such as Seattle, Portland or Austin have been much more successful.

It follows then that cities that are less well placed to attract the creative class and creative activities need to work even harder at sustaining their creative capacity through a focus on their endogenous capabilities. Fostering an indigenous culture of innovation as part of an integrated environment that supports different types of creativity is vital in that 'economic development planning is increasingly emphasizing the human capital side of the growth equation, stressing occupation as well as industry in analysis and policy' (Markusen 2006: 2). This ranges from an arts and culture perspective (Markusen 2006) to activities such as the incorporation of specialist knowledge and expertise into the design of competitive metal products in cities that are still home to traditional heavy industries (Bryson et al. 2008). Even if we can answer questions about what types of cities are creative or whether there is a common set of urban policy initiatives that can be used by creative cities to achieve their objectives (see Wu 2005) there are likely to be obstacles to overcome when the objective is to improve the indigenous capacity of a city to sustain its creative capital. One such obstacle is based on the synergy between the capacity of business and professional service (BPS) firms to provide intermediate inputs of knowledge and expertise and the development of the human resource base of a city region. The tacit knowledge delivered by BPS firms is embodied in people and if the exchange of this, and codified, knowledge between producers and users is to add and stimulate innovation or creativity there needs to be a continuous process of skill upgrading and enhancement (Schleicher 2007). Some of the dimensions of this requirement are explored using the results of a survey conducted in the city of Birmingham, UK.

Sustaining the creative knowledge economy of the city is therefore dominated by demand for more highly skilled workers; for example, the relative ubiquity of information technology ensures that workers need to be better educated, more skilled, and prepared to supplement their skills on a long-term basis in order that they can adapt to ongoing advances in technology.[2] Skills development is shaped by a triumvirate of actors: individuals (or firms), providers of education and of skills training, and public policies that measure and monitor skills needs and develop policies and initiatives that ensure a productive dialogue between the other two parties. Upgrading human capital through access to a range of skills, and especially the capacity to learn, should be at the forefront of public policies.[3]

Rhetoric and Reality

Understanding that the incorporation of knowledge into all forms of creative production is now *de rigueur* and the actual experience of addressing it at national, regional or city level tends to diverge. In the case of the UK the record on the enhancement of workforce skills is patchy; there is a relatively high proportion of people with higher degrees but it is still well behind the US and slightly behind Germany (Her Majesty's Treasury and Department of Trade and Industry 2005). The proportion of the UK population qualified at the intermediate level is also lower than the US and Germany, and there are particular concerns about the quality of vocational qualifications. Critically, the UK has approximately twice the proportion of people with low-level skills as, for example, Germany and the US. The situation is no less encouraging in relation to management skills; international surveys of business executives on the perceived competence and experience of UK managers indicate that they lag behind their colleagues in the US, France and Germany (Her Majesty's Treasury and Department of Trade and Industry 2005).

This is not to suggest that skills improvement is not occurring but progress is far from consistent, especially in relation to intermediate and basic level skills. The skills gap at national level in the UK is inevitably replicated (and even amplified) at city/regional level; employers such as those in BPS require a pool of human resources with the skills that will support the success of their business and employees require an appropriate portfolio of skills that will make them employable, especially in city regions to which BPS tend to gravitate (Department for Education and Skills 2003). Such duality of skills needs requires fine tuning to match those identified by regional development agencies such as Advantage West

---

2   More than 50 per cent of firms in Shanghai and Beijing cited scarcity of a critical skill as hampering their growth and the large majority of firms that lacked a necessary skilled workforce tried to look for a worker with such skills within the local or regional market (see Table 22 in Wang and Tong 2005).

3   Wu (2005) notes that having a readily available and qualified workforce is one of the best investments cities can make (see also Sommers and Carlson 2000).

Midlands (AWM)[4] that have been tasked with creating an environment that links employment and skills in ways that will fulfil the needs of individual workers and firms, as well as enabling innovation and economic growth (Office of the Deputy Prime Minister 2006).

Challenges for Less Advantaged Cities: The Case of Birmingham

The Birmingham city region has been undergoing a painful process of retooling its economic activities to confront the challenges presented by globalization, flexible specialization, or the drive towards a creative economy (Advantage West Midlands and Birmingham City Council 2002).[5] National and regional-level data such as the Annual Business Enquiry (ABI) (National Statistics 2008) or the National Employers Skills Survey (Learning and Skills Council 2005) provide time-series and sector-level information on trends in education levels or some aspects of skills acquisition but more nuanced insights that are also more focused on particular business sectors such as BPS can only be derived using primary data (Daniels and Bryson 2006).[6] A profile of the BPS firms surveyed in Birmingham is provided in Figure 6.1.[7]

With the exception of computer, marketing and design services, more than one-third of the sales/fee income of the firms in the survey originated from clients located within a 10-mile radius (Table 6.1). This helps to underline the importance of endogenous enhancement of employee skills and training if the

---

4   Birmingham is the principal city within the government region covered by Advantage West Midlands.

5   Birmingham is home to a thriving community of creative industries including jewellery, visual and performing arts, craft and design, creative writing, publishing, PR, marketing and advertising, software design and new media, film, television, information and communications technology (ICT) and games, radio, music, music technology and photography. As well as being home to major UK broadcasters such as BBC, ITV Central, and newspaper publisher Trinity Mirror there are a wide range of art galleries and theatres, and the city is home to the Royal Ballet, the Birmingham Opera Company and the City of Birmingham Symphony Orchestra (CBSO). The creative industries are one of the key sectors of growth and expansion in the region, and the many opportunities on offer embrace and enrich the wide diversity of talent located in this unique city. There are major developments planned throughout the Creative City and many are already underway throughout Eastside, the Jewellery Quarter and the Custard Factory, all of which create a focus for further growth and support for the industries in the coming years. See www. birmingham.gov.uk/creative.bcc.

6   The analysis comprises a subsample of a survey of BPS firms conducted in 2005 that included 1196 telephone survey respondents, 206 in-depth interviews and six focus groups at locations across the West Midlands Region.

7   547 telephone interviews along with 93 in-depth face-to-face interviews with a sample of the telephone survey respondents were conducted in Birmingham and Solihull in 2005.

creative knowledge function of BPS can contribute to sustaining the innovation, productivity and competitiveness of other businesses in the city.

The challenge for Birmingham is how to create the conditions conducive to triangulating the learning and skills needs of individuals, organizations (firms and educational institutions), and policy in circumstances where 'people are at the

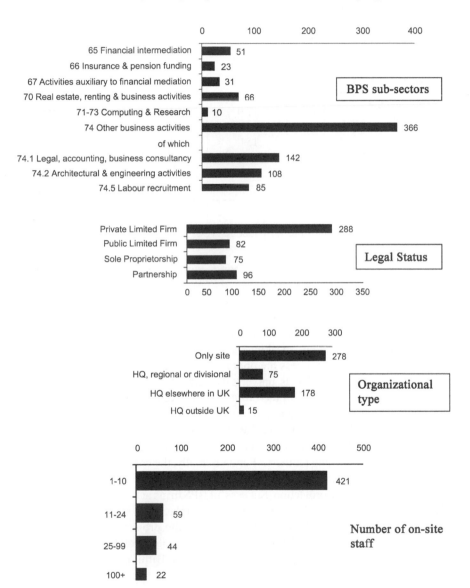

Figure 6.1   Characteristics of BPS firms, Birmingham, 2006

Source: Daniels and Bryson 2006.

Table 6.1   **Location of BPS clients, by share (%) of sales/fees and sector, Birmingham and the West Midlands, 2002**

Location of client	Banking and Insurance	Consultancy and accountancy	Computer, marketing and design	Legal and property services	Total
Local (within 10 miles)	34.9	35.7	18.7	38.3	32.8
Rest of West Midlands	27.8	26.7	25.7	27.7	27.0
London and the South East	4.0	17.0	14.4	7.7	11.3
Rest of the UK	29.8	15.0	30.7	20.1	22.5
EU	0.9	2.1	3.9	2.2	2.3
Rest of the world	2.8	2.1	6.3	1.7	3.0
	100	100	100	100	100

Source: Daniels and Bryson 2002, Telephone Survey.

heart of an economy's competitiveness and, within the right environment, can have a dramatic impact on economic growth' (Accenture 2007, 2). As a minimum this requires the delivery of a mix of knowledge that is specific to an activity (commercial knowledge, languages, engineering, science, mathematics) and enabling skills such as problem solving, adaptability, team working, informational (literacy, numeracy, writing, speaking), and an ability to use existing and new technologies. One way to assess the opportunities and challenges is to undertake a case study of a sector that is universally acknowledged to be critical to the knowledge transfer process and therefore the creative capacity of cities (Daniels and Bryson 2006). The role and quality of BPS as sources of knowledge and expertise for firms located in Birmingham is crucially dependent upon the learning and skills acquisition of their employees as well as the management attributes of their owners and managing partners.[8] Fundamentally, the competitive advantage of all BPS firms rests upon a combination of reputation, expertise, client relationship management and image projection. Successful BPS firms act strategically to ensure that the firm is highly visible in its chosen marketplace; the same is also the case for BPS professionals. Visibility is dependent on the projection of personalities and a professional image for the delivery of commercially relevant expertise. For example, for a medium-sized property consultancy in Birmingham:

---

8   Many of the firms in the BPS sector are organized as partnerships rather than as limited companies.

> Our competitive edge is good communication skills and people that have actually got quite a good outgoing personality. So it's communication skills and personality. It is facility with the written word. So it's being able to put together a good report. It's people that are proactive that will go and suggest ideas to people. … So we are basically looking for good communicators, ideas people, proactive people and what also gives us hopefully an edge is that we have a range of different skill sets and personalities within the office. So we have some people for example that are perhaps slightly introverted, but they just get their head down and they grind through things and we have some people that are terribly, 'hi', 'how are you?', 'great' – maybe not that good at grinding, but the combination of the two together works very well. (Daniels and Bryson 2006)

Assuming that it is important for BPS firms to optimize the use of their existing human capital and achieve high performance, very little is actually known about their demand for skills development, whether firms in the sector actively identify the training and skills requirements of their employees, how these needs are fulfilled by a variety of alternatives, what resources are made available, and how the suppliers of training are identified and evaluated.[9] This docs assume that skills enhancement is an integral part of the business development strategies of BPS firms. However, the majority are small businesses (most employ fewer than five persons) for which employee skills enhancement is but one of several competing priorities. Nor is it not necessarily as high up the list of priorities as, for example, meeting client deadlines or identifying and negotiating new business. There is a correlation between the adoption of high performance work practices (HPWP) by business services firms and the delivery of adequate training, employee development provision, and the management of change but it is only a realistic option for large BPS firms (Sung and Ashton 2004). For the large majority of BPS firms, skills development is achieved on a bespoke basis and is often dependent on individual employees taking responsibility for their own training and education.

Since, in the UK at least, the locally embedded supply of BPS is more prevalent in every region, other than London and the South East, than imports it will be apparent that endogenous skills and training needs will be crucial unless demand can be fulfilled via transfers from elsewhere. The norm that has persisted since at least the middle of the last century is an out-movement of the most talented and skilled (often graduate) employees from the regional cities to London. BPS firms are at least as, if not more, vulnerable to this process. A large property consultancy firm commented that it is:

> … not just Birmingham, compared to London. I think when the economy is in an upswing and the property market is going well then the brighter, younger

---

9   Information about the supply of skills and training, whether provided by the private or the public sector, is more readily available via organizations such the Learning and Skills Council, but the extent to which it is matched to demand has not been mapped.

people are attracted to London. We have a huge number of graduate applications
… the commercial business probably had 700 applicants last year for something
like fourteen or fifteen places in London. We're taking two or three graduates
this year into Birmingham – we've recruited one so far and I didn't count them
up but I suppose the number of graduate applicants to the Birmingham office
would probably be between 75 and 100 but the brighter ones are attracted to go
to London and it is quite difficult to attract the really good people to come and
work in the regional markets. They may do so later on in their career when they
have had a spell in London … those who are brighter and have initiative and are
really prepared to get stuck in as it were are not prepared to settle for second
best – they will be attracted by the bright lights and the more active market in
London. (BS33/Real/25–99).[10] (Daniels and Bryson, 2006)

Other respondents noted that many professionals moved between firms within
Birmingham and that the market could be very competitive; sometimes larger firms
engaged in a sort of 'Dutch auction' to attract professionals (BS46/Leg/100+).

The recruitment and retention in Birmingham of good quality professionals
and graduates is actually a challenge for the minority of large and medium-sized
BPS firms. Most of the micro, and many of the small BPS firms did not mention
the London effect; for many of them recruitment is an unusual and special event
so that '… the last person I took on was seven years ago. The first one before that
was fifteen years ago' (BS81/Acc/1–10).

It is not surprising then that the transformation to a creative, knowledge-based
economy requires, as a minimum, stemming the outflow of skills and expertise.
While initiatives such as Birmingham Future (www.birminghamfuture.co.uk/)[11]
are designed to retain graduates from the city's three universities, it is also
necessary to improve the skill levels and training of those already employed by
BPS firms or those who do not possess the education and skill levels that permit
them to compete for jobs in the sector.[12] Some of the challenges involved will now
be briefly explored.

Skills Shortages

BPS firms face real difficulties recruiting experienced people able to make an
immediate contribution to the business; hard-to-fill vacancies span the full range
of occupational demand (Figure 6.2). As to the reasons why vacancies are difficult
to fill, by far the most important constraint is the low number of applicants with the

---

10   Location and Firm ID No/Firm type/Size group (no of employees)

11   See also Daniels (2003).

12   There are three universities in the city: University of Birmingham, Aston University
and Birmingham City University. There are also several other universities within easy reach:
University of Warwick, Coventry University, University of Wolverhampton, Staffordshire
University and University College Worcester.

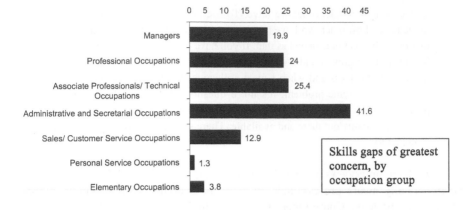

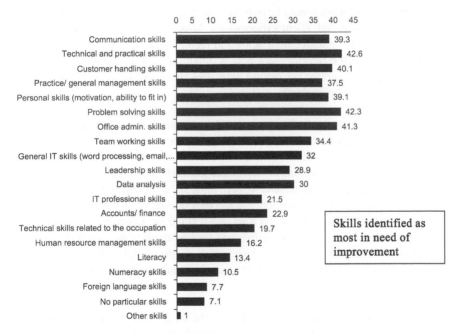

Figure 6.2   Occupations in which the skills deficit is causing concern and the skills most in need of improvement, BPS firms, Birmingham, 2006

Source: Daniels and Bryson 2006.

requisite skills; some 60 per cent of the firms with difficult to fill vacancies cited this reason. Applicants also lacked the qualifications or experience that the firms were looking for with hard to fill vacancies:

> ... the biggest problem that we have. To the point whereby it does have an influence on how much work we take on. So you could say that it's a limitation to growth. It's as fundamental as that. If you have to put it in a strategic context, it is preventing growth and it's not just my office, not just this office it's right through the business and when I meet colleagues in other businesses they've got exactly the same problem. It's not a question of there not being enough people out there who would like to work for us, but there's not enough people of the right quality out there and available. There's a shortage of supply. (BS54/Acc/25–99)

The in-depth interviews revealed that BPS firms found it difficult to recruit commercially aware professionals, as well as customer-focused and/or, work-focused individuals. Commitment to reliably fulfilling client needs was seen as very important, notably a willingness to work after 5 pm or as long as it took if this meant that the work would be delivered in time and at a quality that matched, and preferably exceeded, client expectations. Many of the firms referred to a gap between 'self skills', management skills and technical skills. The latter is of much less concern than the development and acquisition of management and self or softer skills that generally lag behind activities that enhance technical expertise (Figure 6.2). It is not just about:

> ... technical skills, that is first awareness of the market ... it is personality, being able to communicate, ideas, lateral thinking, business acumen. So the ability to see opportunity, go and find it, go and acquire it and at the end of the day we're a business, we're about making money for our shareholders as well as providing a cracking service to our clients and we need those skills in our surveyors. They're not there just to do a technical job; you can only do a technical job once you've got the clients and the jobs to do it on and it is our job to get in high quality work, process it, give the best advice, look for opportunities and maximize our position in the market and make money. That requires people with a skill set which is above just being able to do the technical work and that is where we run into problems. (BS33/Real/25–99)

The small and medium enterprises (SMEs) in the study made frequent reference to multitasking as a common expectation; especially for professional staff working in environments where the pool of support staff is often limited (some firms had no support staff at all). As a result, professionals have no choice but to undertake a range of basic and more advanced support tasks at which they are not necessarily proficient and could benefit from training in. However, time is at a premium in SMEs and any such training can often only take place at the expense of fee earning or income generating activities. It therefore tends to be overlooked.

It seems then that while it may be taken as read that success in the world of BPS firms is attributable to individuals and firms acquiring and continuing to develop technical expertise, there are two additional requirements. The first is an ability to transform technical expertise into something that has commercial value by providing distinctive inputs into the activities of client firms. The second is a requirement for people-focused skills involving interpersonal, presentation and communication techniques; these can be subsumed under the label 'impression management' or 'client relationship building'. The second is relatively easily acquired but time constraints are a major obstacle for SMEs while client relationship building is more elusive if employees do not possess the right personality.

Use of Staff Training

The difficulties associated with acquiring staff with the appropriate skills means that the provision of training for professional and support staff is an important route for BPS firms. Such training can take place in-house with or without inputs from private sector providers or may be sourced externally via the private sector or using public sector providers such as universities or colleges (Figure 6.3). In view of the resource and time constraints, many of the smaller BPS firms only engaged in essential training of the kind required as continuous professional development (CPD) which, if not undertaken, would debar them from practising; for example, lawyers. Most CPD training is obtained via professional bodies such as the Institute of Accountants or from private training providers. Some firms rely on distance learning packages that are delivered using electronic media such as CDs.

Other firms suggested that not only did they not have the time or resources for training but in any event it was best undertaken 'on-the-job' since this is most likely to achieve business- or commercially relevant training.

Investment in staff development and training is mainly made by medium- and larger-sized firms which are also the training grounds for BPS professionals who may ultimately move out to establish independent businesses. There is less investment in the skill needs of support staff in BPS firms; they are assumed to be competent users of key office-based business software so that:

> ... generally speaking we find a lot of secretarial staff now – tend to be fairly experienced at using that sort of package, Word, Excel and presentation packages and things like that. Where we have new accounting packages and staff need to use those to print off accounts we again have in-house training for them, or sometimes the software provider might, you know, come in and spend a morning or afternoon with people depending on what's required. (Bs11/Acc/11–24)

However, important skill deficiencies that hold a business back still occur. The partners in one firm were surprised that the majority of the support staff were unable to take full advantage of the firm's phone system, a deficiency that evidently impacted the effectiveness of the business. The problem was rectified by a telephone training company that was employed to provide intensive telephone training sessions (BS13/Leg/100+).

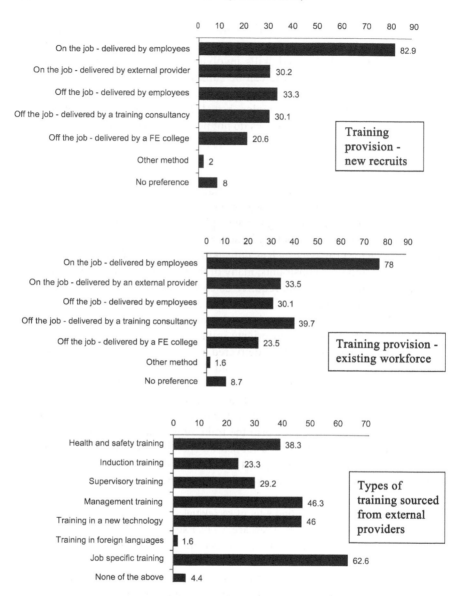

Figure 6.3   On and off the job training providers and principal types of training sourced from external providers, BPS firms, Birmingham, 2006

Source: Daniels and Bryson 2006.

Many of the larger firms have well-developed and evolving staff appraisal systems that are designed to identify training requirements. A branch office of a factoring firm that was part of one of the banks was:

... in the process of trying to implement training plans for each member of the team and we've got a team of about twelve so we tend to look at things like internal training programs and we're members of what is called the FDA (Factors Discount Association) so we try to link in with some of the courses that they might run but we do look at other courses. We are keen for people to develop their own skills and invest in them so if people turn round and say I want to go and learn more of that using spreadsheets, databases, Word and all the rest of it or using computers and we have had people going round the corner to Learn Direct and places like that. (Bs96/Acc/11–24)

Smaller BPS firms are much more likely to adopt much more informal approaches to training that may involve a mix of in-house self training, mentoring by more experienced or knowledgeable employees, and the occasional purchase of training by specialist providers either in-house or elsewhere. Typically:

... we have a sort of continuous training programme, we use outside training organizations, the accountancy training tutors who sort of provide training for professional staff and keeping up with the latest legislation etc. You know we're required by the institutes to make sure all our staff are up to date and sort of by talking to the staff, identifying what areas they feel they need extra help in, you know, I mean we're a small firm so we tend to know what peoples strengths and weaknesses are and try to cover those areas wherever possible. ... the senior staff here, we also help by making our experience available, and also by saying 'this is the way we have to do these things' and explain things. We have in-house reference books etc. which people can use. (Bs11/Acc/11–24)

Training provided by further and higher education institutions is an option for BPS firms located in Birmingham, but the survey respondents indicated that they were discouraged from using these, often cheaper, sources of training because of the rigidity of the course offerings and curricula, and the extended timescales required for attendance and qualification for an award or certificate (several weeks or a year, rather than short, intensive delivery). A particular problem is the mismatch between curricula and the real-world skills needs of staff (especially professionals), a limitation that is exacerbated by the absence of any real dialogue about training needs and opportunities between college-based trainers and BPS firms in the city. While some exceptions were cited, these were outweighed by the references to the absence of firm/employee engagement in course design and content at most colleges. Typically, BPS firms are only interested in using training courses that are:

... short and sharp or they are refreshers so they are half-day sessions or evening sessions or maybe one-day sessions but unless somebody is doing a particular further qualification say in investment or doing an MBA we're not really getting involved in the universities or the college ... . (BS33/Real/25–99)

Although there were difficulties and also misunderstandings about the availability and suitability of training provided by educational institutions, further education colleges engaged more with BPS than universities. Nevertheless, one firm had recently discovered that it could recruit graduates from non-cognate disciplines and train them to fit its own requirements using a course provided by a university in Birmingham:

> ... initially those non-cognate graduates will be less productive because they haven't done the course and they will go off for a day a week for a two-year course – the local course is run at UCE (University of Central England[13]). It would be nice if Birmingham University ran one I should say ... I think it would be a very interesting thing for the university to do. UCE are one of our clients, they are great, but Birmingham University is one of the key red bricks in the UK and having the high quality property real estate course there either as degree or postgraduate course would be I think a very useful adjunct. (BS33/Real/25–99)

There is clearly a good deal of latent potential for further and higher education institutions in the city to provide specialist training and CPD for BPS firms that could also be delivered to clients elsewhere in the country as well as locally. However, there was frequent mention by the survey firms of an absence of any real understanding, and even willingness, on the part of most education institutions to enter into a dialogue intended to generate a better understanding by both parties of specific training needs and delivery issues. Where working relationships did exist the representation of firms on curriculum design or programme management committees in the education institutions, for example, was limited or non-existent. This perpetuated a lack of confidence on the part of BPS firms in the ability of higher and further education institutions to deliver appropriate training or about the value of such training even if it was accessible.

Conclusion

It has been suggested that the process of enhancing the creative capacity of cities in ways that allow them to compete in an increasingly globalized world requires an endogenous approach that complements the exogenous supplementation of the creative economy of the kind advocated by Florida. This is critical for cities such as Birmingham that are located in semi-peripheral or peripheral regions that are reconfiguring the legacies of the Industrial Revolution in ways that require improvements to the supply of locally accessible, good-quality human resources that will facilitate the production of advanced services of the kind typically provided by BPS.

---

13   Now, Birmingham City University (BCU).

The research using a large sample of BPS firms in Birmingham confirms that the competitiveness of BPS firms and therefore their ability to underpin the creative capacity of the city is derived from a combination of technical competence, the personality of individual professional and support staff, and a set of soft skills. This implies that academic excellence and world-class technical competence do not provide the platform for commercial success; it is also crucially founded upon local relationship building, soft skills and personality. At one level, it is difficult and perhaps impossible to develop advanced soft skills in some personality types but the educational system should be designed to develop minimum levels of competence in a set of essential BPS skills. These essential BPS skills are: verbal dexterity; an ability to relate to people from a wide variety of backgrounds and cultures (gender, ethnicity, age, nationality); an ability to listen (to try to understand what the client wants and then to be able to frame an appropriate response); appreciation of body language (its interpretation and projection); and impression management. In addition, much of activity engaged in by BPS is about creative writing. Therefore, advanced literacy skills are required, as well as a level of numeracy.

Although not exclusively an issue for support staff in BPS firms, there is evidence pointing to poorly developed soft skills, including managing relationships – internal and external to firms. This deficiency does not just arise from any inadequacies in the monitoring and training regimes used by employers. It is complicated by a failure on the part of new employees to appreciate some of the survival techniques that are required in the world of employment. This includes an appreciation of the nature of the work environment in BPS, understanding how as individuals they relate to other employees (as well as clients), and the requirements associated with working as part of a team. These are critical skills, as the success of a firm is founded upon the projection of a consistent professional image that is constructed around, and from, the activities of all staff members.

But care also needs to be taken that over-training in non-essential skills is not provided. This can raise expectations on the part of the trainee and undermine the employer perception of the training provider. There are costs associated with external training such as management time spent identifying skills and training needs, time and effort invested in identifying suitable courses, fee earning or support staff time lost as a result of training, and if the training is located 20–30 miles from the user a tired and less motivated labour force in the workplace. On the job training is also not without difficulties that include preoccupation of the internal trainer and the trainee with the task in hand, and a tendency in many instances for the training not to be concluded because everyday business activities have a tendency to distract the participants.

Larger firms have the resources to develop, implement and manage employee appraisal and training schemes The employees in small BPS firms encounter constant demands for multitasking and do not have the spare capacity of time and people that are needed to develop effective training regimes. This means that in cities such as Birmingham, where large firms are more numerous, university and college training providers need to be cognizant of the training schemes of the larger companies while simultaneously identifying and engaging with the advanced training needs of the full cross-section of BPS firms and employees. This will ensure that professional and support staff acquire state-of-the-art skills and expertise that will not only enhance the creative role of BPS firms but also their clients, many of whom are located in and around Birmingham. Sustaining the creativity of cities is not just about fostering dynamic clusters of creative industries, it also requires close attention to optimizing the infrastructure for enabling and delivering the development of human resource skills and training. In many ways these are the building blocks for the creative economy of a city.

Bibliography

Accenture. 2007. *Skills for the Future*. London: Accenture.

Advantage West Midlands and Birmingham City Council. 2002. *Birmingham – Creative City: Analysis of Creative Industries in the City of Birmingham*. Birmingham: Advantage West Midlands and Birmingham City Council.

Brinkley, I. 2008. *The Knowledge Economy: How Knowledge is Reshaping the Economic Life of Nations*. London: The Work Foundation.

Bryson, J.R., Taylor, M. and Cooper, R. 2008. Competing by Design, Specialization and Customization: Manufacturing Locks in the West Midlands (UK). *Geografiska Annaler SeriesB-Human Geography*, 90B(2), 173–86.

Brown, J. et al. 2007. From city of a thousand trades to city of a thousand ideas, Birmingham, West Midlands, UK: Pathways to creative and knowledge-based regions. *ACRE Report WP2.3*. Amsterdam: Amsterdam Institute for Metropolitan and International Development Studies (AMIDSt).

Costa, P. et al. 2008. On 'Creative Cities' Governance Models: A Comparative Approach. *The Service Industries Journal*, 28(3–4), 393–413.

Currid, E. 2006. New York as a Global Creative Hub: A Competitive Analysis of Four Theories on World Cities. *Economic Development Quarterly*, 20(4), 330–50.

Daniels, P.W. 2003. (Re)positioning Birmingham: Old Roles, New Contexts for Advanced Producer Services, paper presented to the International Workshop on Intermediaries of Capital, Knowledge and Labour in Asia and Europe: Comparative and Connected Perspectives, 11–13 December, University San Malaysia, Penang.

Daniels, P.W. and Bryson, J.R. 2002. *Professional Services in the West Midlands: Strengths, Weakness and Opportunities*. Birmingham: Birmingham Forward and Advantage West Midlands.

—— 2006. *The Skill Needs of Business and Professional Services in Objective 2 Areas of the West Midlands, A Sub-Regional and Sub-Sectoral Analysis: Final Report*. Prepared for Birmingham and Solihull Learning and Skills Council. University of Birmingham: Services and Enterprise Research Unit.

Department for Education and Skills. 2003. *Skills Strategy, 21st Century Skills, Realising Our Potential*. London: Department for Education and Skills.

Drucker, P. 1993. *Post-Capitalist Society*. New York: HarperBusiness.

Florida, R. 2002. *The Rise of the Creative Class*. New York: Basic Books.

—— 2005. *Cities and the Creative Class*. New York: Routledge.

—— 2008. *Whose Your City: How the Creative Economy is Making where to Live the Most Important Decision of your Life*. New York: Basic Books.

Her Majesty's Treasury and Department of Trade and Industry. 2005. *Productivity and Competitiveness Indicators*. London: HM Treasury and DTI.

Koch, A. and Stahlecker, T. 2006. Regional Innovation Systems and the Foundation of Knowledge-Intensive Business Services: A Comparative Study in Bremen, Munich and Stuttgart, Germany. *European Planning Studies*, 14(2), 123–46.

Learning and Skills Council. 2005. *National Employers' Skills Survey 2004*. Coventry: Learning and Skills Council.

Levy, F. and Murnane, R.J. 2004. *The New Division of Labour: How Computers are Creating the Next Job Market*. Princeton: Princeton University Press.

Markusen, A. 2006. *Cultural Planning and the Creative City*, paper presented to the annual American Collegiate Schools of Planning meetings, Ft. Worth, Texas, 12 November.

National Statistics. 2008. Available at: www.statistics.gov.uk/abi/.

Oakley, K. 2004. Not So Cool Britannia: The Role of the Creative Industries in Economic Development. *International Journal of Cultural Studies*, 7(1), 67–77.

OECD. 1996. *The Knowledge-Based Economy*. Paris: OECD.

Office of the Deputy Prime Minister. 2006. *Realising the Potential of all our Regions: The Story So Far*. London: Office of the Deputy Prime Minister.

Pratt, A.C. 2008. Creative Cities: The Cultural Industries and the Creative Class. *Geografiska Annaler Series B – Human Geography*, 90B(2), 107–17.

Saxenian, A. 2006. *The New Argonauts: Regional Advantage in a Global Economy*. Cambridge, MA: Harvard University Press.

Schleicher, A. 2007. *Europe's Skills Challenge* (paper presented to Lisbon Council, 15 October). Paris: OECD.

Scott, A.J. 2006. Creative Cities: Conceptual Issues and Policy Questions. *Journal of Urban Affairs*, 28(1), 1–17.

Smedlund, A. and Toivonen, M. 2007. The Role of KIBS in the IC Development of Regional Clusters. *Journal of Intellectual Capital*, 8(1), 159–70.

Sommers, P. and Carlson, D. 2000. *Ten Steps to a High Tech Future: The New Economy in Metropolitan Seattle*. Discussion Paper for the Brooking Institution Center on Urban and Metropolitan Affairs (Washington, DC).

Sung, J. and Ashton, D. 2004. *High Performance Work Practices: Linking Strategy and Skills to Performance Outcomes*. London: Department of Trade and Industry.

Wang, J. and Tong, X. 2005. *Sustaining Urban Growth through Innovative Capacity: Beijing and Shanghai in Comparison.* World Bank Policy Research Working Paper (3545). Washington, DC: World Bank.

Wood, P.A. 2002. Knowledge-intensive Business Services and Urban Innovativeness. *Urban Studies*, 39(5–6), 993–1002.

Wu, W. 2005. *Dynamic Cities and Creative Clusters.* World Bank Policy Research Working Paper (3509). Washington, DC: World Bank.

Chapter 7

Megacities, Regions, and Creativity: Geography of Talent in China

Haifeng Qian and Roger R. Stough

Introduction

Human capital or talent plays an increasingly important role in regional economic development. The new growth theory economists (Lucas 1988, Romer 1986) offer a theoretical explanation of why human capital can drive economic growth. Empirical evidence on this relationship has also been widely provided at both the national and regional levels (Barro 1991, Florida et al. 2008, Glaeser et al. 1995, Rauch 1991). Talent is unevenly distributed across regions, and Moretti (2003) finds a growing disparity of human capital intensity among US cities for the period 1980–2000. In this context, there has been a growing concern among scholars over regional variation in talent, which may help policy-makers to understand how nations, regions or even cities can compete for human capital and further facilitate economic growth.

Geographically bounded factors that may influence the location choice of talent have been extensively investigated in developed countries (e.g. Florida 2002b, Glaeser et al. 2001, Shapiro 2006), addressing both market factors and non-market factors. The traditional market perspective suggests that human capital moves to places where there is market demand (or in other words, job opportunities) for well-educated or skilled workers. However, recent research is dominated by consideration of the effects of non-market local characteristics on talent attraction or retention, such as the quality of life (Glaeser et al. 2001, Shapiro 2006), social diversity or tolerance (Florida 2002b, 2002c), and the presence of universities (Florida et al. 2008).

While most of these empirical studies are performed using data from developed countries, typically the US, there has been little investigation of talent and factors that influence talent attraction within developing countries and their sub-national regions or cities. The primary focus of this study is on talent distribution in China. It describes the geographic distributions of talent and investigates geographically mediated factors that are associated with talent attraction in China. The unit for analysis is the provincial-level region which covers all the provinces in mainland China (except Tibet as an outlier) and four municipalities or megacities directly under the central government (Beijing, Shanghai, Tianjin and Chongqing). This chapter is organized into five sections. A review of the relevant literature follows this introduction. The third section presents data, variables and methods, followed by empirical findings in the fourth section. The last section presents the conclusion.

Literature

The term 'talent' has been used to represent higher-education-based human capital (Florida 2002b), the creative class (Florida et al. 2008), superstars (Rosen 1981) as well as others. While economists have been long studying human capital (e.g. Mincer 1958, Schultz 1961), a geographical perspective has been relatively rare. Lucas's theoretical work (1988) on the role of human capital in economic growth particularly in the context of cities or urban concentrations has spurred numerous empirical studies, in most cases showing positive and significant effects of education-based human capital on city growth (Florida 2002b, Glaeser et al. 1995, Rauch 1991). In recognizing the importance of talent for regional economic development, scholars have been investigating why some regions or cities can better attract talent than others. The world may have become 'flat', but the geographic distribution of talent is becoming 'spiky' (Florida 2006).

Traditional views of regional talent stock are rooted in demand–supply equilibrium thinking about the local labour market. With these views skilled workers tend to stay in places with job opportunities that require knowledge and high skills. Recent research, particularly among geographers and regional scientists, has examined and discovered the important role of geographically mediated non-market factors in attracting talent. To begin with, regions that provide more amenities associated with the quality of life tend to have advantages in attracting talent. Kotkin (2000) and Glaeser et al. (2001) note the importance of regional amenities to attraction and retention of valuable human capital residents. Florida (2002a, 2002c) finds the high correlation between cultural amenities measured by the 'Bohemian index' and human capital. Shapiro's empirical study (2006: 324) shows that 'roughly 60% of the effect of college graduates on employment growth is due to productivity; the rest comes from the relationship between concentrations of skill and growth in the quality of life'.

Social diversity, tolerance or openness is another factor that has drawn attention for its effects on talent intensity, based on the fact that talent is widely distributed and variable across gender, race, country of birth, sexual orientation# and culture. For instance, Saxenian (1999) finds that there are large proportions of foreign-born scientists, engineers and entrepreneurs in the Silicon Valley. According to Florida (2002b, Florida et al. 2008, Mellander and Florida 2007), social diversity signals lower barriers to entry of outside talent with different backgrounds. He reveals a positive relationship between diversity in terms of the gay index and human capital stock. Diversity has also been suggested as associated with high technology industries (Florida and Gates 2001) and growth of rent and wage levels (Ottaviano and Peri 2004), which may further, in a circular causation fashion, attract talent as a result of increased labour market demand.

In addition to the quality of life and social diversity, the presence of universities appears to be critical to the presence and attraction of local talent stock (Berry and Glaeser 2005, Mellander and Florida 2007). Universities are places where higher-education-based human capital is produced. They also create additional demand for human capital – particularly professors and researchers – through the teaching, training and research opportunities they offer.

Although empirical studies on regional variation in talent are not rare, few of them have a focus on China or other emerging or developing national economies. China is experiencing skyrocketing expansion in higher education, and the Chinese government shows growing interest in investing in human capital, research and high technology industries as the basis of its long-term economic development strategy. In this context, it is timely to investigate the geography of human capital in China. Among the few studies, Li and Florida (2006) have tested the effects of non-market factors on human capital production in China and found a positive impact of openness on the number of local universities. In terms of the contribution to regional economic development, talent production, however, is not as important as talent attraction, retention or stock, particularly given the fact that talent production in China, to a large extent, is exogenous of regional characteristics and primarily reliant on national or provincial government policy. As a further development, Qian (2010) and Florida et al. (2008) examine the relationships of both market factors (wage and employment) and non-market factors (universities, amenities and openness) with China's regional talent stock. This work based on regression analysis finds that the presence of universities and the level of openness are significantly associated with the level of talent stock, the level of innovation and regional economic performance. These recent cross-sectional analyses suffer from not only a small number of observations for regression analysis but also an inability to investigate causality when using provincial-level data for one time period.

Methodology

This study is a further development of the work of Qian (2010) and Florida et al. (2008) on the geography of talent in China. Instead of using a cross-provincial dataset for one single time period that only allows for 31 observations,[1] a panel dataset is employed in this research. This dataset covers 30 provincial-level regions in mainland China (Tibet excluded as an outlier) for the period of 2000–2006. With significantly increased number of observations, the empirical results are expected to be more reliable and therefore more definitive. This study also introduces additional explanatory variables to better explain why talent is concentrated in certain regions in China.

---

1   Mainland China has 31 provincial regions.

While it is ideal to study talent in cities, relevant data at the city-level in the case of China are unavailable or unreliable. Even so, our study still sheds light on the role of Chinese megacities in talent attraction since special attention is paid on the four provincial level municipalities directly under the central government (Beijing, Shanghai, Tianjin and Chongqing).

Dependent Variables

The primary focus of this chapter is on the regional variation in talent in China. Two dependent variables are separately used: higher-education-based *human capital* and occupation- and merit-based *professional personnel*. While human capital is often and traditionally measured by the adult literacy rate, the school enrolment rate or the average years of schooling, this study adopts the higher education approach, which is more closely related to talent. Regional human capital level is measured by the proportion of those with a four-year college or above degree among the population 15 years old or older (defined as 'human capital intensity'). These data are available from *China Statistics Yearbook* (National Bureau of Statistics, PRC (2001–07).[2]

Professional personnel (or '*Zhuanye Jishu Renyuan*') is another indicator that has been consistently appearing in China's statistical materials and can be regarded as another type of talent. It is occupation-based and includes such occupations as scientists and engineers, university professors, teachers, agricultural and sanitation specialists, aviators and navigators, economic and statistical specialists, accountants, translators, librarians, journalists, publishers, lawyers, artists, broadcasters, athletes, and so on. It echoes *the creative class* notion of Florida (2002c) in the literature, despite a difference in the inclusiveness of occupations.[3] It is worth noting that not all employees in those occupations are counted as professional personnel in official statistics. Only those in urban units who pass the qualification exam and accordingly are admitted by the government's personnel department belong to this officially recorded category. In this sense it is also a merit-based proxy for talent. Regional professional personnel level is similarly measured by the proportion of those belonging to professional personnel among the population of 15 years old or older (defined as 'professional personnel intensity'). Data can be obtained from *China Labor Statistics Yearbook* (Department of

---

2   Data for the year 2001 are not available. The 2001 human capital intensity is the average of 2000 and 2002.

3   Florida (2002c) defines creative class as consisting of the super-creative core (including scientists and engineers, university professors, poets and novelists, artists, entertainers, actors, designers, architects, nonfiction writers, editors, cultural figures, think-tank researchers, analysts and other opinion-makers) and creative professionals (working in high technology sectors, financial services, the legal and healthcare professions and business management).

Population and Employment, National Bureau of Statistics, PRC and Department of Planning and Finance, Ministry of Labor and Social Security, PRC 2001–07).

Independent Variables

This study investigates the effects of both market factors and non-market factors on the talent distribution among Chinese regions and four megacities. Market factors cover the wage level, the scale of high technology industries, and research and development (R&D). Non-market factors include amenities, openness, the university, and a binominal indicator showing whether a region is a municipality or megacity directly under the central government or not.

Wage
The availability of satisfactory job opportunities may be a major concern when talent selects its work and life location. While there are many ways to describe the quality of job opportunities, the wage level appears to be the most direct measure. High wage positions generally correspond to high value-added work, which is exactly what talented people are seeking. To obtain a high wage job not only enables talent to realize its self value, but also consolidates its financial base, which ensures a good quality of life, *ceteris paribus*. The logarithm of the average wage level is therefore used as an explanatory variable of the talent stock in a region. Data are provided by the *China Labor Statistics Yearbook* (Department of Population and Employment, National Bureau of Statistics, PRC and Department of Planning and Finance, Ministry of Labor and Social Security, PRC 2001–07). The average wage levels of 2001–06 are converted into constant values with the 2000 base year.

High Technology Scale and R&D
More than half of college graduates in China hold a degree in science or engineering and are likely to work for high technology or research-related employers. As a result, a region specializing in the production of high technology products and/ or presenting strong R&D efforts is probably more attractive to talent. In this study, the proportion of value added in high technology industries to gross domestic product (GDP) is employed to measure a region's level or scale of high technology industries. High technology industries in China are officially defined as including electronic and telecommunications, computers and office equipment, pharmaceuticals, medical equipment and meters, and aircraft and spacecraft. High technology value added data are available for 2000–05 from the *China Statistics Yearbook on High Technology Industry* (National Bureau of Statistics, PRC, National Development and Reform Commission, PRC and Ministry of Science and Technology, PRC 2004–07), and GDP data may be obtained from the *China Statistics Yearbook* (National Bureau of Statistics, PRC 2001–07). To measure the relative R&D efforts of a region, the ratio of total R&D capital stock to GDP is employed, both using data with conversion to constant values based on the

year of 2000. While the *China Statistics Yearbook on Science and Technology* (National Bureau of Statistics, PRC and Ministry of Science and Technology, PRC 1995–2007) offers only the amount of new R&D expenditures (or R&D flows) for each year, a capital depreciation rate of 15 percent is assumed to calculate total R&D stock.

Amenities

Talent, with its exceptional financial condition and lifestyle, is more serious to pursue a better quality of life than others, and thus may consider amenities as a major factor when making a location choice. Amenities can be natural, cultural, construction or infrastructure based, or service in nature. This study considers only service amenities, using the location quotient of urban employment in the tertiary sector (researchers and employment in other unknown service industries excluded).[4] Data are available from the *China Labor Statistics Yearbook* (Department of Population and Employment, National Bureau of Statistics, PRC and Department of Planning and Finance, Ministry of Labor and Social Security, PRC 2001–07).

Openness

Diversity, openness or tolerance has been generally measured through the immigration-based melting pot index or the gay index (Florida 2002a, 2002b, Florida and Gates 2001, Florida et al. 2008, Lee et al. 2004, Mellander and Florida 2007). As a result of data unavailability for these two measures in the context of China, this study uses the '*Hukou* index' introduced in Qian (2010) as a proxy for diversity or openness. The *Hukou* system (or the household registration system) is used to manage population mobility by the Chinese government. One's *Hukou* (or resident registration) determines which city or county one 'belongs' to and whether one has rural or urban status. Those with a local *Hukou* or registration are always local permanent residents and can reap local economic, social and political benefits, such as social welfare, health/medical, education and voting rights. Those who live in a jurisdictional area, however, without a local *Hukou* are always temporary residents coming from the outside. They are migrants that may bear different traditions, cultures and knowledge from the local permanent residents. Therefore, *Hukou* appears to be a defensible indicator and thus measure of openness. The *Hukou* index is defined as the proportion of population without local *Hukou* or registration. The higher the *Hukou* index, the more open and diverse the region. Data are obtained from the *China Statistics Yearbook* (National Bureau of Statistics, PRC (2001–07).

---

4   Researchers are excluded since they are related to human capital but unlikely to contribute to service amenities.

University
The university is hypothesized to play an exclusively important role in the geography of human capital in China because of the restriction of population flow imposed by the government. The university measure is defined as the ratio of new college graduates in a particular year to total population. The *China Statistics Yearbook* (National Bureau of Statistics, PRC (2001–07) provides such data for the period of 2000–06.

Dummy Variable for Provincial Municipalities
In China, Beijing, Shanghai, Tianjin and Chongqing are the only four municipalities that are directly under the central government, holding equal or even higher political status than provinces. These four municipalities, generally also considered as provincial level regions, have much higher urbanization rates than the provinces. Most importantly, the resources, preferential policies and treatment they receive from the central government bring them competitive advantages in economic and social development and talent attraction. A dummy variable to control the economic, social and political effects of being a provincial-level municipality is therefore added. With this dummy variable, the role of megacities in attracting talent can also be investigated in the context of China.

Methods

Three types of models are separately used: pooled-OLS (ordinary least squares), fixed effects, and random effects models. The pooled-OLS model, though not as defensible as the other two models in terms of meeting the G-M (Gauss-Markov) assumption, provides additional evidence on what might affect the talent distribution in China. While the fixed effects model does not allow for dummy variables, an interaction term between the university and the dummy variable of municipalities is introduced to manoeuvre around that problem. The random effects model includes the dummy variable for municipalities without using an interaction term. To avoid possible endogeneity problems we use a one-year lag for all the independent variables. Therefore the dependent variables (human capital and professional personnel) are for the years of 2001–06, and independent variables are for 2000–05.

Findings

Geographical Distributions of Talent in China (2000–2006)

Human Capital
Figures 7.1 and 7.2 describe the geographic distributions of human capital in China for 2000 and 2006. In both years human capital is unevenly distributed across regions. In 2000, three out of the four municipalities directly under the

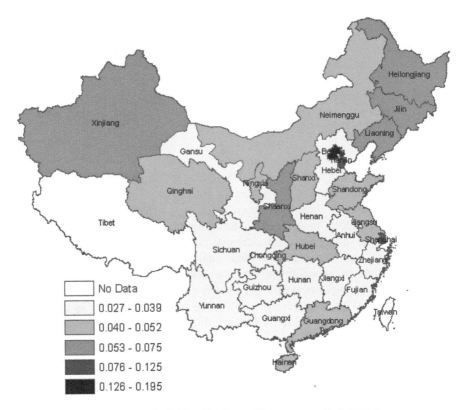

Figure 7.1   Geographical distribution of human capital (2000)

central government – Beijing, Shanghai and Tianjin – took the lead in retaining human capital, showing 19.5 percent, 12.5 percent and 10.8 percent of the local population of 15 years old or older holding a college degree or above respectively. It indicates the power of Chinese megacities in attracting talent. Overall, provinces in Eastern China exhibited higher human capital intensity than inland provinces. One exception was Xinjiang, despite being the farthest province from the coast, presenting 7.0 percent of the local population of 15 years old or older holding a college degree or above and ranking fifth nationally after Liaoning Province. In addition, the regional disparity of human capital intensity was enormous, ranging from 2.7 percent in Yunnan province to 19.5 percent in Beijing.

From 2000 through 2006, all the provincial regions experienced significant growth in both the absolute number of local college-degree holders and human capital intensity, echoing the dramatic expansion of higher education in China during the same period. The geographic pattern of human capital to some extent also changed in 2006 compared to that in 2000. Beijing, Tianjin, Shanghai, Liaoning and Xinjiang maintained their advantages in attracting human capital, all demonstrating human capital intensity of over 10 percent. The increases of the local

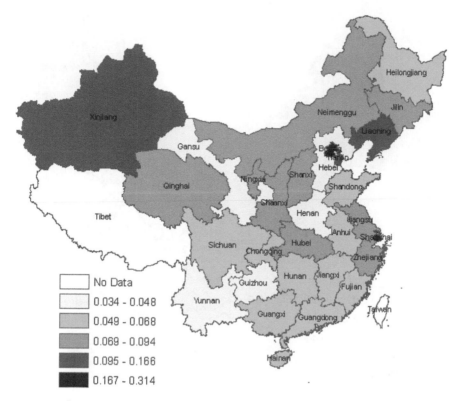

Figure 7.2   Geographical distribution of human capital (2006)

population of 15 years old or older holding a college degree or above in Beijing and Shanghai were particularly remarkable, jumping to 31.4 percent and 23.0 percent respectively in 2006. As a result, the regional disparity of human capital became even larger especially between megacities and other provincial regions.

Professional Personnel

The geographical distributions of professional personnel in 2000 and 2006 are shown in Figure 7.3 and Figure 7.4 respectively. Similarly, Beijing, Tianjin, Shanghai and Xinjiang were among those provincial regions that exhibited the highest professional personnel intensity in both years. Further, for nearly all provinces the scale values of the professional personnel variable were smaller than that of human capital. Beijing was the single region showing more than 10 percent of the local population of 15 years old or older certified as professional personnel (10.2% for 2000 and 10.5% for 2006). Most provinces exhibited a professional personnel intensity of less than 5 percent. As a consequence, regional disparity of professional personnel was not as large as that of human capital, particularly between the coastal provinces and the inland provinces.

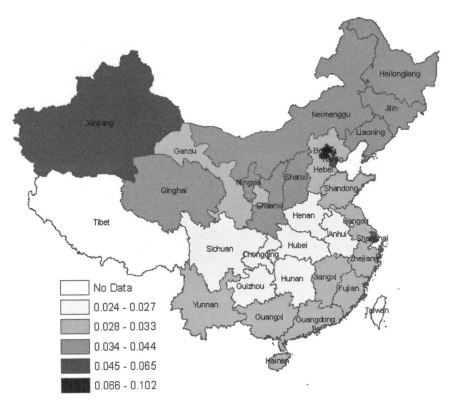

Figure 7.3   Geographical distribution of professional personnel (2000)

The change of professional personnel intensity from 2000 to 2006 was fairly small. It declined slightly for most provinces. The disproportional changes of human capital and professional personnel imply that new college graduates after 2000 might not be so enthusiastic to obtain the certificate of professional personnel. It might also be the case that it often takes years for college graduates to be qualified as professional personnel.

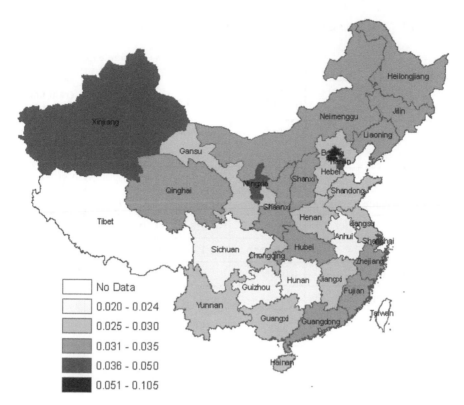

No Data
0.020 - 0.024
0.025 - 0.030
0.031 - 0.035
0.036 - 0.050
0.051 - 0.105

Figure 7.4   Geographical distribution of professional personnel (2006)

Geographically Bounded Determinants of Talent Distribution in China

Determinants of Human Capital Intensity

Table 7.1 presents the regression results when human capital intensity is taken as the dependent variable. Results based on the pooled OLS, the fixed effects and the random effects models are all reported.

Table 7.1 shows that market factors, including the wage level, the scale of high technology industries and R&D expenses may exert limited influences on the human capital distribution in China. The wage level as expected reveals a positive effect on human capital intensity, but only significantly so in the fixed effects model when the effect of being a municipality is controlled through its interaction with the university variable. The effect of high technology on human capital is surprisingly negative, but only significant in the least reliable pooled OLS model. The coefficient of R&D activity is positive and significant in the pooled OLS model and the random effects model but insignificant in either of the fix effects models.

Table 7.1　Regression results – human capital as dependent variable

Explanatory variables (using one-year lag)	Dependent variable: Human capital			
	Model 1: (pooled OLS)	Model 2: (fixed effects)	Model 3: (fixed effects)	Model 4: (random effects)
Wage	0.005	0.007	0.032***	0.010
High-tech scale	-0.295***	-0.057	-0.144	-0.139
R&D	0.351***	-0.103	0.087	0.519***
Service amenities	-0.010	-0.019	-0.029**	-0.012
Openness	0.185***	-0.028	-0.007	0.037
University	17.321***	12.130***	-0.099	10.653***
D_Municipality	0.028***			0.043***
D_Municipality* University			14.227***	
R2 (within for FE and RE models)	0.834	0.501	0.597	0.476
Observations	180	180	180	180

Note: * Significant at 0.1; ** Significant at 0.05; *** Significant at 0.01.

The empirical results suggest that specializing in high technology industries or R&D activity may not be the key driving force of human capital attraction or retention in Chinese regions. This is inconsistent with the finding in studies based on data from western countries (e.g. Mellander and Florida 2007) that high technology is positively associated with human capital. The result of the Chinese analysis might result from data unreliability. China's statistical data especially for economic indicators have been widely criticized for their unreliability. Moreover, statistical standards and categories sometimes change across years, which creates difficulties for panel data analysis. Unfortunately little can be done to improve the officially issued data at this time.

Given the caveat of data unreliability, another possible explanation of the weak relationship between high technology (or R&D) and talent is the restriction of population mobility exerted by the *Hukou* system that likely limits talent migration to those places with job opportunities. Demand for human capital by high technology industries and R&D resources and supply for human capital by talent itself are distorted from labour market equilibriums.

Another possible explanation lies in the characteristics of China's high technology industries. Those high technology industries are primarily based on manufacturing, processing and assembling, not on innovations and services. Compared with developed countries, R&D activity in the Chinese high technology industries is quite limited (Qian 2010). With limited innovative opportunities, the link between human capital and high technology industries is likely weakened. It is very likely that high technology firms in China would rather locate themselves in places with less human capital where total costs of production (including, for instance, land use costs) are generally lower.

Among non-market factors, the university is one of the major driving forces of human capital retention in Chinese regions. Its coefficient is positive and significant across three types of models. This is consistent with the findings by Qian (2010) and Florida et al. (2008) in the Chinese context and Berry and Glaeser (2005) in the western context. In China, however, the effect of the university appears to be driven by what happens in the four large municipalities directly under the central government. Model 3 of Table 7.1 shows that for the remaining provinces alone, the impact of the university is negative and insignificant!

The dummy variable indicating the status of being a municipality or megacity directly under the central government is another major determinant of the human capital distribution in China. Its coefficient (including its interaction term with the university) is consistently positive and significant across all three models. It implies that those with a college degree or above are more willing to stay in those municipalities directly under the central government, even after holding the wage level, high technology, R&D, service amenities, openness and the university variable constant. So why are those municipalities so attractive to the well-educated? The fundamental reason is that those regions with their advantageous political status receive tremendous benefits or preferential policies (i.e. subsidies, direct and indirect) from the central government. For instance, Beijing features the nation's best educational institutions and health systems that serve as talent magnets, and benefits vastly from being the home of the central government. Because of preferential policies granted by the central government, National Economic and Technology Development Zones (NETDZs) in China are the most attractive places for human capital and high technology firms. Tianjin and Beijing each have one of the largest and best NETDZs in China. Shanghai is home to four NETDZs and the only city with more than two. As the economic centre of China, Shanghai receives economic development support from the central government in various forms.

For other non-market factors, neither service amenities nor openness presents a positive and significant coefficient in the fixed effects or random effects models, which is the opposite of findings based on studies in developed countries (e.g. Florida 2002b, Glaeser et al. 2001, Shapiro 2006). This difference might be attributed to different developmental stages between China and western countries.

Determinants of Professional Personnel Intensity
Similar regressions are run simply by replacing human capital with professional personnel. The results are shown in Table 7.2.

Similarly, market factors, including the wage level, high technology and R&D, present weak or ambiguous effects on professional personnel intensity. Among non-market factors, the dummy variable of municipalities is consistently positive and significant, suggesting a similar pattern to human capital that being a municipality directly under the central government is important for talent attraction. The university, while showing a strong and positive effect on human capital intensity, exhibits a negative effect on the presence of professional personnel in one fixed effects model and the random effects model. This is somewhat counterintuitive, and may suggest that the overlap between professional personnel and human capital is not large. Model 3 in Table 7.2 again shows that this negative effect is primarily driven by provincial-level municipalities. In particular, Beijing and Shanghai have experienced skyrocketing expansion in higher education and human capital, a rapid growth in population, but a less significant increase in professional personnel. Different from the human capital case, openness presents a consistently positive effect on professional personnel intensity and significant in two models, echoing the findings uncovered in empirical studies for developed countries (e.g. Florida et al. 2008).

Table 7.2   Regression results – professional personnel as dependent variable

Explanatory variables (using one-year lag)	Dependent variable: Professional personnel			
	Model 1: (pooled OLS)	Model 2: (fixed effects)	Model 3: (fixed effects)	Model 4: (random effects)
Wage	-0.005	0.003**	-0.000	0.002
High-tech scale	-0.142***	-0.021	-0.008	-0.055*
R&D	0.326***	-0.064	-0.092	0.226***
Service amenities	-0.005	-0.003	-0.002	0.001
Openness	0.092***	0.008	0.005	0.012**
University	1.525	-1.595***	0.208	-2.130***
D_Municipality	0.009***			0.023***
D_Municipality* University			-2.097***	
R2 (within for FE and RE models)	0.778	0.230	0.330	0.149
Observations	180	180	180	180

Note: * Significant at 0.1; ** Significant at 0.05; *** Significant at 0.01.

Conclusion

This chapter describes the distributions of talent, including human capital and professional personnel, across Chinese provincial-level regions and megacities.

It also examines the geographically mediated factors that may affect local talent intensity using a panel dataset. While similarly empirical work has been extensively done in the context of developed economies, a focus on China as well as other transitional or developing countries is rare. Given differences in the development stage, the political system, the urban structure and the migration pattern between China and western developed countries, some different empirical findings would be expected in the case of China.

Similar to the developed countries, talent in China is unevenly distributed across regions. Megacities such as Beijing and Shanghai present enormous advantages in attracting human capital or professional personnel, and talent concentrates highly in those municipalities directly under the central government. Multivariate analysis shows that even after controlling for several market factors and non-market factors the effect of being such a municipality directly under the central government on talent attraction is still strong. Moreover, the talent gap between the four municipalities or megacities and non-municipal provinces especially in terms of human capital has been growing. The expansion of higher education initiated by the central government since the late 1990s apparently did not benefit Chinese regions equally.

Among market factors, the wage level and the R&D level have presented unclear effects on human capital or professional personnel. Specializing in high technology production does not lead to the attraction and growth of talent. It is associated with the fact that high technology firms in China mostly focus on standard or more traditional technology-intensive production rather than inventive and innovative production activities. It might also relate to the migration restriction exerted by the central government through the *Hukou* system. Such government interventions have distorted the equilibrium in labour markets and the intrinsic relationship between high technology and talent.

The university as we expect has a major positive effect on regional human capital intensity. While this is also the case for developed countries, it is reasonable to say that the university is more important in China than in the western countries. Florida et al. (2006) point out that US cities with a good university system do not necessarily retain talent, partially because of labour market mobility. In China, by contrast, the government controls population flow through the *Hukou* system. As a result, a college graduate is more likely to stay in the place where she/he obtains her/his higher education.

Among other non-market factors, the effect of service amenities is not significant for either type of talent. Openness presents a consistently positive but not necessarily significant effect on professional personnel. These results may suggest that talented people in China have not been in a situation where their key concerns are to enjoy life and style of work.

Bibliography

Barro, R.J. 1991. Economic Growth in a Cross-Section of Countries. *Quarterly Journal of Economics*, 106, 407–43.

Berry, C.R. and Glaeser, E.L. 2005. The Divergence of Human Capital Levels across Cities. *Papers in Regional Science*, 84(3), 407–44.

Department of Population and Employment, National Bureau of Statistics, PRC and Department of Planning and Finance, Ministry of Labor and Social Security, PRC. 2001–07. *China Labor Statistics Yearbook*. Beijing: China Statistics Press.

Florida, R. 2002a. Bohemia and Economic Geography. *Journal of Economic Geography*, 2, 55–71.

—. 2002b. The Economic Geography of Talent *Annals of the Association of American Geographers*, 92(4), 743–55.

—. 2002c. *The Rise of Creative Class*. New York: Basic Books.

—. 2006. Where the Brains Are. *Atlantic Monthly*, 298, 34.

Florida, R. and Gates, G. 2001. *Technology and Tolerance: The Importance of Diversity to High-Tech Growth*. Washington, DC: Brookings Institution, Center for Urban and Regional Growth.

Florida, R. et al. 2006. *The University and the Creative Economy*. Available at: www.creativeclass.org/rfcgdb/articles/University_andthe_Creative_Economy. pdf.

Florida, R., Mellander, C. and Qian, H. 2008. *Creative China?* CESIS Electronic Working Paper Series, 145, Stockholm, Sweden.

Florida, R., Mellander, C. and Stolarick, K. 2008. Inside the Black Box of Regional Development. *Journal of Economic Geography*, 8, 615–49.

Glaeser, E.L., Kolko, J. and Saiz, A. 2001. Consumer City. *Journal of Economic Geography*, 1, 27–50.

Glaeser, E.L., Scheinkman, J.A. and Shleifer, A. 1995. Economic Growth in a Cross-Section of Cities. *Journal of Monetary Economics*, 36, 117–43.

Kotkin, J. 2000. *The New Geography*. New York: Random House.

Lee, S.Y., Florida, R. and Acs, Z.J. 2004. Creativity and Entrepreneurship: A Regional Analysis of New Firm Formation. *Regional Studies*, 38(8), 879–91.

Li, T. and Florida, R. 2006. *Talent, Technological Innovation, and Economic Growth in China*. Available at: www.creativeclass.org/rfcgdb/articles/ China%20report.pdf.

Lucas, R.E., Jr. 1988. On the Mechanism of Economic Growth. *Journal of Monetary Economics*, 22, 3–42.

Mellander, C. and Florida, R. 2007. *The Creative Class or Human Capital?* CESIS Electronic Working Paper Series, 79, Stockholm, Sweden.

Mincer, J. 1958. Investment in Human Capital and Personal Income Distribution. *Journal of Political Economy*, 66(4), 281–302.

Moretti, E. 2003. *Human Capital Externalities in Cities*. NBER Working Paper (W9641), Cambridge, MA.

National Bureau of Statistics, PRC. 2001–07. *China Statistics Yearbook*. Beijing: China Statistics Press.

National Bureau of Statistics, PRC and Ministry of Science and Technology, PRC. 1995–2007. *China Statistics Yearbook on Science and Technology*. Beijing: China Statistics Press.

National Bureau of Statistics, PRC, National Development and Reform Commission, PRC and Ministry of Science and Technology, PRC. 2004–07. *China Statistics Yearbook on High Technology Industry*. Beijing: China Statistics Press.

Ottaviano, G. and Peri, G. 2004. *The Economic Value of Cultural Diversity: Evidence from US Cities*. NBER Working Paper (10904), Cambridge, MA,

Rauch, J.E. 1991. *Productivity Gains from Geographic Concentration of Human Capital: Evidence from the Cities*. NBER working paper (3905), Cambridge, MA.

Romer, P.M. 1986. Increasing Returns and Long-Run Growth. *Journal of Political Economy*, 94(5), 1002–37.

Rosen, S. 1981. The Economics of Superstars. *American Economic Review*, 71(5), 845–58.

Qian, H. 2010. Talent, creativity and regional economic performance: The case of China. *The Annals of Regional Science*, 45(1), 133–56.

Saxenian, A. 1999. *Silicon Valley's New Immigrant Entrepreneurs*. Berkeley, CA: Public Policy Institute of California.

Schultz, T.W. 1961. Investment in Human Capital. *American Economic Review*, 51(1), 1–17.

Shapiro, J.M. 2006. Smart Cities: Quality of life, Productivity, and the Growth Effects of Human Capital, *Review of Economics and Statistics*, 88, 324–35.

PART III
Urban Cultural Landscape and Creative Milieu

Chapter 8

Creativity, Culture and Urban Milieux

Roberto Camagni

Introduction

The city, intended as an archetype for a specific and historically successful form of social organization, has long since been recognized as the birthplace of innovation and creativity.

This is nowadays almost commonsense in urban economics and geography, as a consequence of the seminal works of Ake Andersson (1985), Edward Glaeser (1998), Peter Hall (1998) and Jane Jacobs (1969, 1984). But also in the literature on innovation and technological change this very fact has been acknowledged for many years (Karlsson 1995, Nijkamp 1986).

More recently, the whole theoretical field has been subject to a growing mediatization, mainly thanks to a normative shift inaugurated by Charles Landry (2000) and to the provocative recipes for urban regeneration proposed by Richard Florida (2005), based on the attraction and development of the 'creative class'. The entire debate entered a new phase, calling for new reflections and possibly clarifications (Scott 2006).

This chapter aims to contribute to this debate in a twofold way. First of all, it will incorporate the creative city phenomenon into a more general and complex theoretical framework with respect to many contributions that straightforwardly underline single explicative elements for urban creativity. The general argument is that not only do many interrelated elements have to be taken into consideration at the same time in order to interpret correctly the long-term creative character of cities and their present manifestation; but also that attention should be addressed not just to traditional functional elements (human capital, externalities, external linkages ...), but mainly to symbolic and cognitive elements (codes, representations, languages, values) replicating the ways in which individuals, groups and communities fully exploit their creative potential through synergy, cooperation and associative thinking.

The second aim of the chapter is to present alternative strategies for relaunching cities' development by enhancing the preconditions for innovation and creativity that emerge from the previous reflections, and reorienting traditional 'vocations' and competencies of the local context towards new and modern activities through the provision of interaction opportunities and places (the French 'lieux magiques' for socialization and interchange), the creative utilization of urban cultural heritage and atmosphere, the enhancement of knowledge-intensive functions and the development of new urban governance styles.

Innovation, Creativity and the City

The reasons for the historical success of the city – the most important creature of mankind according to Lewis Mumford – lies in a series of advantages that the density of infrastructure and the proximity of differentiated activities and people generates almost automatically.

These advantages, synthesized in the concept of agglomeration or urbanization economies, have been indicated in different ways in the urban economic literature looking for single stylized success elements:

- urban size per se (McCann 2004), especially concerning the creation of large human capital pools and wide labour markets (Glaeser 1998, Lucas 1988);
- diversity, concerning the variety of activities and the possibility for specializations in thin subsectors and specific productions, thanks to the size of the overall urban market (Jacobs 1969, 1984, Quigley 1998);
- contacts and interaction, allowing face-to-face encounters reducing transaction costs (Scott and Angel 1987, Storper and Scott 1995);
- synergies, thanks to proximity, complementarity and trust (Camagni 1991, 1999) in more formalized models; these same effects stem from complexity of the urban system and synergetics (Haken 1993);
- reduction of risk of unemployment for households, thanks to the thick and diverse urban labour market (Veltz 1993);
- trans-territorial linkages, emerging from the international gateway role of large cities, particularly crucial in a globalizing world (Sassen 1994).

In a recent paper, this author has argued that the main forces underlying agglomeration economies can be grouped in two main classes (Camagni 2004). First, the provision of physical-functional capital, in the form of diversified and advanced forms of social overhead capital, public goods, transport and communication networks, large urban functions such as fairs, universities, research centres, congress centres, built and cultural heritage. Second, the provision of relational capital (Camagni 1999) in the form of interpersonal and interorganizational networks – the subset of elements encompassed by the social capital concept coming closer to economic performance.[1] These elements are connected with the synergetic action performed by the city, as they enhance: accessibility to information – inherently a cooperative good – through informal, face-to-face and interpersonal contacts; explicit cooperation among actors,

---

1 Relationality presents at the same time costs (psychological, informational and organizational), risks (of rewarding some opportunistic behaviour; of unevenly distributing benefits between partners) and benefits (in terms of enhanced efficiency, effectiveness and innovative character of decisions). For all these reasons, it may be interpreted as a stock, subject to costly accumulation and appreciation/depreciation possibilities, justifying the use of the term 'capital' for it (Camagni, 2004).

stemming from trust, sense of belonging to a community, sharing of similar values; implicit cooperation among actors, in the form of socialized production of skilled labour and human capital for top managerial functions, implicit territorial marketing (economic 'vocation' and success creates local image and trademark), socialized information transcoding.[2]

All the preceding elements underline the complex, systemic nature of urban agglomeration, the necessity of a comprehensive interpretation of the multiple, interrelated roles of the city and, above all, the necessity to consider it not just under a functional approach, but also under a cognitive, symbolic approach.

In this last respect, the multiple roles of the city may be synthesized as in Figure 8.1, distinguishing a functional and a symbolic approach and two spatial logics, a territorial and a network one (Camagni 2001). According to a functional approach (substantive logics), the city appears as a cluster, encompassing the provision of the externalities already mentioned, potential for face-to-face contacts, and diversification of selected and specialized activities. Its main conceptual role resides in the reduction of transaction costs. In a network perspective, the city appears as an 'interconnection', performing the role of node in multiple, interacting networks: transport networks, communication networks, cultural and economic networks, financial networks, headquarter and control networks. All these networks interconnect on the node represented by the city, which also assures the link between the local context ('place') and the external interaction channels.

SPATIAL LOGIC ⇒ COGNITIVE LOGIC ⇓	TERRITORIAL APPROACH	NETWORK APPROACH
FUNCTIONAL (substantive) APPROACH	CITY AS CLUSTER • diversification and specialization of activities • concentration of externalities • density of proximity contacts • reduction of transaction costs	CITY AS INTERCONNECTION • city as a node in multiple and interacting transport, economic and communication networks • city as interconnection between place and node
SYMBOLIC (procedural) APPROACH	CITY AS MILIEU • uncertainty-reducing operator through: ◊ transcoding of information ◊ ex-ante co-ordination of private decisions (collective action) • substratum for collective learning	CITY AS SYMBOL • city overcoming of time and space • city as symbol of territorial control • city as producer of symbols, codes, and languages

Figure 8.1   The roles of global cities: A theoretical taxonomy

Source: Camagni 2001.

---

2   These are the conceptual roles performed by the local milieu (Camagni, 1991, 1999); see Figure 8.1.

According to a symbolic approach (procedural logics) the city performs two other relevant roles. First, the role of a cognitive 'milieu' (Camagni 1991), assuring the reduction of uncertainty to economic actors through (i) the circulation of selected information, its socialized evaluation, selection and transcoding; (ii) the ex-ante coordination of private decisions boosting 'collective actions'; (iii) the enhancement of private/public partnerships thanks to trust and shared behavioural codes; and (iv) the supply of the durable substratum for collective learning processes – embodied in the local labour market and the dense networks of customer–supplier relationships.[3]

Second, the city becomes a 'symbol': symbol of power and territorial control – as 9/11 has paradoxically confirmed – a social device symbolizing human control over time (conserving past heritage and addressing the future) and space (through its global networks); producer of codes, languages and ever renovated symbols.

The above mentioned roles reinforce each other in a cumulative way, generating strong self-sustaining growth, possibly going far beyond the pace and limits that could be manageable by urban planning. This may end up in possible 'contradictions', in the form of scale diseconomies, huge social costs, wide inefficiencies and environmental drawbacks.

The general picture that emerges is evidently an abstract one: the city appears as an archetype of social organization; its role is equated to that of an abstract local 'milieu', characterized by high internal cohesion in spite of its social diversity, high sense of belonging, and shared behavioural and cognitive codes. When transposed into an empirical realm, the picture adapts well to the large, internationalized city, with its local identity and global reach. Does this mean that only the large city, or city-region, can attain sufficient agglomeration economies, international linkages, quality of production factors and in particular of internal human resources?

The answer is no. In fact there are ways by which medium and medium-large cities can reach a sufficient critical mass, in terms of demography and market size, in order to attain comparable (real) incomes, innovation capability and creativity, high international visibility, reputation and lifestyles. The main strategy, pointed out and supported by the European Commission in its territorial cohesion policy concerns networking with other cities of similar size, mainly but not exclusively belonging to the same region, through technological, cultural and logistic

---

3   The initial theorization concerning local innovative milieux made direct reference to the phenomenon of specialized industrial districts. Then, the same concept proved useful in interpreting other kinds of local production systems: tertiary and tourist areas, integrated economic systems organized around natural or cultural resources, and, of course, cities. Cities are considered as more complex systems than the traditional districts: tendentially diverse and de-specialized (when districts and traditional innovative milieux were tendentially specialized), provider of general-purpose infrastructure, characterized by social heterogeneity as opposed to the social homogeneity of the districts. Where the district's identity was defined by its productive vocation, it is urban identity that defines productive vocation in the case of the city (Camagni 1999).

cooperation. By this, network externalities among cities performing similar roles (tourist cities, financial cities: 'synergy' networks) or advantages coming from specialization and division of labour ('complementarity' networks) can provide sufficient scale economies to attain the above mentioned goals (Camagni 1993).

Understanding (Urban) Creativity

Following one of the earliest and most thoughtful works on creative environments (Andersson et al. 1993a), two relevant inspiration elements may be taken from the sociology of science and biology. The first concerns the fact that, in the history of scientific thought, many significant advances have occurred by unexpected mental cross-fertilization between different disciplines. The second concerns the very nature of evolution, being 'a wasteful, fumbling process characterized by sudden mutations of unknown cause, by the slow grinding of selection, and by the dead-ends of over-specialization and loss of adaptability' (Koestler 1964, quoted by Andersson et al. 1993b, p. 9).

Unpredictability and synergy, redundancy and adaptability look to be central elements in the pathway towards creativeness and innovation. But beyond that, a paradigm shift looks necessary with respect to those that traditionally have controlled the evolution of western science and thinking. Simplicity, formalization and prediction have to give way to complexity, qualitative perception and chance, if change, discovery and bifurcation has to substitute for more deterministic scientific and economic development trajectories.

Two elements have to be highlighted in this change of perspective. An associative thinking – like the one used in psycho-analysis, particularly in the recent practice of group analysis called 'social dreaming' (Lawrence 2005) – adds to, or substitutes for causal thinking in revealing new meanings, symbols, psychological constructs or simply in unveiling individual creativity potentials. Second, a crucial convergence looks necessary between formalized, shared and common knowledge, being exchanged on a worldwide dimension, and unencoded, local knowledge that 'creates the semantics which adds complexity to any simpler syntactic entailment' (Andersson et al. 1993b: 9). Local knowledge, with its mainly tacit and implicit character, adds the richness of deeply rooted social competences, tastes, beliefs and values to stylized 'cosmopolitan' components of knowledge, opening the way to potential creativity.

What precedes is synthesized in Figure 8.2, where necessary paradigm shifts in approaching novelty in science, arts, technologies or economic innovation are shown, followed by a series of context characteristics that may help cities translating abstract and potential creativity into actual creativity. These characteristics do not refer only to presence of specific assets or of more or less knowledgeable people, but rather to ways in which the entire urban system works and evolves, thanks to its very identity.

PARADIGM SHIFTS	Complex vs Simple		Associative vs. Causal		Semantical vs. Syntactical
CONTEXT CHARACTERISTICS	Adaptability + Resilience	Leadership (catalyst + vision)	Relational Capital	Social Capital ("bridging")	Tolerance
PROCESSES	Synergetics	Merging global and local knowledge	Cooperation	Interdisciplinarity	Social and Cultural Interaction
POLICY TOOLS & STYLES	Globalized communication networks + Diplomacy	Education & human capital	Identity and cultural heritage	Places of interaction (meeting facilities)	New governance styles (p/p partnership, relationality)
STRATEGY	3K + G COGNITIVE CAPITAL + RELATIONAL CAPITAL + ENVIRONMENTAL CAPITAL + GOVERNANCE				

Figure 8.2　Paradigm shifts, context characteristics and policy strategies for enhancing creativity

The first and probably most important of the context characteristics concerns on the one side its flexibility, open-ness and adaptability, and on the other side and contemporaneously its robustness and resilience: resistances to change and conflicts should be manageable and managed, contacts and interactions among different cultures and classes promoted, transfer of resources from 'sunset' to 'sunrise' functions easily accommodated, but at the same time the system should be sufficiently robust to withstand turbulence, understand and exploit the durable parts of novelties and not be too prone to transient fashions. Often, these difficult processes need a strong political leadership, in order to create the right catalyst for change, speed up citizens' convergence on general societal goals, create new visions for the entire urban system and reinforce or regenerate local identities.

The fuel for a continuing modernization and regeneration of the urban economic fabric lies for sure in the density of external, particularly international, linkages maintained and developed by individuals, groups, associations, firms and institutions: what is increasingly called relational capital (Camagni 1999) or the 'bridging' part of social capital (Gittel and Vidal 1998; Woolcock 1998),[4] responsible for associationism and social networking inside the city.

Tolerance, as emphasized by Richard Florida, may be seen as a last context characteristic conducive to creativity. For sure it links up with such beneficial conditions as diversity, openness, social rewarding of individual talent and

---

4　As opposed to the 'bonding' part − encompassing proximity social networks: family, relatives and close friends ties − which in economic terms may be responsible for parochialism, defensive attitudes, blockage with respect to the non-members of the group, clan or other.

creativity, and deserves consideration in a discourse on urban creativity. More problematic, though, appears accepting Florida's emphasis on tolerance being the *primum mobile* for creativity enhancement, and the following direct causal chain of increased urban attractiveness for the creative class, consequent attraction of advanced firms, technology development, innovation, income growth and so on (Florida 2005: 92).

These context conditions provide the assets – the catalysts or the spatial 'filters' (Rodríguez-Pose 1999, Rodríguez-Pose and Crescenzi 2008) – that help translating potential urban creativity into actual innovation capability. The processes at work are processes of synergetics, synergy and cooperation, social interaction and cultural cross-fertilization.

Once again it looks relevant to underline that, in order to find, maintain and develop a role inside the international division of labour for the local production system, local knowledge and identity, whatever their real visibility and value might be, are crucial in this respect: in order to take full advantage of external linkages; in order to attract new talents and innovative firms; in order to build (or rebuild) a local competitive advantage. Any agglomeration of producers develops in the long term a 'patina of place-specific color', an atmosphere à la Marshall nourished by 'peculiar traditions, sensibilities and norm' (Scott 2006: 7) that creates the 'uniqueness of the place' (as recently D.G. Regio of the EU Commission has pointed out) or of the local milieu. Even the best territorial marketing cannot create a territorial identity, or even image, from scratch, at least of the type required by creative people for defining their locational choice.

The same holds for the provision of urban amenities. Urban policies can help the creation of an urban atmosphere greatly, particularly in some places or districts, which could be attractive for cultivated and creative people. But it is rightly the pre-existent presence of these people that cumulatively adds flavour and real liveliness to these places, not vice-versa. Milan would never have attracted the talents working around the fashion world if it were only thanks to the flavour of the Navigli canals or of the central 'fashion *quadrilatero*', without being the centre of the highly integrated fashion, design and communication industry (magazines, advertising and commercial television); and the fashion industry would not be there without a previous historical specialization in textiles, clothing, advanced textiles machines, followed by CAD-CAM technologies and complemented by international fairs, handicraft clothing production (e.g. for the La Scala opera performances), industrial design and ... Italian taste.[5]

---

5 The pathway towards the present condition of success was by no means a smooth one; in fact, during almost 100 years, it was hit at least four times by huge crisis conditions, mainly linked to international or global trends: the crisis of the cotton and textiles industry in the 1930s, a similar crisis in the 1970s, the ICT revolution in the 1990s and the China challenge of today. The main strategies: focalization, de-localization (to the Milan hinterland, to southern Italy, to overseas), development of specialized industrial districts, enlargement of the *filière* towards complementary, mainly tertiary functions, commercial

Cultural activities are crucial in this picture, in multiple respects. First, as education channels, they help to generate knowledge and boost human capital. Second, they create the atmosphere, the attention, the enjoyment and finally the willingness to pay for the products of human spirit and talent. Third, they provide directly an important part of these 'final' products in which creativity is embedded.

Towards Urban Policies for Enhancing Creativity

As stated at the beginning of this chapter, in history the city has always worked as an innovative environment, attracting the functions that were critical in each historical stage and developing its physical internal structure according to what was needed during each time.

When, after the industrial revolution, industry became the driving force for development, industry was hosted inside cities, even in the city-centre, while peripheries performed the role of assuring a low reproduction cost to a mass labour force. In the subsequent phase, industry disappeared, and was replaced by new headquarter, administrative and control functions located in the city centre, performed by a growing middle class that found new residential locations. In the words of Jean-Paul Lacaze, 'associated segregation', typical of the mill-town residential areas of the working class was replaced by the 'dissociated segregation' of the middle-class suburbs, while the car widely replaced the mass transit system (Camagni and Gibelli 1996) (see Table 8.1, where three archetypes of urban structure and performance are presented: the industrial city, the post-industrial city and the present/future, 'post-modern' city).

When, in the post-industrial city, information and information exchange became crucial, the radio-centric structure of mobility allowed the concentration of face-to-face contacts in the city centre. While in the first, industrial stage internal efficiency was the goal of the planning techno-structure, competitiveness became the primary goal in the second one; large urban schemes and projects were developed through public/private partnerships (with a different power balance between the two partners according to national planning cultures, politics and traditions).

As it is quite evident, creativity is not a distinctive feature of the present/future city: nowadays creativity is only explicitly pursued in forms that are typical of the present stage of development. As is also shown in the last row of Figure 8.2, the main policy goals of the post-modern city are globalization and attractiveness; the leading sectors are higher education, culture and arts, health, new economy, leisure and tourism.

A proactive globalization strategy regards first of all the promotion of international linkages with other cities and the development of an urban diplomacy,

---

linkages with parallel sectors (cars, glassware, perfume production) and utilization of some urban quarters as windows, show-rooms and meeting places for the local *filière*.

Table 8.1   The three development stages of the modern city

	Industrial city	Post-industrial city	Post–modern city
DNA	Competence	Information	Knowledge, creativity, reflexivity
Leading sectors	Industry, headquarter functions, commercial tertiary activities	Advanced tertiary activities, finance, information and communication technologies (ICTs)	*New economy*, R&D, education, leisure, culture, health, arts, global control networks
Urban form	Public transport, dense and continuous urban fabric, metro areas with central gravitation	Car-city, urban *sprawl*, dispersed metro areas ('archipelago')	Multiple-mobility city, local urban villages, polycentric metro areas
Organizzazione sociale	Reduced face-to-face contacts (F2F), 'associate segregation'	F2F in central district, 'dissociated segregation'	Local associations, new F2F places (hubs, edge cities, universities, congress centres)
Urban policies	Efficient urban structure, Keynesian spatial policies (welfare, zoning)	Market oriented urban policies, public entrepreneurship, competitiveness	Urban quality of life, attractiveness, globalization
Leadership in planning	Public techno-structure	Private/public partnership (with differentiated equilibria)	Civic control, participation, new governance forms
Urban policy tools	Plan	Project	Rules + visions
Strategic planning	First generation: structure plans	Second generation: corporate-like strategic planning	Third generation: inclusionary planning
Role of inner city and rings	Low reproduction cost of the labour force	Opportunity for large urban schemes and projects, high-quality suburbs	Places of *mixité* and solidarity; opportunities for cultural enrichment and development of individual capabilities

direct attention to the needs of international firms potentially locating in the area and the provision of high-performing transport and communication networks. Second, it requires attention to human capital formation, not just through high-quality education but also through the organization of international exchange opportunities in culture, arts, science and technology (festivals, forums, fairs, events).

External image and perception of the city, its assets and excellence functions should be carefully managed, not really with a territorial marketing approach but rather trying to interpret and communicate the true identity of the city, of its community of citizens, its specificities at the basis of its competitive advantage. The presence of a rich cultural heritage or of unforgettable natural environments constitutes a crucial advantage in this respect: when coupled with a tradition of hospitality and efficient receptivity structures they may boost the urban and cultural tourism industry and support other forms of tourism (managerial, scientific); when carefully included in the physical urban fabric and form, they enhance quality of life for residents and increase urban visibility and attractiveness; for sure, they provide a charming atmosphere, particularly appreciated by the cultivated and creative 'class'.

In order to increase relational capital, the provision of places of interaction and meeting is crucial. Hubs and airports, edge cities and commercial galleries, congress centres, wharfs, but also just places assigned to public use for concerts, discussions, presentations (as in many new neighbourhoods in some German cities) represent important enhancers of social life, replacing the ancient roman thermal baths, medieval markets and renaissance piazzas. Universities perform more and more this role, both when acting within the scientific international network and when linking up with the urban life hosting events and civic debates.

Citizen participation in urban decision-making looks crucial: diffused imagination and grassroots experience can more easily be given voice and be translated into actual projects. Urban strategic planning can also gain creativity and robustness when it leaves behind old-fashioned corporate-like procedures, typical of the 1990s, and acquires an inclusive character, promoting citizen participation and public/private partnership (Healey 2001). All this requires, of course, new urban governance styles: recognizing the importance of relationality and networking; not really supporting single firms, research institutes or universities but rather new projects launched in partnership among different urban actors, both private and public.

In synthesis (see bottom box in Figure 8.2), the strategy for achieving and enhancing urban creativity cannot be but a complex and integrated one. Utilizing the new concept of 'territorial capital' (Camagni 2008), it should encompass and support three forms of capital – cognitive capital, relational capital and environmental capital – and be managed through new governance styles. In synthesis, a 3K+G strategy.

Cognitive capital brings in not just competence and know how, but especially knowledge, learning to learn capability, serendipity and, as a consequence, adaptability and capability of driving change. Relational capital provides openness and trans-territorial linkages, and also cooperation capability with local and external partners and exposure to novelty. Environmental capital encompasses all manifestations of physical capital going from pure transport and communication infrastructure to urban settlement form and structure, from presence of cultural heritage to quality of the natural and built environment; all these elements impinge on local efficiency and potentials.

In a more traditional view, the preceding elements may be interpreted as supplying different forms of localized externalities, leading to what was called a 'place surplus' (Bolton 2002). But this view has to do with a static efficiency framework, concerned with costs and revenues, while innovation and creativity refer to a dynamic efficiency framework, concerned with change. Therefore, a concept like Scott's 'creative field' (Scott 2006) or GREMI's 'innovative milieu' (Camagni 1991) look more appropriate.

It is interesting to note that in both these concepts a crucial role is attributed to local governance due to a wide presence of market failures in the field of urban policies and the consequent need for ex-ante coordination of actors, collective action and collective sanction against opportunistic behaviour. New governance styles, open to public/private partnership, in fact can allow:

- An easier and creative management of cross externalities between the different competences, disciplines, functions, social groups and classes acting inside the city. The main fields of operation are higher education, security, transport infrastructure, quality of general urban services.
- Control over possible over-exploitation of crucial urban and cultural resources and place congestion.
- Control over possible opportunistic behaviour by some land-owners, which could destroy charm and place atmosphere through selfish decisions.[6]
- Control over land rents (mainly through appropriate taxation) which may jeopardize or destroy some locational advantages, beneficial for the advanced business or art community. In fact, a rise in land prices in the successful city around 'creative' neighbourhoods and cultural sites, thanks to place specificity and scarcity of accessible and suitable places, is a crucial risk for long-term development of the city itself.

It is clear that an 'integrated' strategy is much more difficult to pursue than a direct, straightforward strategy. In increasing order of complexity, other strategies could in fact be adopted:

---

6   The transformation of the city centres of art cities into tourist traps, with the crowding out of the local population, is clear in this respect (see the case of Venice and increasingly of Florence).

- Organization of large (international) events – but no guarantee exists of a successful follow-up: the success of post-olympic Barcelona is one of the rare successful examples while post-Expo Sevilla represents a counter example.
- Territorial marketing, 'city branding', 'experience destination' strategies – but linking up to a pure tourist exploitation may present the risk of the city 'losing its soul'.
- Hosting new developments in special districts in the inner city: successful examples are the urban renewal through modern and advanced activities in Barcelona in a derelict industrial district (project @22); the Navigli area in Milano concentrating architecture and art studios, fashionable restaurants and meeting places, cosy music bars; the Bairro Alto – Chiado area in Lisbon, with similar activities, the relaunching of Bilbao mainly thanks to the visibility of the Guggenheim Museum, new modern internal and external transport infrastructure (metro, airport, ...).
- The attraction of the 'creative class' à la Florida, through the creation of fashionable recreational activities, tolerance and uniqueness of place, to be linked to a (pre-existing) thick local labour market.

This last strategy argues in favour of a discontinuity with respect to previous economic traditions of the city, while the 'integrated' strategy that is advocated here tries to launch new roles and functions, building upon endogenous capabilities, identities (to be reinterpreted and re-oriented), traditional competences and the synergies that can be established among the creative activities (ICTs, high tech, business and financial services, neo-artisanal manufacturing, fashion and communication, health and wellness, cultural products *filière*, media, arts and performing arts) and between them and traditional activities such as building and construction, tourism, the retail trade, craft manufacturing.

The relationship between the strategy suggested by Florida and the integrated strategy parallels the relationship between an exogenous development pattern, driven by foreign investment, and an endogenous development pattern, driven by local traditions and identities, small and medium enterprises (SMEs), urban (rejuvenated) specificities.

A last reflection concerns the relationship between urban creativity and urban sustainability. In an urban context the concept of sustainability cannot refer principally to the preservation of the natural environment, but has to link up with the very nature and the ultimate goals of the city: an artificial environment addressed to human interaction and socialization and to the full development of human attitudes (Camagni 1998). Therefore, defining urban sustainability as 'the co-evolution of the three urban subsystems – the economic, the social, the environmental and built ones – allowing the maximisation of positive cross-externalities among these sub-systems and supplying a non-decreasing level of wellbeing to present and future generations and to local and surrounding spaces' (Camagni et al. 1998), we are allowed by and large to equate the 'integrated' strategy towards urban creativity to a sustainability pathway. In fact, enhancing (social) knowledge, (social and economic) relationality and quality of the local natural/physical/cultural environment, and building on all the potential synergies that the local context can provide, this strategy is likely to help to provide a forward-looking, cohesive and attractive urban society.

On the other hand, more straightforward and single-dimension strategies for enhancing urban creativity, with their emphasis either on large urban regeneration projects or on the artificial blowing up of a creative class could perhaps look more concrete and effective, but would probably lead to a less balanced urban realm and to a lower exploitation of true local potentials.

Towards a Conclusion

In abstract terms, cities are inherently creative environments, thanks to the agglomeration economies and the external linkages that they provide, enhancing communication and interaction. But not all cities in practice are able to exploit these potentialities, or do so in a sense that could be coherent with respect to the needs of present modern societies.

Other disciplines such as sociology of science and social psychology show us that some general context characteristics, having to do with cognitive processes, are crucial in enhancing creativity and innovation: associative thinking, diversity and interdisciplinarity, openess and relationality, semantical vs. syntactical reasoning, complexity vs. simplicity.

In the present development stage, three main elements that constitute the local 'territorial capital' look indispensable: cognitive capital, in the form of competence and knowledge, adaptability and capability of driving change; relational capital, assuring openess, external interaction, cooperation capability, internal synergies (and Richard Florida's tolerance); and environmental capital, in the form of high-performance communication and transport infrastructure, quality of the working and living context, presence of cultural heritage and charming urban neighbourhoods and 'villages'. These 3Ks have to be managed and combined through new governance styles, open to citizen participation and inclusionary planning, supportive of public/private partnerships, attentive to relationality and the provision of places devoted to social interaction.

All the mentioned elements have to merge and be considered together by conscious and reflexive urban policies. An integrated strategy for relaunching urban creativity has necessarily to start with present 'vocations' and competencies of the local context, reorienting them towards new and modern activities through the enhancement of knowledge-intensive functions, the creation of international city-networks, the creative utilization of urban cultural heritage and atmosphere, and the provision of interaction opportunities and places. This strategy works mainly bottom-up, making the most of existing potentialities; the attraction of external 'creative' firms, professionals and people could follow and cumulatively reinforce the initial necessary endogenous steps.

Bibliography

Andersson, A.E. 1985. Creativity and regional development. *Papers of the Regional Science Association*, 56, 5–20.

Andersson, A. et al. (eds) 1993a. *The Cosmo-creative Society: Logistical networks in a Dynamic Economy*. Berlin: Springer-Verlag.

—— 1993b. Logistical dynamics, creativity and infrastructure, in *The Cosmo-creative Society: Logistical Networks in a Dynamic Economy*, edited by A. Andersson et al. Berlin: Springer-Verlag, 1–16.

Bolton, R. 2002. Place surplus, exit, voice and loyalty, in *Regional policies and comparative advantage*, edited by B. Johannson, C. Karlsson and R. Stough. Cheltenham: Edward Elgar.

Camagni, R. 1991. Technological change, uncertainty and innovation networks: Towards dynamic theory of economic space, in *Innovation Networks: Spatial Perspectives*, edited by R. Camagni. London: Belhaven-Pinter, 121–44.

—— 1993. From city hierarchy to city network: Reflections about an emerging paradigm, in *Structure and Change in the Space Economy*, edited by T.R. Lakshmanan and P. Nijkamp, Festschrift in honour of Martin Beckmann. Berlin: Springer Verlag, 66–87.

—— 1998. Sustainable urban development: Definition and reasons for a research programme. *International Journal of Environment and Pollution*, 1, 6–26.

—— 1999. The city as a Milieu: Applying the Gremi approach to urban development. *Revue d'Economie Régionale et Urbaine*, 3, 591–606.

—— 2001. The economic role and spatial contradictions of global city-regions: The functional, cognitive and evolutionary context, in *Global City-regions: Trends, Theory, Policy*, edited by A.J. Scott. Oxford: Oxford University Press, 96–118.

—— 2004. Uncertainty, Social Capital and Community Governance: The City as a Milieu, in *Urban Dynamics and Growth: Advances in Urban Economics*, edited by R. Capello and P. Nijkamp. Amsterdam: Elsevier, 121–52.

—— 2008. Regional competitiveness: Towards a concept of territorial capital, in *Modelling Regional Scenarios for the Enlarged Europe*, edited by R. Capello et al. Berlin: Springer Verlag, 33–48.

Camagni, R. and Gibelli, M.C. 1996. Cities in Europe: Globalisation sustainability and cohesion, in *European Spatial Planning*, edited by Presidenza del Consiglio dei Ministri. Roma: Il Poligrafico dello Stato.

Camagni, R., Capello, R. and Nijkamp, P. 1998. Towards sustainable city policy: An economy-environment-technology nexus. *Ecological Economics*, 24, 103–18.

Florida R. 2005. *Cities and the Creative Class*. New York: Routledge.

Gittell, R. and Vidal, A. 1998. *Community Organising: Building Social Capital as a Development Strategy*. Thousands Oaks, CA: Sage Publications.

Glaeser, E. 1998. Are cities dying? *Journal of Economic Perspectives*, 12, 139–60.

Haken, H. 1993. Synergetics as a theory of creativity and its planning, in *The Cosmo-creative Society: Logistical Networks in a Dynamic Economy*, edited by A. Andersson et al. Berlin: Springer-Verlag, 45–52.

Hall, P. 1998. *Cities in Civilization*. New York: Pantheon.

Healey, P. 2001. New approaches to the content and process of spatial development frameworks, in *Towards a New Role for Spatial Planning*, edited by OECD. Paris: OECD.

Jacobs, J. 1969. *The Economy of Cities*. New York: Random House.

—— 1984. *Cities and the Wealth of Nations*. New York: Random House.

Karlsson, C. 1995. Innovation adoption, innovation networks and agglomeration economies, in *Technological Change, Economic Development and Space*, edited by C.S. Beruglia, M.M. Fischer and G. Preto. Berlin: Springer, 184–206.

Koestler, A. 1964. *The Act of Creation*. New York: Pan Books.

Landry, C. 2000. *The Creative City: A Toolkit for Urban Innovators*. London: Earthscan.

Lawrence, G.W. 2005. *Introduction to Social Dreaming*. London: Karnac Books.

Lucas, R. 1988. On the mechanics of economic development. *Journal of Monetary Economics*, 22, 38–9.

McCann, P. 2004. Urban scale economies: Statics and dynamics, in *Urban Dynamics and Growth: Advances in Urban Economics*, edited by R. Capello and P. Nijkamp. Amsterdam: Elsevier, 31–56.

Nijkamp, P. 1986. *Technological Change, Employment and Spatial Dynamics*. Berlin: Springer-Verlag.

Quigley, J.M. 1998. Urban diversity and economic growth. *Journal of Economic Perspectives*, 12(2), 127–38.

Rodríguez-Pose, A. 1999. Innovation prone and innovation averse societies: Economic performance in Europe. *Growth and Change*, 30, 75–105.

Rodríguez-Pose, A. and Crescenzi, R. 2008. Research and Development, Spillovers, Innovation Systems, and the Genesis of Regional Growth in Europe. *Regional Studies*, 42(1), 51–67.

Sassen, S. 1994. *Cities in a World Economy*. Thousands Oaks, CA: Pine Forge Press.

Scott, A.J. 2006. Creative cities: Conceptual issues and policy questions. *Journal of Urban Affairs*, 28(1), 1–17.

Scott, A.J. and Angel, D.P. 1987. The US semiconductor industry: A locational analysis. *Environment and Planning A*, 19, 875–912.

Storper, M. and Scott, A.J. 1995. The wealth of regions: Market forces and policy imperatives in local and global context. *Futures*, 27, 505–26.

Veltz, P. 1993. D'une géographie des coûts à une géographie de l'organisation: quelques thèses sur l'évolution des rapports entreprises/territoires. *Révue Economique*, 4, 671–84.

Woolcock, M. 1998. Social capital and economic development: Towards a theoretical synthesis and policy framework. *Theory and Society*, 27(2), 151–208.

Chapter 9
Creativity: The Strategic Role of Cultural Landscapes

Xavier Greffe

Introduction

A landscape comprises the visible features of an area, including its physical elements, living elements and human elements such as human activity and the built-up environment. Since a landscape is shaped by human activity, we may use the expression 'cultural landscape'. A priori this does not say anything about the quality/state of preservation or the level of preservation it demands.

Traditionally, a landscape is considered as an expanse of natural scenery that people come to see and enjoy. Defining and protecting a landscape is not a new issue. The study of the landscape as a form of visualization has begun to shed light on the processes through which a landscape can be used as a cultural and political instrument. This approach has been criticized for placing too strong an emphasis on representation at the expense of considering our material interactions with the world (Rose 2002). Taken to a post-modern extreme, such theorizing has, in some cases, led to a totally immaterial conceptualization of the landscape denying the 'connectivity' of representations with the 'world outside' and downplaying the importance of the relationship between the material world and its representations. The challenge at the heart of Western scientific thought is that it is generally based on the separation of man from nature: a position that may eventually undermine our concerns about ecology and sustainability. Contesting the understanding of the world that divides it into subjects (minds) and objects, Ingold advocates 'an alternative mode of understanding based on the premise of our engagement with the world, rather than our detachment from it' (see Greffe and Maurel 2009: 212). In a similar vein, the geographer Mitch Rose argues that 'the engine for the landscape's being is practice: everyday agents calling the landscape into being as they make it relevant for their own lives, strategies and projects' (Rose 2002).

But this problem has become important today for three reasons:

- The ecological movement has convinced city-dwellers and communities of the need to look after the environment and prevent its deterioration, which may not always be immediately visible.
- Cultural tourism: the landscape becomes a lever of economic development through tourism.

- The role of cultural atmosphere for creating a creative city. This last point is very important now and can be considered from a twin perspective. A good cultural landscape can favour creativity by generating new activities, jobs and revenues. This very common idea is usually backed by cultural tourism or the cultural attractiveness of the creative class. But another perspective is to start with the fact that a creative city is a city where people have values and behaviours that distil and disseminate new activities and new jobs. Then creativity can be defined as a skill for designing and implementing new projects. The relationship between cultural landscape and creativity is clarified: as far as a cultural landscapes can integrate architecture, public art, arts centers that demonstrate how to produce new projects, cultural landscapes can be a lever for more creative cities.

Indeed, once the conservation of cultural landscapes is recognized as an important element for reinforcing the economic base of a territory rather than as a simple expression of an aesthetic need, the nature of the debate changes. It is no longer a question of compromising the growth of employment and income to protect a few old stones, fish or obscure plants, but to take action for the area's sustainable economic development by avoiding useless or irreversible damage to the natural, cultural and, therefore, human environment. The long-term economic health of a community may demand that the cultural landscape not be sacrificed by blindly pursuing unregulated development. Promoting it to attract visitors to the area can largely compensate for the earnings expected as a result of the destruction of the cultural landscape. These motivations – ecological, tourist or cultural – will determine the type of actors who will play a role in the formulation of such policies. Some will intervene in the name of safeguarding the quality of the biosphere and the living conditions of the local inhabitants. Others will intervene in the name of preserving culture as an intangible element, while still others may invoke the beauty and integrity of a landscape. The latter will intervene in the name of tourism to facilitate access to landscapes regarded as capable of attracting visitors.

In order to understand this challenge, we shall consider:

- the consequences of considering landscapes as a form of visualization, in terms of criteria for assessing the quality of an urban cultural landscape and instruments for implementing and managing these urban cultural landscapes;
- the consequences of considering landscapes as a form of experiences, in terms of comprehensive criteria and components, focusing on cultural assets.

Urban Cultural Landscapes as a Form of Visualization

Which Criteria to Use to Assess the Quality of an Urban Cultural Landscape?

How traditionally do we appreciate and assess landscapes? The criteria used may be both economic and aesthetic, and they are probably related because the economic value of a place is enhanced by its aesthetic value. But the aesthetic value is elusive and ambiguous. Moreover these criteria are mainly 'negative' (Berleant 1997):

- Ugliness: this is the easiest criterion to define since it is the antithesis of beauty. But it seems to be an extreme criterion because things that are not beautiful are not necessarily ugly. Moreover, some things may be both ugly and aesthetic such as the gargoyles in the Notre Dame Cathedral in Paris.
- Offensiveness: this criterion is less exceptional. Although the ugly may not necessarily be unbeautiful, it may still offend us. The offence occurs in this case because it is believed that a rule pertaining to good taste or proportion has been transgressed. So when a commercial interrupts an interesting programme on radio or television, we can say that there is an offence because of the insensitivity of the interruption.
- Banality: this may also be considered as offensive, but only by default. A landscape is banal because its possibilities have not been fully exploited.
- Dullness: this is generally a consequence of banality, but is not confined to it. The landscape may be dull because of an absence of inventiveness. In fact, this criterion is easier to apply to the arts and to monuments than to the environment. When a landscape looks dull, it is because of human intervention, for example, when houses or gardens are planned unimaginatively so that they give an impression of dullness.
- Lack of fulfilment: an object is not appreciated because it does not live up to our expectations. This criterion seems difficult to apply to landscapes too. In fact, lack of fulfilment is frequently associated with another criterion.
- Inappropriateness: ethnic designs used without regard to local building traditions can have a disastrous effect.
- Trivialization: this is a form of inappropriateness. Things that are serious, such as housing, are treated in a very casual manner. Trivializing the past is often a cause of such inappropriateness.
- Deceptiveness: this intensifies the negativity of the trivial by making evident a voluntary dimension. Much of the vernacular heritage has been created on such a basis, which simultaneously induces inappropriateness, trivialization and deceptiveness. But the insensitivity of this criterion is linked to its moral roots.
- Destructiveness: this is the most extreme criterion, much more so than ugliness.

Which Instruments?

Let us start by stressing two points:

- As far as cultural landscapes are concerned, there is both a positive and a negative demand for their protection: some actors such as ecological movements demand more conservation; others such as urban developers or promoters are less concerned about conservation.
- There are many stakeholders. But who are they? There are those who have a direct interest in the ownership or utilization of environmental resources; those who are passionately committed to nature, especially nature that is unimproved, undefiled, inviolate; gentlemen farmers and owners of estates lying beyond the fringe of suburbia; farmers who are more or less prompted by aesthetic concerns; citizens who are concerned in different ways according to their preference for specific ownership or localization; tour operators who want minimal protection to create a demand for their services, but not too much to oblige them have to manage their flow of tourists; environmental non-governmental organizations (NGOs), and so on. Similarly, there are many instruments: education for environmental protection is probably one of the most important and it should be announced at the very beginning. But here we shall focus on the more traditional economic instruments. Anyway, it is necessary to increase incentives for people to participate with the landscape through active perception, and this signifies creativity for cultural landscapes as well as physical investment. By understanding the interplay of so many elements in a landscape, it is possible to enhance its 'beauty' and emphasize its cultural dimension (Whyte 2002 [1968]).

The Police

The police force is employed to prevent people from using their land or property in a way that is harmful to public welfare or, to be more exact, in a way that alters the cultural landscape in a negative manner. Usually this type of policing consists of zoning an area to indicate the type of estate or resources that can be developed and the principles governing this development. As far as cultural landscapes are concerned, the police define zoning laws for restricting the built-up area on an acre of land. This policy is not always effective since developers tend to violate laws related to the built-up area. Thus extending the areas for construction compensates for what is lost in intensiveness. Another issue is the parochial nature of local zoning, which requires extensive monitoring by governments and law courts. Finally, another difficulty is that owners of properties have to be paid compensation if the new zoning laws affect their properties. When regulated by the police, there is a possibility of their wealth being undervalued and it is therefore legitimate to compensate them for the loss. But the cost can increase dramatically in such cases.

'Fee Simple'

The best way to prevent any harmful use of real estate or land is to acquire absolute ownership with unrestricted rights of disposal (fee simple). The possibility of the government exercising this power is bound to influence owners and make them use their properties more prudently. But the problem is that when a local government wants to buy land, it has to indicate its objective clearly and the court that assesses the property and decides the price must accept this objective. 'Public purpose' is one condition that permits alteration of private property, but the point is whether the court will recognize conservation of a cultural landscape as a 'public purpose'. At times, the time needed for these legal proceedings may have negative effects on the final agreement by raising the expectations of the seller forced to part with their property. Further, the cost of an outright purchase is very high, and many local governments do not have the necessary resources. Naturally, this does not apply to governments who have ample resources at their disposal and are in a position to offer a part of these resources in exchange for the property in question.

Another solution is to create a system of revolving resources where the purchase of a new piece of land or an urban estate is ensured through the sale of other resources. Another system is the lease-back system. The local government buys the land but allows the previous owner to use it only for specific purposes in accordance with the demands of landscape preservation. Another system is preventive buying. But whatever the choice of system, the cost of the fee simple policy is bound to be high, much higher than the use of policing powers.

It has to be said that in many European countries, the use of this instrument is not dramatic since local governments traditionally own important amounts of land. Naturally the problem is to know how they use these resources. During the last decades local governments have been more sensitive to the employment issue than to the landscape one, by offering significant amounts of land to new builders, superstores, more or less interesting entertainment parks, and so on.

Easement

Instead of buying a property, it may be much more economical to resort to the easement law that gives a person certain rights or privileges in another's land. When a property is bought, the buyer acquires a whole bunch of rights tied to the property, but the buyer is basically interested only in one type of right. To achieve this purpose, the buyer needs to acquire just one or a few rights in the property. In such a case, it is better to resort to an easement and leave the remaining rights with the owner. Some easements are positive while others are negative. In the first case, it is possible to buy the right to do something with a part of the property. For example, it is possible to create a public footpath or a bicycle trail. In the second situation, the owner is told not to use their property rights fully, for example, to desist from adding extensions to the main building.

One advantage of easement is that it can be used no matter who owns the land. Moreover, if it is necessary to pay compensation to the owner who suffers as a result of the easement, their successors are not entitled to any payment because the price of the property includes the losses caused by the easement.

One difficulty with easement is the assessment of the compensation amount to be paid to the owner. The rule of the thumb for assessing the easement is to figure out the 'before and after' value of the property. This depends not only on the nature of the easement but also on the future prospects of the property, which are always debatable. If the easement deals with the essential property rights, then the property will lose its value and the easement will be as costly as the fee simple.

Another issue concerns employment. If the easement leads to the destruction of jobs, its social cost may be much higher than its financial cost, and this should to taken into account when dealing with welfare.

Development Rights
In an urban environment restrictions can take a different form. When a limit is imposed on the height of the buildings to safeguard an urban landscape, it is possible to limit the constructible height in one area while allowing its increase in another area. The advantage for the authorities is that they do not have to spend anything because the loss of a right in one area is compensated by its availability in another area where it can be sold to owners and developers. This attractive solution was developed in the United States. But although it holds the possibility of earning profits from the new rights, this is far from obvious, except for promoters.

Moreover, such cases are often referred to the court to ensure that the new rights have the same value as those that have been suppressed. So it is necessary to wait for the court's verdict and there is a good chance that the local government may be obliged to pay additional financial compensation.

Taxation
Another means of controlling the use of various natural spaces that form a cultural landscape is to treat them differently through taxation. Thus uses needing preferential treatment because they contribute to the conservation of the cultural landscape can be taxed lower or even exempted from tax, whereas uses considered to be harmful for the conservation of the cultural landscape can be very heavily taxed. This can be done by defining the basis for taxation as also through the differentiation of tax rates. There should be prompt approval of such fiscal expenditure since it is directly intended to protect the general interest. Thus, if farmers are asked to become gardeners by preserving certain species of trees or crops that convey a positive image of the landscape, but which make them incur losses as compared to other uses, this solution appears normal.

Another advantage of this instrument is that its cost for the government is relatively low. For example, if the owners participating in the protection of the cultural landscape are granted significant advantages. Those whose activities are detrimental to the preservation of the cultural landscape could finance this measure.

The main disadvantage of this instrument is that its long-term consequences can be very unfavourable. For example, if the use of land for constructing dwellings is highly taxed, these dwellings will be built in some other place where they will contribute to the over-use of land and raise the corresponding costs of congestion for the local government. The tax instrument must therefore be a part of long-term planning and associated with other instruments.

Urban Cultural Landscapes as a Form of Experience

A New View: Starting from the Aesthetics of Engagement

Human intervention does not always harm nature. In fact, to understand what type of criteria should be applied, it is necessary ask oneself how aesthetics operate in a landscape. The underlying idea is that adopting a purely contemplative approach cannot enhance the aesthetics of a landscape, as it tends to exacerbate subjectivism and conflicts between various actors. It should be based on the aesthetics of engagement, which calls for an active involvement in the aesthetic field (Chang Chun Yuan 1963).

Positive aesthetic values function in four ways: mechanically, organically, practically and humanly:

- The mechanical function: The most efficient function is that of an object adapted to a specific task that it performs with the greatest economy of movement and a minimum of wasted effort. Here, the paradigm is the machine that produces results with maximum efficiency. Normally we cannot deduce any aesthetic value from it, but some machines have managed to penetrate the field of arts and demand aesthetic attention. From Russian constructivists to Italian Futurists via Leger's industrialized human forms, the machine has found its place in the world of artistic aesthetics. The mechanical function has two positive characteristics: it is highly practical and it is eminently pleasing. An aesthetic function emerges and overlaps our perception of the landscape, mainly through design, architecture and urban planning.
- The organic function: a machine calls for the arrangement of various parts in a specific order to fulfil a predetermined task. The organic function goes beyond this simple interrelatedness by demanding cohesion and mutual responsiveness so that the function of the whole is more than the action of the individual parts. The organic function generates its own purpose whereas a mechanical function receives its purpose from its external environment. The paradigm here is the human body or, if we were to consider the arts, it would be dance. There is another difference here between the mechanical and the organic functions. In the case of dance, the external audience tends to participate somatically, joining the dancer in a common activity. Then

the person appreciating the dance is involved in a synthesis. There should be unity between the perceiver and the object in order to have a full organic function. Finally, the organic function adds elements of vital harmony and self-generation to the austere efficiency of the mechanical function. When considering a cultural landscape, this organic function stresses the coherence of the landscape so that each of its elements is fully merged with the others.

- The practical function: this refers to a context of use in which an object is associated with a person in a relationship involving a means to an end. It is this practical function that differentiates fine arts from applied arts: Does beauty depend on need? In fact, this distinction between fine and applied arts is more philosophical than artistic in nature. The activity of the artist has always been a synthesis of the practical and the aesthetic, an activity where practical skills are used to create objects of art with the continuous involvement of both in a mutually responsive manner. Both practical and aesthetic aspects are usually present in a work of art. In the case of a landscape, it is difficult to deny the relevance of architecture from this practical perspective. The interest of the object of art fuses with the interest of the user or perceiver. A building that is considered successful achieves both aesthetic and practical success. Architecture joins hands with the landscape and urban planning to perform these functions coherently. The practical function thus embraces both mechanical and organic functions in a fuller context of interrelation and dependence where an object of art and the aesthetic subject engage in a creative exchange.

- The humanistic function: the cultural landscape becomes a kind of conjunction between the landscape content and its perception bringing them together in a mutually fulfilling transactional relationship. A landscape does not fit in with the traditional models of disinterestedness, isolation and permanence. It fits in much more with the machine, dance and architecture linking functional, organic and practical dimensions considered from the perceiver's point of view. It incorporates the practical, but goes even further by making the receiver experience a relationship with the landscape. The landscape brings together its previous creators, its actual content and its receivers to form a whole. A cultural landscape creates (and has to create) a synthesis between aesthetic perception, social relevance and human fulfilment in which these three elements become inseparable from one another.

Urban Cultural Landscapes as an Ecosystem

A Simple Economic Approach
Whether the stakeholders want to invest in the conservation of a cultural landscape will depend on its initial quality. If this cultural landscape is in a poor condition, these actors are more than likely to neglect it, which will only accelerate its further

deterioration. Conversely, if a cultural landscape is in a good condition, it will be appreciated and attract more attention and, consequently, increased resources for its conservation. In Figure 9.1 (Greffe 2004a), the X-coordinates measure the attention paid to the cultural landscape by stakeholders (positive on the right of the origin and negative on the left). The Y-coordinates measure the quality of the cultural landscape (which is supposed to be directly related to the amount of resources devoted to the maintenance of the landscape). Curve AA'A'' thus represents the attention paid to the cultural landscape according to its quality or, conversely, the improvement of the quality of the landscape due to the attention paid by the stakeholders. There is a limiting threshold (A'), which determines a break-even point in terms of the behaviour of the stakeholders, and then in terms of resources allocated for conservation. Below this threshold, the attention paid to the cultural landscape is too feeble to guarantee the support of the users and the corresponding effort for its maintenance. Above this threshold, the attention is sufficiently strong to develop greater attention and the corresponding effort for maintenance.

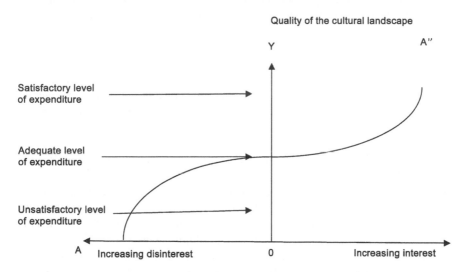

Figure 9.1   The dynamics of a cultural landscape

An amount of resources (OB on the axis OY) is allocated by the actors for the conservation of the cultural landscape in the form of both public and private expenditures. This amount depends on the spontaneous efforts made by various agents to protect the cultural landscape minus the efforts of others contributing to its deterioration because the latter want to put the land resources to a different economic use. The issue then is to know whether the amount of resources (OB) guarantees the minimal threshold of the quality of conservation (OA'), which will create a positive interest among the stakeholders in favour of the cultural landscape.

- If this condition is fulfilled, local actors will consider their cultural landscape positively (OB > OA'). They will then make the necessary efforts for is conservation. This tends to raise OB and generate an enhanced quality, D'', followed by increased attention and so on.
- If this condition is not fulfilled (OB < OA'), local actors will consider their cultural landscape in a negative manner (OB > OA'). They will then reduce their efforts. This tends to decrease OB and to generate a lower quality, D'', followed by reduced attention and so on.

There are two ways of enhancing the process of protection:

- The first is to reinforce the public or private instruments for raising funds for conservation by introducing tax incentives or by increasing the effectiveness of these expenditures by improving their productivity.
- The second is to raise interest in the landscape, even if the initial protection level is very low, by disseminating information and training or by organizing events. Consequently, the sustainability threshold will be lower.

The development of a cultural landscape thus goes beyond the traditional economic framework of the production and consumption of a given good. The plurality of values and the interdependence of different types of behaviour encourage us to consider the cultural landscape as an ecosystem, i.e. a system likely to generate positive or negative dynamics depending on whether certain types of behaviour or budgets are satisfactory.

What Type of Local Community will Protect the Urban Cultural Landscape? A New Set of Criteria

Which are the social factors that could support and enrich this approach. To be more precise, it is important to identify the type of social organization that will endorse the protection of cultural landscapes (Berleant 1997, Greffe 2007). This issue concerns the community in charge of a cultural landscape.

We have to begin in the most realistic manner from a *utilitarian perspective*. Utilitarianism believes in the ultimacy of the individual, a being located in a rational universe, separated from other beings and whose intelligence is largely calculative. In order to win support for the preservation of cultural landscapes, it is possible to raise issues such as the ethics of prudence or the morality of conscience, but these are articles of faith when we take into account the utilitarian interests of individuals. But we cannot consider individuals as totally autonomous since they are interdependent as long as they face collective constraints such as the scarcity of resources or the external effects of technological development. Persons do not exist singly and it is necessary to consider conflicts between egoism and altruism, between the self and the other. It is also necessary to consider the possible relationship between alternative social complexes and the quality of cultural landscapes. Three views are thus possible:

- The *rational community*: this is a community of individuals who view society as an artificial construct and the government as a dark mechanism to be tolerated at best as an unwelcome necessity. Central to this approach is the individual motivated by self-interest, guided by reason and protected by rights. In this community individuals are guided by prudential motives and they agree to collective actions when they can identify their common interests. This rational community is more of a social order than a community. This model can be seen in the political process: it is always seen as satisfying special interests through a process of political competition. Some alterations may interfere with this model: self-sufficiency is frequently recognized as a utopian ideal in a way that the self is more a folk category than an entity; mutuality is recognized, but it does not mean dependence and it is for the individual to make the choice.

  This is where landscapes come into the picture: in the face of pollution and negative external effects, collective action may be undertaken to protect and maintain cultural landscapes. But this is a very minimalist view inspired by the principle of prevention when the problem is clearly identified, analysed and recognized as legitimate. In such a community cultural landscapes will be marginally protected in the face of severe or even irreversible challenges.

- The *moral community*: Multiple bonds hold the members of a community together. Their interdependence stems from the morality of conscience: a moral obligation is a binding force that goes beyond desire or usefulness. Sometimes the moral community is so powerful that it becomes an organic community: an authoritarian centre of power regulates relations between its members who are subsumed in the whole. These communities may foster some common values, but the brittleness of the links between individuals can make these objectives very superficial.

  In the case of cultural landscapes, there is a fear that opposing forces that can determine the quality of the landscapes may destabilize such a community. If on the one hand a certain amount of solidarity within the community can be expected, on the other hand there is always the fear that some private real estate company with sufficient clout may undermine the basis for their collective and coordinated maintenance.

- The *aesthetic community*: this community does not signify just an external relationship between different things. The community is also prepared to take on external challenges. The relationship between individuals is not based on autonomy or internal control but with reference to an external dimension. Individuals may differ from one another and have contrasting attitudes but this is of secondary importance when dealing with the environment where the community is located. Fusion takes place at a more fundamental level. The link with nature and the external environment is probably the most important factor of internal connectedness. An aesthetic community is thus a community where the difference between the exterior and the interior is blurred, a community where the observer is also a participant.

Opposition in rational and moral communities preserves the difference between the observer and the participant in such a way that judgements are made independently of actions and can lead to personal strategies. In the aesthetic community, the participant is primarily a committed person and this encourages others. Self-awareness of observation is secondary and dependent.

Once a cultural landscape is identified, there is a relationship of undivided reciprocity with natural elements. A person identifies and experiences a personal relationship, a personal connection that unites individuals with one another. They do not objectify and control but enter into an intimate association. There is more than a simple connection; there is continuity and finally a community. By recognizing the multidimensional reciprocity of a cultural landscape, we recognize its social dimension and the aesthetic conditions of human fulfilment. We protect cultural landscapes not as an external treasure to be transmitted to posterity but as a part of our revolving identity and life.

Which Cultural Landscape for Creative Cities?

Whatever the perspectives and sets of assessment criteria, it is important to screen the elements that can make cultural landscape more proactive for implementing creative cities. Reconverting brownfields, implementing cultural districts and quarters appear as very seminal elements to distillate and disseminate creativity into the city. Then it may be useful to see that this public policy is not that easy, taking as an example the case of France.

Reconverting Brownfields

The conversion of former industrial sites into art facilities can contribute to local development in various ways. It can rehabilitate old buildings, improve the quality of life by offering new facilities in often underserved areas, and can offer local groups and communities the chance to rebuild their identities, to become part of creative culture, and to undertake projects that will have positive fallout for the entire city (Greffe 2004b).

A 'retrofit' of this kind typically occupies a former industrial, commercial or military site that has deteriorated through its former use to the point where it can no longer be used without a thorough overhaul and cleanup. The proliferation of abandoned industrial, commercial and military facilities has created gaping voids in the larger cities, particularly, witnesses to a bygone era. Activists promoting the revival of cultural and associative life set out to meet the challenge posed by these derelict urban spaces and to invest in them. The movement was born in the late 1970s, with the emergence of the counterculture, squatter invasions, and the growing awareness that new generations of artists and audiences had entirely new aspirations. Little by little, the sponsors of these conversions attempted to transform the surrounding districts and to make them a force for integration and education, where artistic activities represented a means as well as an end. The 'retrofit' thus produced a new urban territory.

These converted facilities contribute to both better urban cultural landscapes and increase local social development in four ways:

- By re-invoking the emotional and symbolic significance of a place that was once used for industrial production, the retrofit suggests that creation is inherent to the territory and it invites us to explore those aspects: Derelict urban sites show how ephemeral any social organization is, and at the same time they contain the seeds of a possible future. These wastelands, where ordinary life is suspended, speak to us of the unspeakable, the unnameable, the dark aspects of society. Because they suspend the unconstrained process of production and consumption, of use and wear, they blur the frontier between what is thinkable and what is not. For people who have spent all or part of their working life in these old warehouses, forts and factories, to return to them now is quite a surprise, because they can see there traces of new hopes and uses rather than ruins. The members of cultural associations that choose to set up shop there see these places as covering all possibilities, from rebirth to ultimate physical destruction. An abandoned building thus highlights two related concepts, the ephemeral and the creative. By testifying to successive modes of production in the past, the derelict suggests that development can continue, but in different forms.
- Moreover, when factory sites are used for amateur performances, such as practical theatre, rap or hip-hop workshops, they allow individuals and groups to become actors in their own development. Other cultural institutions have been doing this for a long time, but one of the key features of the retrofits is to push these practices further by enlisting amateurs from outside their walls, in disadvantaged urban neighbourhoods. By reaching out to a new potential audience and bringing it into a practical theatre workshop, and showing that theatre is not the preserve of the rich and privileged, these recycled facilities are fulfilling one of their key social callings.
- Retrofits help to popularize a forward-looking 'project' culture by instilling the principle of creativity in an environment that is often degraded and disadvantaged, and from which that principle has long since disappeared. This creative activity brings a new life to the territory, one that goes far beyond evening concerts or theatre productions. The Friche Belle de Mai in Marseilles is not limited to helping artists with their tangible work, for example by providing them with studio space: it takes account of all the conditions and skills needed to bring a work to life. Experts in the new technologies provide coaching and artistic support for radio or newspaper

ventures, for example. The combination it offers of housing for artists, visual art workshops, training in the practical arts, recording studios, discussion forums, sporting activities and restoration makes it a place of creativity not only for artists but for the entire district. Spectators and fans are in close contact with the artists, and so they can better understand the creative impulse in which they are now involved. What is special about these retrofits is that 'they are built at the meeting point between professional and less professional artists, young and not so young, they attract different audiences, and they keep cultural, social and symbolic resources circulating in a flexible space where many artistic disciplines interact'. Artists can act at all levels; they can join forces with others to create a new artistic and social endeavour according to redefined standards. In assessing the Belle de Mai, F. Raffin resorts to the image of DNA (Raffin 1994): 'in the DNA chain, some elements appear more relevant than others …. As with DNA, the Belle de Mai brings together at one place and at one time a host of players involved in the creative process, in this case artistic. Artistic collaborations appear, disappear, and interact continuously with each other on site, whenever there is a presentation or an event. Internal solidarity is constantly being reshaped as a function of cultural endeavours'.

- Converted brownfields have sparked new activities that can contribute to social integration. Take the case of the Belle de Mai. It began initially with the 'Système Friche Théâtre' or 'factory theatre' association that moved into part of the premises made available by the closure of the old tobacco factory located in the Belle de Mai neighbourhood of Marseilles in 1992. Its basic principle was to give artists a centre where they could produce works aimed at a wide range of audiences. Little by little, the Belle de Mai has extended its activities by reaching out to the surrounding city and using its networks of associations and activists to establish many cultural and social interaction points such as information centres for marginalized groups, boxing clubs for young people, and artistic or musical workshops (Greffe 2002). Artistic activities are not the only means of promoting social integration, but they can be powerful tools because they disseminate symbols and communication techniques that are used in all fields of activity. In the urban areas enlivened by Belle de Mai, there are now small businesses producing for market or offering personal services, properties are being maintained, and restaurants have appeared. These efforts to reach out to people in their daily lives are gradually giving them the skills and experience that open up jobs and access to the mainstream. By participating in these activities, individuals contribute to the emergence of an evolving, multifaceted culture that disseminates creative skills. Outcomes are varied, as some participants go on from these small urban productions to mainstream venues, while others will remain in these local enclaves that use only a portion of their skills (Roulleau Berger 1993). These retrofits thus perform a particularly effective socialization function: they achieve artistic socialization by bringing artists and their works together, and they achieve cultural socialization through the dissemination of standards and values. Some of the people active in these centres may however be suspicious of the term 'social', as smacking of utilitarianism.

What does it take to convert a brownfield site for cultural uses? Whether these effects are being achieved is far from clear. The establishment of such facilities did not initially stem from any ethical choices, nor was it even voluntary. Artists and other people engaged in cultural pursuits were looking for urban spaces that were wide, unencumbered and as inexpensive as possible, and this led them to set up in peripheral areas devastated by industrial collapse and poverty associated with social exclusion. Aware of these inequalities and feeling a common cause with the excluded, the groups that took over these derelict buildings tried to develop an entirely separate system, based on a new approach to cultural activism and creativity, and in the end helping to inspire local development. Then, too, they are quickly faced with a number of constraints (Greffe 2004a, 2010):

1. Physical: moving into a derelict industrial site generally provides significant and cheap space that can be used very flexibly, with its high ceilings, big windows, and solid structure. While these elements favour reuse, they also imply various costs of refurbishing the building, cleaning up any pollution, and bringing it up to municipal standards.
2. Territorial: the districts in which these buildings are found are generally disadvantaged. The buildings often lack utility services, the potential users often have no money, and they may be indifferent to what they see as elitist cultural pursuits. The environment, then, is not very attractive. In the midst of such isolation, it is not always easy to strike up local partnerships, including with the authorities, who will not always be convinced that a retrofit can help upgrade the neighbourhood. Even if these facilities are generally well received by the local populace, they are often the targets of vandalism that can threaten their survival. The public does not always respond as hoped, and people in the poorer neighbourhoods may stubbornly refuse to frequent these places despite the efforts to entice them and make them feel welcome. People who feel themselves excluded from cultural activities may put initiatives of this kind down to the whim of trendy folk who think it is fun to go slumming. To overcome the effects of such isolation, they have to offer low or even symbolic ticket prices or user fees. But this can pose a risk for their management.
3. Management: these converted art centres often have trouble covering their capital and operating expenses from their own resources. These include box office receipts (which will be low, given ticket prices), revenues from the bar or restaurant, or proceeds from activities conducted at the centre (workshops, housing, training programmes). But these revenues will at most cover artists' fees and overhead. There will be little left to pursue an

outreach policy or to buy expensive equipment. Artists, coordinators and administrators are rarely paid a salary, and most of them will be unable to rely exclusively on their artistic activities for their livelihood. Most often, any employees will have been hired under some kind of government works programme.

To make up for these shortcomings, the retrofits have several resources:

- They may pursue external growth strategies.
- They may provide training for young people who can then become independent or even go commercial to support themselves, and perhaps to turn a profit, while continuing to collaborate with the 'parent association'.
- They may provide and bill for services to other businesses and thereby put to profitable use their specific skills in such areas as electronic music, new technologies, or the organization of events or performances. Revenues from activities of this type can finance new projects in the centres themselves.
- In the end, they can also ask for public subsidies, which will immerse them in another set of constraints, this time political. In fact, the retrofits that have been created have often sprung from alternative cultures, and they will be disinclined to cooperate with the world of politics and business. Their participants feel that they embody certain attributes that those worlds do not, and they devote themselves to undertakings in which both their political outlook and their local roots are evident. But they will soon be forced to apply for financial or technical assistance from the local municipal authorities. Quite apart from the technical problems in obtaining such funding, subsidies are likely to be seen as opening the way to the kind of manipulation against which participants had revolted in the first place. In France, the *friches culturelles*, as they are called, have survived in large part thanks to the subsidies that cover 50 per cent of their financing.
- A last resort, then, is for the retrofits to organize themselves in a network. Does this networking, or even internationalization, of retrofits mean that their contribution to landscapes and local development will be compromised? The risk of manipulation of these centres will be all the greater if the authorities can exert subtle pressure through the network leader or via subsidies. The best way for these centres to remain faithful to themselves is still to work at transforming the local setting, by reinforcing identities and publicizing their projects and skills. The retrofit constitutes a subtle and important link between arts, landscapes and creative cities, for the arts will be more creative if they rely on local development, and at the same time the territory's development will be more sustainable if it can benefit from the constant creative renewal that artists bring.

Cultural Districts

For a long time, the insufficiency or volatility of the demand for cultural goods was used to explain the difficulties faced by new cultural companies. Initially described as superior goods by Veblen, cultural goods were thought of as being supported by very small markets. Furthermore, the difficulty of recognizing the quality of cultural goods with oftentimes intangible attributes was emphasized by economists (from Marshall to Stigler) and sociologists (Bourdieu) as an excuse for expressing doubts about the sustainability of their markets. Nowadays, these difficulties are more readily associated with supply than with demand. Baumol and Bowen stress the negative consequences of absence in productivity gains. In a recently updated article, Schumpeter claims that cultural goods change the traditional risk paradigm. Indeed, using Mantegna's paintings to illustrate his argument, Schumpeter shows how artistic creativity passes from one standard to another, such that 'this transition can be brought back from nothing to successive movements towards the margin' (Schumpeter 1933, cited in Greffe 2010: 12). This idea of uncertainty, which goes beyond the simple probability distribution perspective, is dealt with at length by Richard Caves (2000) in his work on creative industries, which dwells on the leitmotiv: 'Nobody knows'. De Vany (2004, cited in Greffe 2010: 16) develops this idea by underlining how in an economy of culture, which is an economy of extremes, there is a possibility for success as well as failure, so that a result cannot be anticipated when a new venture is conceived and planned (Greffe and Simmonet 2008).

Certain other elements support these interpretations on the specificity of cultural goods in comparison to other goods. There is greater emphasis nowadays on the intangible dimension of cultural goods, undoubtedly due to the way in which the advent of digitalization, along with increased copying capabilities, has upset the cultural sector's functioning. Given this intangible dimension, increasing numbers of cultural assets are characterized by high fixed production costs along with very low duplication and distribution costs, thereby conferring to cultural goods the attributes of semi-collective goods. Producers must thus adjust by studying markets according to their ability to pay (*windowing* audio-visual programs or *versioning* between *hardback* and paperback books); by reducing the lifespan of their products; by resorting to a frequent renewal of their components (adapting or editing news); by linking the use of their products to dedicated support mechanisms and then raising the cost of duplication to prevent copying (videogames); and by incorporating technical devices in order to further block copying (Digital Rights Management).

These two characteristics – uncertainty and intangibility – oblige cultural companies to frequently change their products and their corresponding production functions. Cultural companies are consequently forced to change the specific skills they need.

The first consequence affects the lifespan of companies. Since these companies are often created for a specific project and must mobilize specific skills required for that purpose, once the project is over they quickly need to consider other projects and mobilize any required skills if they want to survive. This represents a strong challenge, and many observers have commented on the fragility of such companies. One difficulty is to define a new project which can satisfy the new demand (artisan risk), while another challenge is to handle the constant reorganization required in moving from one project to another (industrial risk), albeit that the boundaries between these two difficulties are sometimes blurred.

Another consequence ensues from exposure to more or less favourable environments through which one might overcome the constraints of variety. These past few years a number of studies have argued in favour of the assumption that cultural companies can benefit from being close to one another. To justify this claim one must consider not only the industrial district approach, but also the idiosyncratic nature of cultural goods.

The chapter from Marshall's *Principles of Economy* on the geographical concentration of specific companies has shown that such cultural companies could profit from both external and internal economies (Marshall, 1890). Indeed these companies benefit from external economies thanks to the general development of their industrial sector, while also benefiting from internal economies due to the way in which they are organized. This being said external economies particularly benefit companies that are clustered together within a same area. Their local contiguity allows for free exchange of ideas and information, a rapid recognition of the need to innovate their equipment and structure, the development of ancillary and intermediate services, and the creation of a local market of relevant skills. Although these arguments are mainly focused on the production side, Marshall does not forget the consumption side: he admits that such concentrations attract consumers, since high travelling costs are compensated by the variety of goods offered.

For a long time this theory remained ignored or underestimated. Powerful factors had undoubtedly come into play at the time Marshall wrote, reducing transportation costs and widening labour markets. Over the last quarter of the twentieth century, however, doubts expressed by 'fordist organizations' focused more on the district, which was considered to be a more flexible industrial organization, as well as on the concept of industrial atmosphere which allows those who belong to it to understand and assimilate others' experiences much faster. Arguments such as abstract knowledge, mutual confidence or social capital were invoked to justify their interest in the district (Bagnasco and Sabel 1995, Becattini 1992, Piore and Sabel 1984, Storper 1997).

At the same time, this underscores the idiosyncratic nature of cultural goods. Several cultural production companies often exist side by side under local and historical conditions (Santagata 2007). This is usually justified by claiming that intangible components, such as a specific know-how or organization, can play an important role along with tangible components (Limoges, Murano or Hollywood). However, this does not imply any absence of creativity or innovation, and such areas have witnessed successive waves of creativity, as in the Parisian fashion industry for instance, which embodies this idiosyncratic characteristic (OECD 2004: 55–69).

The rather logical conjunction of these two approaches, namely those of idiosyncratic nature and industrial district, has led present-day observers to bring forth the concept of cultural districts (Greffe 2010, Santagata 2006, Scott 2000). The cultural district brings together companies producing cultural goods and services, as well as companies that manufacture required equipment and deal with the distribution of cultural goods. Moreover, these districts highlight the specificity of a labour market named *ad hoc labour market,* where very specific skills are exchanged, and in a flexible way which makes artists as well as producers see an advantage in becoming part of it (Greffe 1999). Artists must constantly remain present in this market since they must frequently move from one project or employer to another. Producers too must be present since they must continuously redefine the nature of the skills they need to mobilize and employ.

While it is one thing to stress the advantages of such clustering and synergies, it is another to scientifically demonstrate them. These new companies are also in competition with one another. In order to influence the net final effect, it is thus advisable to assess the signs as well as the extent of the synergy effect – which should act in a positive way – along with the signs and the extent of the competition effect – which should act in a negative way. The statistical analysis of new cultural companies is thus useful for understanding such important stakes in the development of contemporary economies. In particular, it should enable us to clarify at least two questions: Are cultural companies born and developed like other companies? Is their geographical clustering a lever of sustainability?

Based on data that were collected by the French SINE national survey (Système d'Information sur les Nouvelles Entreprises*),* which was conducted by the INSEE (Institut National de la Statistique et des Etudes Economiques) and covers a representative sample of companies created in 1998 over a five-year period, we have drawn two central conclusions:

- The rate of disappearance of cultural companies (which is comparable to that of other companies) is not stable, and it differs according to the subsector taken into consideration. One can indeed observe that traditional artistic companies exhibited a very high rate of disappearance during the first three years, while this rate fell substantially during the last two years of the survey. On the other hand, audio-visual companies reflected a constant rate of disappearance in comparison to cultural-product companies, which registered growth after two years of existence. These results indicate that within the cultural field the 'artisan risk' remains very strong throughout the first three years, while the industrial risk becomes relatively significant only after two years.
- The rate of survival of cultural companies is to a large extent dependent on their geographical clustering. On one hand a new cultural company can suffer from the proximity of companies pursuing an identical activity (the 'competition effect'), yet on another hand it can benefit considerably from the presence of a large number of new cultural companies with

diverse cultural activities (the 'synergy effect'). In absolute value, we have shown that the second effect tends to exceed the first (the net effect on the life expectancy was +70%), thus testifying to the productivity of cultural clusters or districts. This result explains both the tendency towards geographical concentration of many cultural activities which has been observed during the past few years; and the need for local governments to support such clustering that may be useful as well for economic activity than for bettering landscapes.

Cultural Quarters

Cultural quarters offer a new approach to urban planning. In the place of cumbersome, top-down programming, they introduce a more flexible and independent perspective, one that can directly engage the energies of consumers and producers. In any urban redevelopment programme, there are two possible sets of linkages: there is the vertical chain, that of the official public sector, and there is the horizontal chain, made up of small, interdependent units generating, and benefiting from, external economies. When flagship cultural projects are undertaken, the vertical chain wins out over the horizontal. When strategies for rehabilitating or creating cultural quarters are implemented, the two chains intersect, and the project's success will depend on their ability to create proper synergy.

Cultural districts are seen as serving many purposes:

- Reinforcing a city's identity, attractiveness and competitiveness. But in this field no success is permanent, and it must constantly be renewed, for example by seeking to attract cultural tourists in the wake of more traditional tourists. Moreover, as the authorities seek to extract themselves from cultural or intangible investments, this capacity will come to rely increasingly on private sector players, commercial or not.
- Stimulating an entrepreneurial approach to the arts and culture. Culture is the basis of the new economy, and is giving rise to many creative activities that produce high value-added. The city becomes creative by using culture as one of its possible levers. We may introduce here the 'bohemian' model, as reflected in the neighbourhoods of Montmartre, Montparnasse, or SoHo, but giving it today a business-oriented side.
- Fostering cultural democracy and cultural diversity. This is the urban doctrine developed by Bianchini, where culture becomes a political rather than an economic instrument (Bianchini and Parkinson 1993). With it, new channels and new means of expression can be created that will allow traditionally marginalized or even excluded groups or communities to express themselves and enrich public spaces.

Today we are witness to an impressive number of local development initiatives seeking to institute cultural quarters. To clarify the various forms and the effects they may produce, several classification criteria are available:

1. Cultural quarters distinguished by the range of activities they embrace: some are horizontal (where different cultural sectors coexist), and others are vertical (where cultural activities in the strict sense are accompanied by consumer or entertainment activities).
2. Cultural quarters distinguished by the way they are financed: government funding is usually more important than private, but the two forms often succeed each other in cycles.
3. Cultural quarters distinguished by their accessibility: if the quarter is physically shut off, and squeezed into a single site, it will be a ghetto that will be hard to change. If the quarter is more open and spread across the urban space, its atmosphere and its identity will be altered, and opportunists will move in to capture economic rents, ultimately denaturing its contents.
4. Cultural quarters distinguished by their organizational approach: some are top-down, while others are grassroots inspired. Generally speaking, consumer-oriented quarters are often organized by government fiat, while those organized around production represent slower and more decentralized approaches.
5. Cultural quarters distinguished by their location within the city: heritage or museum quarters are usually found in the city centre, redeploying a classic urban hierarchy, while others will be located on the margins of traditional spaces, thereby altering those urban hierarchies. Inter-quarter linkages will then appear, and they will reduce transaction costs, speed up the circulation of capital and information, and reinforce social intercourse and solidarity (Landry 2000).

Things do not always take their expected course. Public programmes can often spark a host of private initiatives that will take a completely different tack from that planned. Culture in this case can turn city centres back into pure consumption centres. This is quickly relayed by the marketing of values, spaces and relationships. Culture will be condemned, then, to walk a tightrope. In her work on (New York City's – as confirmed from internet sources) SoHo district, Sharon Zukin showed how the district progressed from a depressed rag-trade area into a zone for the expression of ethnic cultures, which was indeed its desired state, and then on to a zone for the consumption of works of art, an activity that led to 'ghettoization' and gentrification of the quarter (Zukin 1991). Many such experiments show that anything that can be reopened or renovated – warehouses, convents, quays, monasteries, gas plants or military barracks – can also become a source of exclusion. Art becomes a pretext, a kind of lure to which commercial and political interests flock. Experience shows that when a cultural quarter is created and promoted, this can spark a jump in property prices and the rejection of young artists. Where some had hoped to see new movements emerge that would clearly track progress, ruptures and occupational refinements, they find instead that entertainment, fun and commerce have been telescoped together (Hannigan 1998). The people managing these quarters have trouble themselves in arbitrating between public and private, entertainment and creativity, respect for traditional cultures and immersion in a global culture. These balances become ends in themselves, where they should be only the means. These situations are not always happy, because many 'post-modern' consumers will only accept economic values if the underlying cultural value is respected.

The cultural quarter can thus lead to antagonism instead of the desired synthesis. Some cultural quarters that have welcomed big retail chains selling products that are 'cultural' to some degree have been promptly transformed into commercial districts. This outcome was not bad from the viewpoint of renovation and enlivening the urban landscape, but it came at the price of banishing the cultural quarter's role as a lever of creativity and sustainable development. This trend also raises questions about the governance of cultural quarters. They are often based on a few projects that are considered worthy, but the sum of these projects does not guarantee the hoped-for result, because other interests – real estate, politics or business – may be pushing in other directions. The situation is made worse by the fact that the national authorities do not recognize the originality of such quarters and continue to hand out subsidies without looking at the specific projects (Bilton 1999). While past cultural policies were often compromised by an approach that was too vertical, a purely horizontal approach based on lining up projects does not always produce the desired result.

Some Public Policies Issues: A French Illustration

From the Monument to its Environment, from Centralization to Community Development
The protection of cultural assets is affected by the pressures exerted by the environment on the quality of a monument. Hence it is not surprising that the first initiatives in this field were taken in the name of improving the 'visibility' of the monument (Greffe 2007). The law of 1913 stipulated a protective perimeter of a radius of 500 meters around the listed monument, in order to prevent the reduction or the alteration of its visibility from different approaches. Within this zone, it was impossible to carry out any changes or new construction without prior permission from the government. The law of 1943 reinforced this system by defining the protective perimeter in stricter terms and by extending this protection to all monuments situated in the area. These measures were not enough because the development of this protective zone was not carried out in an organized manner. In 1962, the notion of a 'safeguarded sector' came into existence. Unlike the previous systems, it involved the definition and implementation of a plan to safeguard the architectural unit in question. Finally in the 1980s, the concept of Zone for the Protection of Architectural, Urban and Landscape Heritage (ZPPAUP) was

defined. In fact, the Ministry of the Environment introduced this new regulation for the protection and conservation of exceptional elements in urban areas. The constraints resulting from this system have to be included in the urban planning charter. The permission for building is not given unless the building plans are in conformity with the urban planning charter. An important difference between this instrument and the previous one is that the local government can take decisions, whereas in the previous case it was necessary to obtain permission from the national government.

Though this system was successful, it faced some problems in the initial stages:

- It was necessary to find a new balance between the central and local governments, given the trend towards devolution. The ZPPAUP made it possible to define the perimeter as well as the corresponding clauses regarding incentives and fiscal expenditures in more flexible terms. The request for permission has to come from local communities, but in the absence of such initiatives, the old legislation remains effective.
- Are there not more regulations related to the protection of monuments than to the protection of cultural landscapes? There are many landscapes affecting our day-to-day life as well as a vernacular heritage existing alongside protected monuments. Instead of formulating an approach that would benefit from the synergy between these two types of assets, there is a tendency to consider vernacular heritage as being subsidiary to protected monuments. This aberration reaches absurd heights in areas developed to promote cultural tourism: local residents start leaving the area gradually so that it is soon devoid of its daily activities and deprived of the day-to-day conservation efforts carried out by them.
- This shortcoming of the conservation policies was pointed out by a powerful NGO: the League for Protecting Cultural Assets and Landscapes. This League was set up with the idea of mobilizing all the assets instead of focusing on the aesthetic aspects of listed cultural assets. It claimed to be 'an innovator who respects rather than a conservative who destroys. It tried to approach town planning on a human scale by integrating the respect for traditions and classified cultural assets without endangering their overall economic viability.

The landscapes policy is no longer the exclusive preserve of experts; it has become an important political issue. A landscape is not just the indifferent scenery of a place and the life of its community. As pointed out in the European Landscape Convention, the 'landscape is an important part of the quality of life for people everywhere: in urban areas and in the countryside, in degraded areas as well as in areas of high quality, in areas recognized as being of outstanding beauty as well as everyday areas' (UNESCO 1977, 1985).

The Labelization: 'City and Country of Art and History'
In 1985, the National Network of Historic Buildings and Sites created this label
and a new network of *Cities and Countries of Art and History*, which currently
includes more than 60 towns and cities and more than 20 countries. It has
been remarkably successful. In the early 1960s, many of the old cultural urban
areas were dilapidated and on the verge of ruin. Confronted with this situation,
André Malraux, then Minister of Cultural Affairs, put to vote in 1962 the law
on *safeguarded sectors*. However, the major issue regarding these cultural assets
was not just conservation but getting the public interested. The first new policy
included guided tours conducted by freelance guides during the summer months.
These visits were primarily meant for tourists. But tourists are not really the best
persons to defend cultural assets that they do not see every day. This gave rise
to the idea that the safeguard and protection of these cultural assets should be
entrusted to the inhabitants of the area instead of tourists.

In accordance with this principle, the label of Towns of Art and History was
devised. To obtain the label, a town or city must *have an important cultural
landscape and wish to emphasize it by animation* and it must sign a convention
with the Ministry of Culture. *Cultural assets as economic clusters* are areas that
possess cultural and natural, tangible and intangible assets. The idea underlying this
new policy was that a good collective management of such assets could generate
and fuel local development. In this case, a cultural asset clearly goes beyond
the concept of a monument to include natural landscapes, vernacular heritage,
intangible know-how and crafts. This perception is generally strengthened by the
proper projection of cultural assets. This projection gives coherence to the territory
through the choice of proper actions and since the points to be developed are
directly related to the specific features of a particular territory, they vary from one
cluster to another.

This policy was defined and managed by an organism directly under the Prime
Minister's Office. The two ministries involved, Culture and Ecology, have been
moving in exactly the same direction.

There are approximately 60 clusters of this type and there is a variation in the
themes. Some have been organized taking the cultural landscape as their starting
point, while others take monument as their starting point. Finally, some revolve
around a traditional craft or other local resources. But whatever the focal point,
the three dimensions must be present. Moreover, the combination of these three
dimensions must be shown as having a potential for future development (Greffe
2004a, 2007).

Bibliography

Bagnasco, F. and Sabel, C. (eds) 1995. *Small and Medium Size Enterprises*.
London: Pinter.

Becattini, G. 1992. Le district marshallien: une notion socio-economique, in *Les régions qui gagnent*, edited by G. Benko and A. Lipietz (eds). Paris: PUF, 35–56.

Berleant, A. 1997. *Living in the Landscape: Towards an Aesthetics of Environment*. Kansas: The University Press of Kansas.

Bianchini, P. and Parkinson, M. 1993. *Cultural Policy and Urban Generation: The West European Experience*. Manchester: Manchester University Press.

Bilton, C. 1999. *Management and Creativity: From Creation to Creative Management*. New York: Wiley.

Caves, R. 2000. *Creative Industries: Contracts between Arts and Commerce*. Cambridge: Cambridge University Press.

Chang Chung Yuan 1963. *Creativity and Taoism*. New York: Julian Press.

Greffe, X. 1999. *L'emploi culturel à âge du numérique*. Paris: Economica.

—— 2002. *Arts and Artists from an Economic Perspective*. Paris: UNESCO Publishing.

—— 2004a. *La valorisation économique du patrimoine*. Paris: La documentation française.

—— 2004b. *Culture and Local Development*. Paris: OECD.

—— 2007. *French Cultural Policy*. Tokyo: Bookdom (In Japanese).

—— 2010. Introduction: L'économie de la culture est-elle particulière? *Revue d'économie politique*, 120(1), 1–34.

Greffe, X. and Maurel M. 2009. *Economie globale*. Paris: Dalloz.

Greffe, X. and Simmonet, V. 2008. La survie des entreprises culturelles: le rôle de la concentration géographique. *Recherches Economique de Louvain*, 74(3), 327–57.

Haanigan, J. 1998. *Fantasy City*. New York: Routledge.

Landry, C. 2000. *The Creative City*. London: Comedia.

Marshall, A. 1890. *Principles of Economics*. London: Macmillan.

OECD. 2004. *Culture and Local Development*. Paris: OECD.

Piore, M. and Sabel C. 1984. *The Second Industrial Divide: Possibilities for Prosperity*. New York: Basic Books.

Raffin F. 1994. *La Friche Belle de Mai*. Marseille: Friche Belle de Mai.

Rose, M. 2002. Landscape and labyrinths. *Geoforum*, 33(4), 455–67.

Roulleau Berger, C. 1993. *La ville intervalle*. Collection Klienksieck. Paris: Gallimard.

Santagata, W. 2007. *Creativita culturale*. Milano: Franco Angeli.

Scott, A.J. 2000. *The Cultural Economy of Cities: Essays on the Geography of Image-producing Industry*. London: Sage.

Storper, M. 1997. *The Regional World: Territorial Development in a Global Economy*. New York: Guilford.

UNESCO. 1977. *World Heritage Convention*. Paris: UNESCO.

Whyte Willima, H. 2002 [1968]. *The Last Landscape*. Philadelphia: University of Pennsylvania Press.

Zukin, S. 1991. *Landscapes of Power: From Detroit To Disney World*. Berkeley: University of California Press.

Chapter 10

Cultural Heritage and Creative Cities: An Economic Evaluation Perspective[1]

Faroek Lazrak, Peter Nijkamp, Piet Rietveld and Jan Rouwendal

Cities: The Place to Be

In contrast to a century ago, the majority of the current world population is living in cities (or urban areas). This statistical fact is not just a cool number, but tells us that cities have become the home of man (Ward 1976) who expect – and are entitled to find – in modern urban settlements an identifiable, attractive, sustainable and financially viable living environment.

According to modern planning theory (see, for instance, Batey et al. 2008) land use planning gives direction to urban architecture and the identity of new cities. This changes the view on modern cities and calls for visionary ideas with inspirational practice that may guide our appreciation for the emerging urban settlement pattern of our planet, as is exemplified by cities such as Beijing or Dubai.

Linking the cultural heritage of cities to the urban planning gives people the opportunity to connect their own identity to the city. Clearly, cities are not just densely populated areas, but enjoy an enormous innovative and cultural potential that positions them nowadays at the crossroads of a transformation of civilization, sustainable development and societal imagination by seeking a balance between business, arts and culture, liveable environment and human well-being. In this vein, we see the emergence of the concept of the 'creative city' as a new mantra for urban architecture and urban planning in which the urban way of life forms a breeding place for (inter)cultural interaction of residents, businessmen and visitors who engage themselves in shaping a place transformation marked by imagination and future orientation without a blueprint in mind. It goes without saying that the urban creativity movement prompts a tour de force from all actors and stakeholders involved (businessmen, planners, economists, artists, environmentalists, non-governmental organizations (NGOs), policy-makers) to assemble original concepts that will shape – or perhaps lead to the discovery of – a new creative and vital urban society with a distinct place identity (see Florida

1 This chapter was written in the context of the CLUE cultural heritage research programme at VU University, and the NICIS project on 'Economic valuation of cultural heritage'.

2002, Scott 2003). One of the important drivers of that distinctive place identity is the cultural heritage, more specifically the cultural building heritage.

The creative city idea captures two elements: creative creatures shape a creative city, and an enchantment of imaginative cities will create and attract innovative people aligned to an open and global society. City planning goes through evolutionary cycles based on learning from successes and failures. In the history of our world, numerous examples can be found on the close connection between socio-economic progress and artistic and cultural expressions (for example the Golden Age in Europe). And also the growing economics literature with regard to cultural aspects in economics witnesses the importance attached by many scholars to culture and the artistic sector as a cornerstone for innovation, urban vitality and wealth creation (Throsby 2001). However, the broad recognition of the economic significance of the creative industry for cities really took place with Florida's publication *The Rise of the Creative Class* (2002). In his view, a modern – often urbanized – society with a broadly differentiated work force and economy needs not only high-knowledge intensity, but also creative minds. Such a creative class does not have a '9 to 5' work rhythm, but is a mobile and floating group of well-educated, often young, innovative people who need an inspiring work environment and a high-quality living environment. The 'sense of place' (place of identity) is then decisive for their job career and work involvement. The arts, media, sports, fashion, research and cultural sectors usually offer such inspiring urban environments for creativity, which are generally characterized by the availability of talents, the presence of high-tech sectors, and the existence or acceptance of a tolerant lifestyle.

From a different perspective, in his book *The Art of City Making* Landry (2006) makes an alternative plea for a creative urban movement that forms a balance between passion and compassion: the ultimate goal is not to be the most creative city *in* the world, but the best and most imaginative city *for* the world, so that individual wants and collective and planetary needs are served in parallel. Consequently, the city is also an emotional, cultural and artistic experience; a city offers a human cultural narrative about the past, the presence and the future of our civilization. And therefore, cultural heritage – not as a fixed asset firmly rooted in the history of a place, but as a transparent collective property aligning the past to an open future – is the soul – and hence one of the critical infrastructures – of modern cities. The value of the heritage is linked to the dynamic opportunities it will give to the future.

For an economist, the overarching emphasis on seemingly intangible and visionary dimensions of creative urban environments may sound somewhat awkward, as creativity and imagination do not belong to the standard jargon of mainstream economists, and are not often monetarized. But even when cultural expressions and artistic profession may seem to take the lead in contemporaneous urbanistic thinking, it ought to be recognized that cities are dynamic spatial entities that go through difficult and often unanticipated stages as a result of human acts and man-made decisions, even in an evolutionary economic perspective which should be taken into account in the monetarization process. Schumpeter has emphasized the importance of 'creative destruction' in which innovative entrepreneurs conquer new markets by developing new products. Such behaviour may create economic cycles that are closely related to city lifecycles. If cultural heritage attracts creative people to cities, it may well provide a city with some long-lasting and sustainable anchor points that helps it to survive to the next upswing and benefit more from it. Against this background cultural built heritage may be regarded as one of the driving forces for future development of cities.

In addition to the changing views on and perceptions of the position of the city as a core determinant and signpost of modern civilization and innovation, it ought to be recognized that changing lifestyles – in particular, the unprecedented rise in geographical mobility (which stimulates both global tourism and short-haul recreation) and the emerging welfare society in many countries (which produces more discretionary income and leisure time) – most likely prompt a global appreciation for the attractiveness of modern cities as places to live, work, enjoy or visit. From that perspective, cultural capital is a powerful asset of any city. In the next section we address the significance of cultural heritage in a modern urban economy in particular.

Cultural Heritage: The Appreciative Economy of Cities

Cities such as Venice, Jerusalem, Florence, Amsterdam, Istanbul, Kyoto, Bangkok or Acapulco house a wealth of cultural assets that make them distinct from other places and determine their place identity (see Coccossis and Nijkamp 1995). It is noteworthy that such cultural capital is not present in the form of petrified lava, but offers an appealing productive asset to the urban economy. The tourism sector depends even to a large extent on the appreciation by millions of people for the artistic, aesthetic, historical, cultural or emotional sense of cultural assets. Thus, a significant part of modern urban economies – especially centres of culture and arts performance – is based on collective emotion from millions of visitors. This is also the type of environment that inspires and attracts creative minds, such as artists, designers, scientists, and so forth. And there, the 'appreciative economy' of cities offers a socio-economic added value that supports and enriches the traditional urban economy and offers a stepping stone for attracting innovative talents in an imaginative environment. 'Made in the city' has become almost a paradigmatic belief of the creative class.

Creative classes are usually found in inspirational, cultural environments. From that perspective, the presence of an attractive cultural heritage that is actively enjoyed by many people is a major asset. Cultural heritage has many appearances and takes the form of many artefacts, such as monuments, historic buildings, museum collections, 'cityscapes', archaeological sites, and also historic environments in cities, artists' cafés, and so on. Throsby (2001) mentions three criteria to identify cultural goods: a form (or sense) of creativity in the production;

the provision and communication of symbolic meaning; the output embodies some form of intellectual or artistic property. To qualify as cultural heritage, such goods should of course have a historic significance (intergenerational meaning) and a sense of local identity (Coccossis and Nijkamp 1995).

Cultural heritage is interpreted in this chapter in a broad sense and comprises those parts of cultural capital that have an explicit and distinct historical connotation to the past of a place and that may be seen as a self-identifying landmark for that place.[2] Cultural heritage and place identity are thus closely interrelated concepts.

It is interesting that UNESCO (1972) makes a distinction between cultural, natural and intangible heritage (see also Arizpe et al. 2000, Klamer and Zuidhof 1999). Cultural heritage is next subdivided into monuments, groups of buildings and sites. Clearly, cultural heritage is a broad and heterogeneous concept, so that a comparative study of cultural heritage or an unambiguous economic evaluation is fraught with many problems, of both a methodological and empirical nature.

It is noteworthy that in recent years not only cultural cities have made an attempt to be more visible in a globally mobile and competitive tourism world (witness the 'cultural capital cities of Europe' programme of the European Community), but also villages have joined hands to be more visible on the tourism stage – witness the emergence of the 'beautiful villages association' (see Gülümser et al. 2009). Culture, creativity and economic progress appear to be well connected in current debates on urban revitalization and urban futures for the high-tech and service sector.

In general, cultural heritage may produce many market benefits (such as tourism revenues, spillovers to the hospitality and service sector) as well as non-market benefits (externalities through the appreciative economy in the form of local self-esteem, attraction of creative minds, open-mindedness of urban population etc.) (see also Navrud and Ready, 2002).

We may thus conclude that cultural heritage has become an asset of both historic-cultural and socio-economic significance in a modern society. The needs of a leisure society as well as the needs of those who want to relax from their daily labour race are often met in places with a local identity and a specific built environment. Consequently, cultural heritage is highly appreciated for its externality value for the broader community (see Hubbard 1993, Riganti and Nijkamp 2007).

The question may be raised whether cultural goods may be seen as 'normal' economic goods (cf. Throsby 2001). Clearly, the answer is affirmative, in the sense that they contribute to our well-being under conditions of scarcity, while the means to acquire or maintain them are limited and can be used for alternative purposes. It is equally clear, however, that cultural goods have specific characteristics that

---

2   Cultural capital is a more general concept and may be seen as those forms of public or quasi-public goods that comprise the stock of cultural value embodied in an asset, while this stock may give rise to a flow of goods and services over time in the form of commodities that themselves may have both cultural and economic value (Throsby 2001).

may render their socio-economic evaluation slightly more complicated (Riganti and Nijkamp 2007) than that of conventional market goods. In the first place, they are usually subjected to many (often positive) externalities, as the market value is often absent or does not incorporate at full scale the variety of user preferences. In this context, stated preference methods such as contingent valuation methods or conjoint analysis may be helpful or revealed preference methods such as travel cost methods or hedonic price methods. A second special feature of cultural goods is the fact that very often they are specific, authentic or unique in the sense that they cannot easily be substituted or replaced by an identical good. If this is the case, one might have to resort to a Hicksian economic compensation analysis in order to find out in an experimental way which value would be foregone if a certain good would be no longer available. Valorization schemes on cultural potentialities may also be helpful here, by including next to direct and indirect use values also option values, legacy values, existence or bequest values. Also with the help of hedonic price methods it may be possible to value cultural heritage in an indirect way. Although valuation of cultural heritage is often difficult and existing methods are not always able to uncover their value completely, the discussion in the present section already makes it clear that the presence of cultural heritage contributes to the creation of an appreciative urban economy that in turn may favour the rise or reinforcement of a creative class. The sections that follow provide a discussion of the application of some economic evaluation methods to cultural heritage and then a more detailed review of the hedonic price method. We conclude with some remarks on policy implications.

Economic Assessment of Urban Cultural Heritage

A Variety of Methods

To develop a rational basis for smart and in a sense creative urban policy, a systematic assessment of various choice options has to be made. The economic literature offers a wide array of evaluation methodologies. Examples are stated preference methods (such as contingent valuation techniques or conjoint analyses), travel cost methods or multi-attribute utility methods (see Mitchell and Carson 1989). Some of these methods have been fruitfully applied to the valuation of unpriced goods with external effects (Schuster 2003). Cultural heritage has usually been approached with the so-called stated preference methods. However, the economic elicitation of cultural heritage valuation based on market-oriented methods, in particular hedonic prices, has received far less attention. In this section we will briefly review such evaluation methods and their applications to cultural heritage. In the next section we take a closer look at the hedonic price method and the challenge it offers for valuing cultural heritage.

Physical Compensation

If a physical good is (threatened to be) lost, one may try to compensate for this loss by reconstructing the same asset preferably at the same place. If the physical asset is not really rebuilt, the estimated economic costs involved in reconstructing the same good might be interpreted – in a Hicksian sense – as the financial compensation for the loss of the good concerned. The essence is that compensation takes place for a loss in welfare. Thus, the physical compensation may be virtual. Although this is often impossible, cultural heritage sometimes has the possibility of being reconstructed. There are numerous examples of old buildings (castles, mansions) or even small towns (for instance, Heusden in the Netherlands, Willemstad in Curacao) that have been restored in their former glory after a period of decay. In such cases the amount of money necessary to rebuild or restore a physical cultural heritage good in its historical state provides a shadow price for that good that offers useful information for compensation costs in project evaluation. It will then be worthwhile to rebuild or restore an asset when its social value is at least equal to its shadow price. Clearly in this case we do not need an exact assessment of the value of the good in question: it suffices to know that its value exceeds the particular threshold given by the shadow price. However, in many cases, also in valuing cultural heritage, it is hard to determine the shadow price or it is uncertain whether the social value will exceed the shadow price, unless further investigation into the valuation of the good in question is undertaken.

Project Evaluation Methods

Project evaluation methods are usually deployed to assess the economic benefits and sacrifices of projects that may have significant socio-economic and other (e.g. social or ecological) implications. In the cost–benefit analysis tradition a sophisticated toolbox has been developed to deal with complex project evaluation, sometimes with large spillover effects. In the past decade, many attempts have also been made to incorporate intangible (e.g. environmental) effects in these calculation schemes. In this subsection we will briefly discuss two general methods: economic impact analysis and multicriteria analysis. In the two subsections that follow we focus on two instruments from the toolbox that aim specifically at the determination of the willingness to pay, which is the key concept of evaluation from the perspective of welfare economics.

In the context of cultural goods, Tyrrell and Johnston (2006: 3) describe economic impact analysis as seeking 'to estimate changes in regional spending, output, income and/or employment associated with tourism policy, events, facilities of destinations'. Economic impact studies are used for valuing various types of cultural heritage, especially cultural heritage that attracts large numbers of tourists who spend money from outside the impact area (Snowball 2008: 33). Impact studies try to monetize the direct and indirect effects of an event on an impact area. Snowball (2008) points out that impact studies focus mainly on the private good character of the arts that is captured by market transactions instead of merit or public good characteristics.

To measure direct net impacts of cultural heritage provisions on user groups, it is important to identify the main spending groups in the impact region affected by the cultural event. Spending groups that otherwise would spend their money in another way in the impact area should be identified. It is important to take only the spending into account that otherwise would not be realized. Indirect net impacts depend on chain effects or induced effects of direct net impacts for the impact area. Clearly the amount of leakage in a multiplier sense depends on the size and nature of the impact area (Snowball 2008). Baaijens and Nijkamp (2000) offer an empirical meta-analysis approach with regard to those leakages and present a rough set analysis approach to estimate income multipliers for different characteristics of impact areas (see also Van Leeuwen et al. 2006).

The criticism on the validity of impact studies focuses in particular on methodological issues and conceptual problems (Snowball 2008). Conceptually, there are a number of problems that impact studies have to address and to solve. First, the definition of the impact area influences the outcomes of the study. Because there are alternative spending opportunities, the size of the impact area influences the size of those alternatives. Another caveat is that redistribution issues usually remain implicit. Especially with regard to cultural events, rich residents usually profit more than poor residents (Richards 1996). Impact studies are also plagued by methodological issues in valuing the public good characteristics of cultural events. Costs are relatively easily measured, but when cultural events are free of charge the benefits are hard to quantify. In conclusion, economic impact analysis is a prerequisite for a solid evaluation of cultural goods, but its implementation is still fraught with uncertainties.

Another strand of the project evaluation literature focuses on methods that do not require a monetary translation of project impacts, but are able to capture in principle all relevant intangible effects. These methods are usually captured under the heading of multicriteria analysis (see e.g. Nijkamp et al. 1991). An example can be found in Coccossis and Nijkamp (1995), who show how multicriteria analysis in the cultural heritage field may be applied. Multicriteria analysis offers an opportunity to assess and weight qualitative and quantitative effects. With the broad range of value-generating aspects of cultural heritage, multicriteria analysis makes it possible to deal with categorical information in economic evaluation and to address policy trade-offs by assigning weight to the different attributes of cultural heritage. Often multicriteria analysis is pursued on an item-by-item stated preference evaluation.

Stated Preference Methods

The core concept of economic evaluation methods is the marginal willingness to pay of a consumer for a particular commodity. For market goods, the marginal willingness to pay is equal to the price, which is clearly convenient for welfare analysis. However, many valuable commodities – cultural heritage offers important examples such as historical city districts – are not traded on a market. Optimal design of public policies with respect to such goods calls for an estimate of the willingness to pay and mainly for this reason economists have worked for decades already on the development of techniques to uncover this value, also for non-market goods. In the present subsection we will review one family of such methods that exploit the idea that one may approximate the willingness to pay for a non-market good by asking consumers about their preferences. Such methods are known as stated preference methods. In the next section we pay attention to a second important class of economic evaluation techniques, which are based on market outcomes.

Stated preference valuation techniques try to discover what individuals are willing to pay or are willing to accept, through the use of survey questionnaires. Contingent valuation methods form an important sub-class of preference elicitation methods and focus directly on willingness to pay by using open-ended questions (for an overview see Mitchell and Carson 1989). A second subset of stated preference techniques are choice experiments. That is, one tries to investigate the preferences of people from the choices they make between bundles of attributes that describe the good to be valued at different levels (Noonan 2003, Snowball 2008). Although choice experiments extract the willingness to pay in a more indirect way than contingent valuation methods, their focus on concrete choices is generally regarded as an advantage, because it reduces the risk that respondents indicate a willingness to pay on the basis of a superficial impression of the value of the good in question.

The strength of stated preference techniques is that they can be applied in circumstances in which consumers do not reveal their preferences in other ways, for instance, by buying goods at a price on the market. However, this strength is also a considerable potential weakness of these techniques. The hypothetical character of the statements made by consumers prompts questions on reliability for investigating the willingness to pay of the consumers in actual cases (Arrow et al. 1993, Hoevenagel 1994, Murphy et al. 2005, Snowball 2008). Indeed, a number of biases in stated preference methods has been identified in the literature (see, among others, Kahneman and Knetsch 1992, Snowball 2008). One problem with the implementation of these methods is that the stated willingness to pay often differs significantly from the willingness to accept, while differences should be negligible (Morrison 1997a, and 1997b Kahneman et al. 1990, 1991). However, there exist a number of potential advantages that reduce the risk of some biases. It is noteworthy, however, that, Snowball (2008) identifies also a number of potential problems in the literature with regard to the conjoint method. These problems relate to complexity and choice consistency, and individual valuation and summation (Snowball 2008).

In the context of valuing cultural assets, Snowball (2008) mentions two reasons why a mixed good[3] could cause a bias. In the first place, there is an incentive for users to overstate its non-use value (see Throsby 1984). Users who experience use values of cultural heritage have an incentive to overstate their willingness to pay over users who only experience the public aspects of cultural heritage and therefore only experience non-use value. This incentive arises because non-users also benefit from the use value so they overstate the non-use value and then drive up the price for the non-users. Second, as Seaman (2003) suggests, willingness-to-pay studies may also capture expected economic benefits that not only reflect present earnings but also bequest earnings. Because mixed goods could have positive employment effects, people may take this into account and thus come up with higher willingness-to-pay figures.

There is evidence that research conducted according to the NOAA recommendations[4] (Arrow et al. 1993) is more valid, more reliable and reduces the size of a number of biases (Noonan 2003, Snowball 2008). Benefits transfers can be used if estimations in one context can be generalized to indicate values in other contexts. Nevertheless, it remains a problem to realize benefit transfers because every situation is very bounded to a specific context and the adaption to the new context stays troublesome. Because validity and reliability of the contingent valuation method is still a source of debate (see for example, Diamond and Hausman 1994) benefit transfers are harder to realize.

One of the first contingent valuations of cultural heritage is of Navrud and Strand (1992), which values the Nidaros Cathedral in Norway. Stated preference techniques have been applied to the evaluation of cultural heritage in numerous evaluation studies. Noonan (2003) offers a meta-analysis of this rich literature. Snowball (2008: Chapters 4 and 6) provides an update of the contingent valuation literature, while she also reviews the application of choice experiments in the cultural economics field. In a study undertaken by Alberini et al. (2003), the value of the cultural and historic dimensions of a square in a city is determined by comparing the actual square with a hypothetical square. The hypothetical square is made as close as possible to the real square, except for its cultural and historic dimensions. The authors conclude that aesthetic and use attributes contribute to the explanation of the hypothetical choices individuals made. Also, they found significant differences between the actual and hypothetical square. They conclude that with the exception of the height attribute, the attributes are valued differently for the actual and abstract squares (Alberini et al. 2003).

---

3   Goods that have both private and public good characteristics.

4   The National Oceanic and Atmospheric Administration (NOAA) reviewed the use of contingent valuation and concluded that, if its guidelines and recommendations were followed, 'CV studies can produce estimates reliable enough to be the starting point of a judicial process of damage assessment, including lost passive-use values' (Arrow et al. 1993: 24).

An example of a choice experiment recently undertaken by Willis (2009) is a case study to investigate the preferences of visitors in the management of Hadrian's Roman Wall. The study results can be used in managing an archaeological site and heritage sites. In his study interaction effects between attributes are examined and the study concludes that visitors to Vindolanda were clearly able to state their preferences for the future management of the archaeological site.

Revealed Preference Methods

Introduction

In the subsection 'Project Evaluation Methods' above, we observed that the stated preference method is sometimes criticized for the fact that it uses hypothetical alternatives. Contrary to the stated preference methods, revealed preference valuation focuses on market outcomes. There are a number of ways for an economist to pick up the clues that real market transactions give about the economic value of cultural heritage. One is the travel cost people are prepared to pay to visit particular cultural locations. Another is the price actors are willing to pay for real-estate objects that can be considered as cultural heritage or are located in the proximity of such objects. The travel cost and hedonic price methods use these two pieces of information to investigate the willingness to pay. These will be discussed in the next subsection.

Travel Cost Method

Visiting cultural heritage means that one has to travel to its location. The associated travel cost acts (i.e. the financial sacrifice to get there) as a price for the visit in much the same way as a ticket price does, and indicates the consumer's willingness to pay for the cultural heritage good. The costs of visiting the cultural heritage good not only refer to monetary outlays but also to the time spent at the site and all other costs that stem from that visit (Navrud and Ready 2002, Snowball 2008). The demand curve for the cultural heritage good can be derived from the differences in travel costs of different visitors. In this way one can use demand analysis, even if there is, strictly speaking, not an unambiguous (ticket) price associated with the visit.

Although it is based on revealed preferences, and therefore avoids some of the problems associated with stated preference techniques, the travel cost method has some other problems. In the first place, travel cost methods are faced with the problem of multipurpose trips. A tourist making a trip and visiting several cultural heritage goods will find it hard to distinguish which part of the costs of the trip are related to a particular cultural heritage good. They buy essentially a non-separable bundle of goods. A related problem is that the visitor of a cultural heritage good can derive utility from the trip itself or from the company in which it is made (social externalities). Second, the opportunity costs of a visitor are hard to estimate; currently the visitor's wage is often used to value the opportunity cost (Navrud and Ready 2002). Third, with travel cost methods substitutes of cultural heritage can cause disturbances and provide difficulties to address the direct effect of cultural heritage. Especially for those who live near cultural heritage this is a problem.[5] When people who chose to live in the vicinity of cultural heritage value cultural heritage most, the distance to cultural heritage itself is then a residential factor that may cause complications in estimating the related demand function.

There are two types of travel cost models which use differences in total travel cost between visitors. First, there are *visitation frequency models* in which the survey of the travel cost method asks visitors at different distances from a cultural heritage good how often they visited it in a specified period. This 'trip generation function' is estimated where the number of visits is taken as a function of total costs corrected for other explanatory variables. The 'trip generation' function is then the demand curve from which the consumer surplus can be derived. Because actually most people visit most sites only zero visitors are usually aggregated geographically into groups. This zonal travel cost model approach simplifies sampling, because it makes it possible to take a random sample at the cultural heritage good site (Navrud and Ready 2002). The second travel cost model is the *site choice model*, which surveys potential visitors and asks them which cultural heritage goods they have visited. The utility function is then estimated based on the characteristics of each heritage good and the total costs for that good. According to Navrud and Ready (2002), site choice models may become a promising variant of the travel cost method, because they are able to measure site quality of cultural heritage, which is often most policy relevant (Navrud and Ready 2002).

There are various examples of cultural heritage studies that estimate the values with the travel cost method. Boter et al. (2005) use a site choice model to estimate the value of different Dutch museums, while Bedate et al. (2004) estimate consumer surplus of four cultural heritage goods in the Castilla y Leon region in Spain on the basis of a travel cost method. The travel cost method was also used by Poor and Smith (2004) to value the historic St. Mary's City, which is considered to be one of the United States' most significant archaeological sites. They use a zonal travel cost model to estimate consumer surplus. Depending on the functional form, individual consumer surplus ranges from $8.00 to $19.26. Finally, it is interesting to note that Alberini and Longo (2006) combine the travel cost method with contingent valuation to value cultural heritage in Armenia. This combination offers them also interesting opportunities to separate use and non-use values.

Hedonic Price Method
Contrary to the rather direct valuation of cultural heritage by the travel cost method, the hedonic price method measures the value of cultural heritage in an indirect way. The hedonic price method is based on the observation – generally attributed to Lancaster (1966, 1979) – that:

---

5   Cultural heritage goods are often abundant at densely populated city centres.

... goods are valued for their utility bearing attributes characteristics. (Rosen 1974)

This leads straightforwardly to the idea that prices of heterogeneous goods are related to the characteristics of the varieties. Griliches (1971) and Rosen (1974) developed the idea of implicit prices for characteristics, which can be estimated by regressing prices on these characteristics. Like ordinary prices, these implicit prices reveal the marginal willingness to pay of consumers. Although Rosen's (1974) original analyses were developed for a market with perfect competition the method is applicable under alternative market conditions (Bajari and Benkard 2005, Rouwendal and van der Straaten 2008).

Like the other methods, hedonic price methods have some intrinsic weaknesses and problems. For example, Jones and Dunse (1996) point out that the measurement of different attributes of the hedonic price method raises questions about the correct model specification. In another paper they criticize the fact that the method assumes equilibrium throughout the property market and no interrelationship between the price of attributes (Dunse and Jones 1998).

An important problem for hedonic price analyses is that, in principle, there can be many variables that influence the value of real estate. In a conventional cross-section, limited information about potentially relevant characteristics implies the risk of omitted variable bias. On the other hand, there is the possibility that some other determinants of value are strongly correlated with the variable of interest (for instance, an architectural feature that is typical for a particular period or style), which makes it difficult to pin down its effect. Moreover, economic theory offers little guidance for the specification of a hedonic price function.

Nevertheless, a further development and use of hedonic price analysis may offer a considerable promise for a better understanding of the value of cultural heritage. In particular the recent frequent availability of large databases – constructed, for instance, by land registry or cadastral offices – containing often detailed information about transactions in the real asset market offers large opportunities. These data are especially useful if they compose disaggregated data about the characteristics of the properties sold. In this context geographic information system (GIS) techniques often offer the possibility of further enriching such data with information about geographic neighbourhood characteristics. With such data, the problem of omitted variables can be mitigated considerably, while the large number of observations enables the analyst to incorporate a satisfactory number of regressors. This, therefore, leads to the following concluding remarks.

In the available literature various methods are used to value cultural heritage. Most of the existing studies use stated preference methods. A significant disadvantage of these methods is the presence of a number of serious potential biases. The increasing use of revealed preference techniques may provide alternative information about the value of cultural heritage. Because of the increasing availability of rich databases about real-estate transactions, application of the hedonic method seems to offer a promising avenue for further research. For this reason we will in the next section offer a concise and selective review of a few such studies that offer applications of to the valuation of cultural heritage.

Hedonic Pricing Studies on Cultural Heritage: Some Examples

Although the existing literature on valuation of cultural heritage often uses stated preference techniques, applications of revealed preference methods are not completely absent from the literature. From the various applications in the literature, we have selected a few studies that use the hedonic price method to explicitly value cultural heritage or aspects of cultural heritage. Clearly, the literature covers various aspects of cultural heritage. Some early studies concentrate on the effect of designation of a building as cultural heritage. Designation is supposed to have various use effects, both negative and positive. An important negative aspect of designation of buildings is that it restricts the owner's property rights. An important positive aspect of listing is being eligible for various forms of tax deductions. Leichenko et al. (2001) offer a table that surveys the estimated impacts of such designation on property values. Their summary table starts with a study of Heudorfer (1975) which uses difference-in-difference[6] to evaluate the impact of designation. Of the seven difference-in-difference studies, three found a positive impact and four failed to find any effect. The question is what a difference-in-difference study measures as treatment. For example is it the cultural heritage status or is it the tax deduction or restrictions caused by the designation?

The first study estimating a full hedonic price function mentioned in that table is Ford (1989) who reports a positive impact of designation on property values. Three other hedonic studies measure a positive impact, two a negative impact and one measures no impact of designation on property values.

More recently, Deodhar (2004) used a hedonic price function to estimate the market price difference between heritage listed and regular, unlisted houses in Sydney's upper north shore. She finds a 12 per cent premium of listed over unlisted houses in Ku-ring-gai after controlling for other property attributes.

Ruijgrok (2006) used a hedonic pricing method to monetize housing comfort value with respect to cultural heritage in the old Hanseatic town of Tiel in the Netherlands. In this study, houses with historic characteristics were selected from heritage listed in national and municipal lists. She finds a positive effect of historical characteristics of almost 15 per cent (Ruijgrok 2006). Her study is a good starting point for further exploration of the positive effects of cultural heritage on housing prices in the Netherlands.

Noonan (2007) offers further insight into the different effects property designation and district designation has on property. He estimates a hedonic price

---

6   Difference-in-difference is an econometric technique that examines the effect of one type of treatment. In our case this is the effect of designation on real-estate values.

function on data from the Multiple Listing Service of Northern Illinois, which includes Chicago. As explanatory dummy variables, an indicator for allocation in a designated district (*district*) and an indicator for designation of an individual property (*landmark*) are included. Prices of landmarks are higher than those of otherwise comparable houses, while for districts a smaller premium is estimated. Noonan also conducts a repeat-sales analysis[7] with the dataset. The repeat-sales approach offers evidence for proximity effects. The more landmarks in the block group are designated, the higher the cultural heritage premium is; this shows that the external effects of designation are stronger when more cultural heritage gets designated (Noonan 2007). The repeat sales approach can value this because it follows neighbourhood trough time.

One concern raised by these studies is that it is not always clear whether there is a causal effect of designation itself (for instance, because it protects the valuable characteristics of a building or a district) or whether the listing merely signals the presence of valuable characteristics that are already recognized by the market. Even with a repeat sales approach a positive coefficient might be interpreted as the effect of listing or as the effect of increased appreciation of specific aspects of cultural heritage (both effects can be present simultaneously).

There are several hedonic studies thast evaluate architecture and architectural quality (Hough and Kratz 1983, Moorhouse and Smith 1994, Ruijgrok 2006, Vandell and Lane 1989). The studies focus on different measurable aspects of architecture or architectural quality in a city. For example, the authors focus on architectural style, number of façade features, and historical or architectural quality. In 'Can "good" architecture meet the market test?', Hough and Kratz (1983) investigate the way the office market of downtown Chicago values 'good' architecture. Their results indicate that a considerable rent premium is paid for 'good' new architecture, but not for 'good' old architecture. In another study, Moorhouse and Smith (1994) regress the original purchase price as dependent variable on relevant architectural characteristics collected by Smith through visual inspections of houses that were built between 1850 and 1873.

Finally, we refer here to some studies addressing the effect of churches on neighbourhood quality. Regression of church amenities on transaction prices of neighbourhood property measures the effect of churches on the neighbourhood (examples can be found in Carroll et al. 1996, Do et al. 1994). The study of Carroll et al. indeed finds a significant impact of the proximity to churches on house values. Our overall conclusion on the hedonic price approach in the cultural heritage domain is that the literature on this subject is still limited and fragmented. There is clearly a scope for an intensification of research efforts in this domain.

---

7   A technique that uses time series data to estimate the effect of listing on house price.

Cultural Heritage and Creative Urban Policy

Cities are not just geographical settlements of people; they are also the 'home of man' (Ward 1976). They reflect the varied history of mankind and are at the same time contemporaneous expressions of the diversity of human responses to future challenges.

Actually, modern urban planning shows an avalanche of varying initiatives focussed on creative urban development, in particular by centring on culture, and acts as a multifaceted cornerstone for innovative development of the city. Consequently, it has become fashionable to regard cultural expressions such as arts, festivals, exhibitions, media, communication and advertising, design, sports, digital expression and research as signposts for urban individuality and identity and departures for a new urban cultural industry (see Florida 2002, Scott 2003). 'Old' cities such as London, Liverpool, Amsterdam, Berlin, Barcelona, New York, San Francisco, Sydney or Hong Kong witness a profound transformation based on creative cultures. This new orientation not only provides a new dynamism for the city, it also has a symbolic value by showing the historical strength of these places as foundation stones for a new and open future.

Undoubtedly, a main challenge of the modern creativeness fashion is to translate creative and cultural assets and expressions into commercial values (value added, employment, visitors, etc.), which means that private-sector initiatives are a sine qua non for effective and successful urban creativeness strategies. Consequently, an orientation towards local identity and local roots ('the sense of place'), a prominent commitment of economic stakeholders (in particular, the private sector), and the creation of a balanced and appealing portfolio of mutually complementary urban activities are critical success conditions for a flourishing urban creativeness strategy.

The revitalization of modern cities is anchored in two drivers: their historic-cultural heritage and their future economic dynamism. The real-estate market is undoubtedly influenced by these two force fields. It seems plausible that a significant part of the creative opportunities in the urban economy is rooted in the sense of place, including its cultural history. From that perspective, it is of great importance to assess the impacts of the past (e.g. the urban cultural heritage and history) on the value of real-estate properties in relevant urban districts. Various empirical studies using a hedonic price approach highlight the importance of architectural, cultural, historic or artistic factors for real-estate transactions, but the overall evidence is fragmented and at times ad hoc, or partial. There is, therefore, a clear need for solid empirical studies, in the form of both quantitative comparative studies and systematic case studies based on solid and extensive datasets, so that more reliable insight can be obtained on the potential of using hedonic price studies for estimating the impact of cultural heritage on real-estate values.

Bibliography

Alberini, A. and Longo, A. 2006. Combining the Travel Cost and Contingent Behavior Methods to Value Cultural Heritage Sites: Evidence from Armenia. *Journal of Cultural Economics,* 30, 287–304.

Alberini, A., Riganti, P. and Longo, A. 2003. Can People Value the Aesthetic and Use Services of Urban Sites? Evidence from a Survey of Belfast Residents. *Journal of Cultural Economics*, 27, 193–213.

Arizpe, L., Preis, A.B. and Taurus, M. 2000. *World Culture Report 2000: Cultural Diversity, Conflict and Pluralism.* Paris: UNESCO.

Arrow, K., Solow, R., Portney, P.R., Leamer, E.E., Radner, R. and Schuman, H. 1993. Report of the NOAA Panel on Contingent Valuation. *Federal Register*, 58, 4601–14.

Baaijens, S. and Nijkamp, P. 2000. Meta-Analytic Methods for Comparative and Exploratory Policy Research An Application to the Assessment of Regional Tourist Multipliers. *Journal of Policy Modeling*, 22, 821–58.

Bajari, P. and Benkard, C.L. 2005. Demand Estimation with Heterogeneous Consumers and Unobserved Product Characteristics: A Hedonic Approach. *Journal of Political Economy*, 113, 1239–76.

Batey, P., Baycan Levent, T., Button, K. and Nijkamp, P. (eds) (2008). *Urban Planning.* Cheltenham: Edward Elgar.

Bedate, A., Herrero, L.C. and Sanz, J. 2004. Economic Valuation of the Cultural Heritage: Application to Four Case Studies in Spain. *Journal of Cultural Heritage*, 5, 101–11.

Boter, J., Rouwendal, J. and Wedel, M. 2005. Employing Travel Time to Compare the Value of Competing Cultural Organizations. *Journal of Cultural Economics*, 29, 19–33.

Carroll, T.M., Clauretie, T.M. and Jensen, J. 1996. Living Next to Godliness: Residential Property Values and Churches. *The Journal of Real Estate Finance and Economics*, 12, 319–330.

Coccossis, H. and P. Nijkamp (eds) 1995. *Planning for Our Cultural Heritage.* Aldershot: Ashgate.

Deodhar, V. 2004. *Does the Housing Market Value Heritage? Some Empirical Evidence.* Australia Economic Research Paper, 403. Sydney: Macquarie University.

Diamond, P. and Hausman J.A. 1994. Contingent Valuation: Is Some Number Better than No Number? *Journal of Economic Perspectives*, 8(4), 45–65.

Do, A.Q., Wilbur, R.W. and Short, J.L. 1994. An Empirical Examination of the Externalities of Neighbourhood Churches on Housing Values. *The Journal of Real Estate Finance and Economics*, 9, 127–36.

Dunse, N. and Jones, C. 1998. A Hedonic Price Model of Office Rents. *Journal of Property Valuation and Investment,* 16, 297–312.

Florida, R. 2002. *The Rise of the Creative Class.* New York: Basic Books.

Ford, D.A. 1989. The Effect of Historic District Designation on Single-Family Home Prices. *Real Estate Economics*, 17, 353–62.

Griliches, Z. 1971. Introduction: Hedonic Price Indexes Revisited, in *Price Indexes and Quality Change*, edited by Z. Griliches. Cambridge, MA: Harvard University Press, 3–15.

Gülümser, A., Baycan Levent, T. and Nijkamp, P. 2009. Beauty is in the Eyes of the Beholder: A Logistic Regression Analysis of Sustainability and Locality as Competitive Vehicles for Human Settlements. *International Journal of Sustainable Development*, 12(1), 95–110.

Heudorfer, B.S. 1975. *A Quantitative Analysis of the Economic Impact of Historic District Designation*. MA thesis. Brooklyn, NY: Pratt Institute.

Hoevenagel, R. 1994. *The Contingent Valuation Method: Scope and Validity*. Amsterdam: Vrije Universiteit.

Hough, D.E. and Kratz, C.G. 1983. Can 'Good' Architecture Meet the Market Test? *Journal of Urban Economics*, 14, 40–54.

Hubbard, P. 1993. The Value of Conservation. *Town Planning Review*, 64(4) 359–73.

Jones, C. and Dunse, N. 1996. *The Cutting Edge 1996*. London: Royal Institute of Chartered Surveyors.

Kahneman, D. and Knetsch, J.L. 1992. Valuing Public Goods: The Purchase of Moral Satisfaction. *Journal of Environmental Economics and Management*, 22, 57–70.

Kahneman, D., Knetsch, J.L. and Thaler, R.H. 1990. Experimental Tests of the Endowment Effect and the Coase Theorem. *Journal of Political Economy*, 98, 13–25.

—— 1991. Anomalies: The Endowment Effect, Loss Aversion, and Status Quo Bias. *Journal of Economic Perspectives*, 5, 193–206.

Klamer, A. and Zuidhof, P.W. 1999. The Values of Cultural Heritage: Merging Economic and Cultural Appraisals, in *Economics and Heritage Conservation: A Meeting Organized by the Getty Conservation Institute, December 1998*, edited by R. Mason. Los Angeles: The Getty Institute, 23–61.

Lancaster, K.J. 1966. A New Approach to Consumer Theory. *The Journal of Political Economy*, 74, 132.

—— 1979. *Author Variety, Equity, and Efficiency: Product Variety in an Industrial Society*. New York: Columbia University Press.

Landry, C. 2006. *The Art of City Making*. London: Earthscan.

Leichenko, R.M., Coulson, N.E. and Listokin, D. 2001. Historic Preservation and Residential Property Values: An Analysis of Texas Cities. *Urban Studies*, 38, 19–73.

Mitchell, R.C. and Carson, R.T. 1989. *Using Surveys to Value Public Goods: The Contingent Valuation Method*. Washington, DC: Resources for the Future.

Moorhouse, J.C. and M.S. Smith 1994. The Market for Residential Architecture: 19th Century Row Houses in Boston's South End. *Journal of Urban Economics*, 35, 267–77.

Morrison, G.C. 1997a. Resolving Differences in Willingness to Pay and Willingness to Accept: Comment. *American Economic Review*, 87, 236–240.

Morrison, G.C. 1997b. Willingness to Pay and Willingness to Accept: Some Evidence of an Endowment Effect. *Applied Economics*, 29, 411–17.

Murphy, J.J., Allen, P.G., Stevens, T.H. and Weatherhead, D. 2005. A Meta-analysis of Hypothetical Bias in Stated Preference Valuation. *Environmental and Resource Economics*, 30, 313–25.

Navrud, S. and Ready, R.C. 2002. *Valuing Cultural Heritage: Applying Environmental Valuation Techniques to Historic Buildings, Monuments and Artifacts*. Cheltenham: Edward Elgar.

Navrud, S. and Strand, J. 1992. The Preservation Value of Nidaros Cathedral, in *Pricing the European Environment*, edited by S. Navrud. Oxford: Oxford University Press.

Noonan, D.S. 2003. Contingent Valuation and Cultural Resources: A Meta-Analytic Review of the Literature. *Journal of Cultural Economics*, 27, 159–76.

—— 2007. Finding an Impact of Preservation Policies: Price Effects of Historic Landmarks on Attached Homes in Chicago, 1990–1999. *Economic Development Quarterly*, 21, 17–33.

Nijkamp, P., Rietveld, P. and Voogd, H. 1991. *Multicriteria Evaluation in Physical Planning*. Amsterdam: Elsevier.

Poor, P.J. and Smith, J.M. 2004. Travel Cost Analysis of a Cultural Heritage Site: The Case of Historic St. Mary's City of Maryland. *Journal of Cultural Economics*, 28, 217–29.

Richards, G. 1996. Production and Consumption of European Cultural Tourism. *Annals of Tourism Research*, 23, 261–83.

Riganti, P. and Nijkamp, P. 2007. Benefit Transfer of Cultural Values: Lessons from Environmental Economics. *Journal of Environmental Policy and Law*, 2, 135–48.

Rosen, S. 1974. Hedonic Prices and Implicit Markets: Product Differentiation in Pure Competition. *The Journal of Political Economy*, 82, 34–55.

Rouwendal, J. and van der Straaten, W. 2008. *The Costs and Benefits of Providing Open Space in Cities*. CPB discussion paper, 98.

Ruijgrok, E.C.M. 2006. The Three Economic Values of Cultural Heritage: A Case Study in The Netherlands. *Journal of Cultural Heritage*, 7, 206–13.

Schuster, T. 2003. *News Events and Price Movements: Price Effects of Economic and Non-Economic Publications in the News Media*. Finance 0305009, EconWPA.

Scott, A.J. 2003. *The Cultural Economy of Cities*. London: Sage.

Seaman, B. 2003. *Contingent Valuation vs. Economic Impact: Substitutes or Complements?*, paper presented to the Regional Science Association International Conference, Philadelphia.

Snowball, J.D. 2008. *Measuring the Value of Culture: Methods and Examples in Cultural Economics*. Berlin: Springer Verlag.

Throsby, D.C. 1984. The Measurement of Willingness-to-Pay for Mixed Goods. *Oxford Bulletin of Economics and Statistics*, 46, 279–89.

—— 2001. *Economics and Culture*. New York: Cambridge University Press.

Tyrrell, T.J. and Johnston, R.J. 2006. The Economic Impacts of Tourism: A Special Issue. *Journal of Travel Research*, 45, 3–7.

UNESCO 1972. *Convention Concerning the Protection of the World Cultural and Natural Heritage*. Paris: UNESCO.

Vandell, K.D. and Lane, J.S. 1989. The Economics of Architecture and Urban Design: Some Preliminary Findings. *Real Estate Economics*, 17, 235–60.

Van Leeuwen, E.S., Nijkamp, P. and Rietveld, P. 2006. Economic Impacts of Tourism: A Meta analytic Comparison of Regional Output Multipliers, in *Tourism and Regional Development New Pathways*, edited by M. Giaoutzi and P. Nijkamp. London: Ashgate.

Ward, B. 1976. *The Home of Man*. New York: Norton.

Willis, K.G. 2009. Assessing Visitors' Preferences in the Management of Archaeological and Heritage Attractions: A Case Study of Hadrians's Roman Wall. *International Journal of Tourism Research*, 33, 85–108.

Chapter 11

Cities as Creative Hubs:
From Instrumental to Functional Values
of Culture-led Local Development[1]

Guido Ferilli, Pier Luigi Sacco and Giorgio Tavano Blessi

Introduction

One of the trends that has characterized the recent (in historical time) turn of the century has been a massive increase of interest in culture[2] as a major policy leverage for urban change (Miles and Paddison 2005). In face of the difficulties experienced by conventional approaches to strategic urban planning, culture has appeared to many as an attractive bypass in many different respects (Evans 2001, 2003): the construction of urban identity (Gimeno Martinez 2007), the involvement of the local community (Goldbard 2006), and the attraction of resources and talent (Florida 2002a) are just examples of issues where traditional tools seemed to fall short of credible recipes, and where instead culture sounded promising and was likely to introduce refreshing elements of novelty, both at the conceptual and at the policy practice levels (Landry 2000, 2006). In the breakthrough of the post-industrial transition that is deeply changing the ways in which economic and social value are produced and circulated, moreover, culture is admittedly playing a new, major role, and is by many no longer regarded as an economically marginal sector, but as one of the potential engines of the emerging experience economy (Scott 1997, 2000) and a major factor of local competitiveness (Currid 2007). There is, however, a far from unanimous feeling that this emphasis on culture- and creativity-led urban policies is warranted enough by the available evidence and by a convincing background of case studies (Bontje and Musterd 2009, Peck 2005).

The issue is, in fact, much more complex than one could expect at first sight. What seems certain – and this is little surprise if one filters out the cheesy rhetoric that usually accompanies marketing-oriented accounts of the developmental

---

1 We thank two anonymous referees for useful comments and suggestions, while keeping full responsibility for the chapter.

2 Throughout the chapter, by 'culture' we will mean all those expressions of meaning that are the result of a purposeful, professionally organized creative activity *and* that are conventionally classified into the spectrum of cultural and creative activities as distilled by the recent theoretical debate (KEA European Affairs 2006).

benefits of culture – is that culture is not a panacaea that somewhat magically 'works' to make cities cooler, richer and trendy, in a word: cosmopolitan (Young et al. 2006). The idea that culture may bring about substantial effects on the aforementioned dimensions and possibly many others is in principle not without a rationale, but why, how and under what conditions this happens is an issue that has to be tackled seriously and systematically for each specific point. What is fundamentally put into question is not the idea of designing, implementing and testing culture-based local development strategies, but rather the somewhat irrational 'belief' in culture as 'the' solution to several currently outstanding issues in urban policy theory and research.

The most disappointing consequence of adhering to a fideistic attitude towards the beneficial influence of culture is giving up a serious, intellectually compelling analysis of the various aspects to indulge in an apologetic form of argumentation; this trend is particularly evident in creative class-based policy recipes being developed by an uncountable number of consultants in the wake of the media success of Richard Florida's books (2002a, 2005, 2008, see also Florida 2002b). At the end of the day, nothing harms more the credibility of culture as a sustainable source of useful concepts and policy tools for the urban designer than this sort of 'fashionability'. In particular, the most dangerous aspect that emerges is the idea that culture is an *instrument* to pursue other social or economic goals that are intrinsically unrelated to culture itself – in other words, culture runs the risk of being *instrumentalized* by the followers of these approaches, and to be turned upside down in terms of its very psychological, social and even economic sense. The aim of this chapter is to develop a careful discussion of this issue, and to provide a few guidelines for a logically sounder and empirically grounded approach to culture-led local development.

Creativity, Creative Sectors and the Economy

There are many different ways in which culture may have a relevant impact on the local economy. Let us briefly discuss some of the most significant ones.

Attraction of Resources

Culture may be a factor of attraction for firms and human capital insofar as there is a sound understanding of the links that exist between a thriving cultural milieu and economic opportunities. From the point of view of firms, different local conditions as to individual and social attitudes towards the production, circulation and use of knowledge[3] may make a big difference, irrespectively of the firm's

---

3   The social dynamics of knowledge is clearly not limited to the production and circulation of culture, but it is important to remark that culture plays a prominent and often somewhat overlooked role – which is often strategically complementary to other forms of

size. A culturally active milieu, if adequately supported by a large enough level of social capital (Swart and Kinnie 2003), certainly favours the emergence of open-minded, ready-for-change organizational environments, and makes room for incentive schemes based on non-monetary forms of compensation that involve training, workshops, and other forms of knowledge-intensive activities (Harris 1999). This makes room for the individual and collective accumulation of cultural capital (Throsby 1999), thereby driving a positive feedback growth dynamics, as more skilled workers interested in further investment in their intangible capitals, but also more financial investments and venture capital, are attracted.

It is important to stress that this particular attraction-driven growth dynamics is not necessarily linked to 'creative class' factors. Different urban environments may be attractive for a variety of different reasons in a knowledge-intensive economic and social scenario: There is a large number of factors concurrently at work, and they may interact in very complex ways (van Winden et al. 2007). For example, issues of knowledge base, quality of life and diversity may be weighed against those of accessibility, social equity, industry structure or scale: a far more intricate conundrum than the simple '3T' formula (Technology–Talent–Tolerance) that has won a massive popularity for Richard Florida (2002a). Perhaps the most compelling (and convincing) critical appraisal of the limitations of the 'creative class' way of thinking of urban attraction dynamics is that of Peck (2005). He takes up former criticisms by Glaeser (2004) and Malanga (2004) questioning the association between cities faring well in Florida's '3T league tables' and actual growth and occupational performance, but goes on to stress that the major shortcoming of 'creative class' explanations of local development – and this is an irony if one thinks of the sociological origin of the term 'class' – is their basic neglect of issues of social inequality, marginality and poverty, which is easily conducive to new forms of dualistic societies based on the contraposition between the 'creative haves' and the 'creative have-nots' (Bourdieu 1984). And thus, in terms of social sustainability, there is room to believe that the long-term attractiveness potential of a city basically depends on its ability *not* to encourage such dualistic dynamics, but rather to develop a socially inclusive milieu (on the actual social architecture of a local 'creative community', see e.g. the careful fieldwork of Lloyd 2006).

The main tension between 'creative class' vs. local community values-focused accounts of attractiveness lies, as already anticipated, in the issue of instrumentality. From the creative class point of view, the emphasis on creativity[4] and the creative class (creatives) stems from the fact that the latter (according to Florida) 'bring the money' in the current scenario, that is to say, they are a means to pursue a social goal (to make cities thrive), whereas from the local community point of view, the

---

knowledge – in such dynamics, and especially with reference to that particular aspect of the knowledge economy that is the social dynamics of innovation.

4 A discussion on the definitions and meanings of creativity lies well beyond the possibilities of this chapter; but see the fundamental contribution of Garroni (2010).

emphasis is on the social practices of creativity and their *meaning*: Having the opportunity to live a creative life is an end in itself, and all of its further implications come as a second thought (Markusen 2006, Markusen and Schrock 2006). But the paradox is that only when one reasons in non-instrumental terms is there a real possibility that economic and social value can be generated in a sustainable way. And the reason is simple, as is learnt quickly when, say, one is willing to develop a network of friends *because* one needs friends to do good business: perhaps they will grab a few occasional, nice business opportunities, but certainly they will not really make friends with many people, thus self-defeating their very basic premise for looking for 'friends' (warranted or not, it does not matter).

The latter is a point whose implications go far beyond anedoctic wisdom: It has to do with some basic properties of intangible assets that are crucial to define, and develop, suitable norms of substantial and procedural rationality in a post-industrial context. From the point of view of both substantial and procedural rationality, the optimal accumulation of intangible assets (such as human, social or cultural capital) requires an intrinsic motivational drive – in the absence of which, people will not do what is meaningful to acquire knowledge and skills, relationships or cultural experience, but only what is instrumentally functional to their stated goals – and, as a consequence, their conduct will be too narrowly focused and lacking that 'serendipity dimension' that easily allows telling the difference between the truly committed and the opportunist.

The literature on the relationships between culture and local development processes has given so far relatively little relevance, both at the theoretical and policy levels, to issues related to the effect of culture on human capital (Castello and Domenech 2002, Judoson 2002) and social capital (Glaeser et al. 2002, Putnam 2001), which proves to be instead a rather fundamental connection to understand the developmental role of culture in post-industrial societies. This is at least partly due to the fact that a proper conceptualization of the role of culture in shaping the codes of rationality for the production and accumulation of intangible assets is still somewhat lacking. As we will see, the (lack of) instrumentality issue will be a sort of red thread across our whole analysis, and we will argue that the crucial condition for a viable culture-led[5] local development is the existence of social governance mechanisms that encourage individuals and groups to give importance to intrinsic motivation and to link social approval and recognition to commitment towards knowledge-intensive activities and experiences. This is a basic pillar of the emerging knowledge society and urban policy-makers have to learn to take this apparently subtle and impalpable dimension into due account when designing their social mechanisms and regulations.

---

5   By 'culture-led' local development we will mean a local development process where cultural production and participation (both directly and in their strategic complementarity with other forms of knowledge-based productive activities) play a driving role similar to that of education in classical endogenous growth models (see e.g. Aghion and Howitt 1997).

Innovation

Needless-to-say, innovation is the true name of the game in today's global competition. Innovation, however, is an extremely subtle and complex notion that escapes understanding when tackled through a narrow disciplinary focus. To make sense of the phenomenon, multiple perspectives have to be employed: economical, sociological, anthropological, socio-psychological, geographical, technological, to stick to a very simple, short list. And yet, despite this manifold complexity, that calls into question a number of heterogeneous factors and therefore paves the way to 'anything (nothing?) goes' approaches, there has been – even here – an upheaval of interest in the specific role of culture in fostering innovation. Not later than July 2009, the European Community (KEA European Affairs 2009) presented a document that illustrates systematically the many, different connections between innovation and culture, arguing that they occur at several logical levels, and putting forward some preliminary indications for the future mainstreaming of policies aimed at developing those connections.

It seems, therefore, that innovation-related policies may be an important part of a culture-led local development model. Clearly, the effect of culture on innovation is hardly rationalized in terms of the same, almost tautological links that connect to science and technology. The impact of culture is more subtle, but no less interesting, nor important: Culture acts as a sort of global platform that creates the social conditions for the onset of innovative waves, by providing people with opportunities to revise and even reshape their cognitive background through contact with unexpected, and even upsetting, situations and contexts – that is to say, with the essence of cultural experience (Sacco, Ferilli and Tavano Blessi 2009). This effect, in particular, lays the foundations of a new model of clusterized local development that no longer depends on the exploitation of vertical integration – as was customary of classical 'industrial districts (or clusters)' models, but of a new form of horizontal integration where the common ground among firms does not lie in belonging to a same value chain, but rather in the common stakes in the development of an effective local innovation strategy – this is what we call a system-wide cultural district[6] (Sacco et al. 2008, Sacco, Tavano Blessi and Nuccio 2009). Seen in this perspective, culture may become the engine of a new model of endogenous growth where the acquisition of cognitive capabilities through cultural attendance and their socialization across culturally mediated interaction may

---

6 A comprehensive discussion of the system-wide cultural district model is beyond the scope and space limitations of this chapter. For the sake of the present argument, it is useful to remark that in actual local development processes the dynamics of innovation proceeds by alternating phases of horizontal and vertical integration – i.e. by successive waves of search and optimization at the system level (see e.g. Peretto and Connolly 2007). In this particular phase, the search dimension gets a particular relevance in view of the strong emphasis that in the recent past has been given to the competitive potential of vertically integrated, specialized clusters.

cause an innovation-based productivity shift that can be dynamically reinforced (see Sacco and Segre 2009). It is interesting to remark that this particular type of innovation-based growth dynamics does not overlap with the attraction-based one, but is complementary to the latter: therefore, a culture-led local development dynamics may be the effect of the confluence of several distinct factors that are in principle open to strategic complementarity and spillovers.

The implications of this exciting new field of theoretical research and policy design are not easily dismissed. In particular, it is important to stress how the dynamic mechanism briefly presented above crucially hinges upon the *actual* acquisition and development of cognitive skills and upon their effective socialization: In other words, it depends on the success of intrinsic motivational factors in driving individual and collective behaviours. In the case of innovation, the contradictions lying in the instrumental rationality dimension become outstandingly clear: on the one side, there is the need to move across unknown territories to develop new ideas that can have a successful business translation, but on the other, there is a fundamental suspicion and lack of interest towards everything that does not fit nicely into the existing schemes – and can therefore be safely evaluated in terms of familiar cost–benefit trade-offs. And when faced with this dilemma, instrumental rationality inevitably tends to prefer to stay within the circumscribed territory where things 'make sense' and are therefore amenable to evaluation, thereby omitting to deal with most of the things and issues that are potentially conducive to real innovation. This is why, in his still seminal work about the nature and economic effects of innovation, Schumpeter (1934) argues that the entrepreneur is following a peculiar norm of rationality that is escaping the narrow calculations of the ordinary *homo oeconomicus*.

The links between cultural experience, creativity and innovation are therefore much more substantial than one could expect at first sight, and are rooted in the very foundations of the rationality norms that govern non-instrumental behaviours. And this deep link does not lend itself to mechanistic recipes that sound like 'do this, and that will happen', which are the daily bread of instrumental rationality and that are clearly reflected in approaches such as the 'creative class' one. The formation of an innovative wave cannot but be the effect of self-organization – a product of that 'spontaneous order' that is the hallmark of the Austrian school, that is to say, the core of Schumpeter's background as an economist. One can operate upon the contextual conditions for the emergence of innovation, in terms of resources, social attitudes, infrastructure, education, and so on – but it is a matter of fact that similar initial contexts will typically generate much different outcomes, and not only in terms of performance. The socio-psychological micro-structure that prevails in the different contexts will have a very significant bearing on the outcome, and in ways that cannot easily be anticipated. Therefore, it is hard to justify – and implement – a culture-based innovation policy if there is no interest in the intrinsic effects that it produces, that is to say, if one does not give value to the very fact that people and groups may acquire culturally transmitted cognitive competences. Judged from the standpoint of instrumental rationality, there may be safer, less vague ways of pursuing the goal of excellence in innovation, and certainly among them there are ways that are more easily evaluated and can thus command more political and social consensus.

One can then expect to see that the route of culture-based innovation will be likely to be taken by these local communities where the socialization of knowledge as a goal with an intrinsic value has been successfully accomplished, at least to a degree – and it is there that one can expect, *on average*, to witness the onset of new, powerful innovation waves. In this vein, the already cited fresh commitment of the European Community on the issue will create an interesting field for experimentation and empirical validation, and may possibly become one of the most practical and convincing ways to give substance to the ambitious goals of the Lisbon strategy.

Cultural Production: Cultural and Creative Industries

A third fundamental pillar of a culture-based local development model cannot but be cultural production itself, in view of its weight in the overall level of activity of the economic system. As a matter of fact, cultural production per se is one of the most ancient human activities, with roots digging deep in mankind's pre-industrial past. But what makes a difference, and in particular what makes culture so relevant in the late industrial and post-industrial phases of socio-economic development, is the volume and economic value of cultural production. To appreciate this point fully, it is advisable to resume briefly the main stages of the process.

In the pre-industrial era, cultural demand was basically traceable back to cultural commissioning, and in particular to patronage: A situation in which the commissioner makes use of his political and economic power to offer to cultural producers a 'free zone' to work and operate, shielded against the necessities of daily life. It is important to stress that, in the pre-industrial context, the channels of creation of economic value and of political power are basically disconnected from cultural commissioning, which rather appears as a way to make use of otherwise accumulated resources to self-celebrate one's own power and prestige, in ways and forms not accessible to the majority of others. But cultural production occurs at the bottom of economic production, and of all of the social relationships that underlie the latter.

With the industrial revolution, we witness the beginning of a process, initially very slow but, in its most advanced phases, quickly accelerating: the development of cultural industries, where ever-expanding markets centred on the creation and circulation of cultural products rely on the new possibilities unleashed by technological change to devise more and more sophisticated interfaces to address new audiences (Sassoon 2006). Here come printed books, magazines and newspapers, vinyl records and subsequently CDs, cinema on film and then digital, radio-television broadcasting, video-games and recreational software, and so on. A truly astounding broadening of possibilities of cultural experience – one that could have never been imagined still at the middle of the nineteenth century, when the industrial revolution was, in other fields of production, already well on its way. In this new epoch, cultural production is no longer a pre-economic kind of activity, based more on the economics of gift than on market exchange, as was customary of pre-industrialism. On the contrary, it becomes one of the many ways to create economic value: culture, therefore, does not lie any longer at the bottom of the value chain, but becomes one component of the chain itself among many, and the real issue becomes that of its relative dimension (in terms of turnover) with respect to that of more traditional productive sectors. An issue that, in Europe at least, has been taken seriously only recently, whereas in North America it has been long credited an appropriate status in the economic (Caves 2000) and even in the political discourse (Nye 2004). The reasons for the European neglect are manifold. On the one side, the fact that in Europe culture is abundantly subsidized in an explicit way leads quite naturally to think of it as a meta-sector that *absorbs* economic resources more than it is able to create them – and the explicit nature of the subsidy makes a difference here: in the United States and partly elsewhere (mostly in Commonwealth nations), culture (or at least certain cultural sectors) is being subsidized implicitly under the form of tax benefits (Wu 2002) or at least commands forms of social reward such as reputation or public image of donors, but the indirect character of this form of support causes the emphasis to lie on the cultural producers' capacity to attract or to generate (private) economic resources. On the other side, in the European tradition industrially organized cultural production has been seen, especially through the legacy of the Frankfurt (Horkheimer and Adorno 1972[1944], Marcuse 1964) and French (Debord 1967) schools, as a form of mass manipulation, laden with conflictual socio-political implications. These positions have maintained a significant influence in the public orientations towards culture in Europe, providing an important background for continued, explicit public support to cultural production as a way to defend pluralism and to prevent at least partially the commodification of culture.

On the basis of these premises, the 'rediscovery' of cultural industry in Europe as a primary engine of economic and social development is a relatively recent fact. The urge to find workable ways to re-activate de-industrialized, stagnating local economies left behind by the post-industrial transition has allowed the burden of doubt and controversy that hung over the topic to be bypassed rather quickly, once it became clear that culture was likely to be one of the sectors of the future post-industrial economy for which demand was very likely to rise, because of the fact that technological progress would leave people with more free time to spend (Howkins 2001). To be fair, both the North American and the European approaches have their own merits and shortcomings. In North America, cultural industries have struggled to conquer their own spaces of survival and growth on the market, thereby creating the most formidable 'mass seduction machine' of the history of mankind, as Adorno/Horkheimer/Marcuse and Debord would put it (and with some reason). In Europe, the same cultural industries that have benefited, at least partially, from substantial public support have had the possibility of experimenting in ways, and to degrees, that were often impossible in the American context where the verdict of the market (and thus of the audience) was the ultimate one. Thus, on the one side we find a context where the economic value of culture is clear and well recognized although there is a tendency to confine culture almost entirely within the realm of entertainment; on the other side, there is a context where culture is widely thought to be a 'resource drain', but that at the same time the fundamental role that culture has, and ought to have, in the shaping of collective values and in building European civilization is recognized: a task that goes well beyond the dimension of entertainment, to root itself deeply into the very foundations of civil society. Not surprisingly, then, the audience-oriented North American cultural industries have conquered global markets, whereas the European cultural industry has chosen for itself the less spectacular role of an 'R&D lab', an experimental playground somewhat parallel to that of basic research in the science and technology fields. It is this difference that explains why and how cultural influence has been such an important part of America's soft power, as the already cited Joseph Nye puts it.

On the basis of these premises, it is understandable that when, at the end of 2006, the European Community published a document on the levels of economic turnover of European cultural industries (KEA European Affairs 2006), the results have been received mainly as a shock. Some of the most widely reported highlights were the fact that the turnover of the whole cultural and creative sector in Europe was roughly twice as big as that of the automobile industry, and that the relative weight of the sector was similar to giants such as information and communications technologies (ICT), real estate, and food–beverages–tobacco. This was a surprise, especially for continental Europe; however, as in the UK the cultural industries issue had been taken quite seriously long before. The first intuition of the potential of creative industries as an economic development arena was sparked in the late 1990s, at the dawning of former Prime Minister Tony Blair's New Labour era. The Department for Culture, Media and Sport (DCMS) set up its Creative Industries Unit and Task Force, identifying creative industries as 'those industries that have their origin in individual creativity, skill and talent and that have a potential for wealth and job creation through the generation and exploitation of intellectual properties' (DCMS 2004). The UK has thus begun a systematic mapping of its cultural industry system, with two rounds in 1998 and 2001 that included: advertising, art and antiques markets, crafts, design, designer fashion, film and video, interactive leisure software, music, performing arts, publishing, software and computer games, television and radio. The list is significantly broader than cultural industries properly meant, and makes significant additions towards the somewhat blurred realm of creative industries, with their edge on traditional, non-cultural consumer markets.

This extension of the domain of culturally relevant productions to creative industries is far from innocuous. The most important point is that, unlike cultural industries, for creative industries such as design or advertising, the establishment and formalization of a certain sphere of meaning is no longer the end, but rather a means that improves the product's marketability: the creative content is a component of a multi-layered production chain that operates on goods markets *outside* the cultural field. In other words, the use of cultural content in the creative realm is programmatically *instrumental* to goals that are fixed by those (firms, generally) that require the creative input service. And, therefore, in the case of creative industries, the instrumentality/non-instrumentality dialectics necessarily unfolds differently: creative producers are much more dependent on the commissioners' briefing than cultural producers are. Not incidentally, the development of an entrepreneurial attitude towards creative industries and its trespassing from an 'arts and crafts' to an industrial mode of production mostly occurs between the very last stages of the industrial era, and the beginning of the new post-industrial one, with all of its 'new economy' implications.

This complex articulation of the cultural and creative sectors (in which we also find a 'core' that comprises cultural productions that are not amenable to a fully fledged industrial organization form, see below) basically conforms to the concentric circles model of Throsby (2001), which assumes that it is the *cultural* value of *cultural* goods that is the ultimate source of value for these sectors. The core of the concentric circles consists of 'unconstrained' creative ideas and content, whereas moving outwards one finds a progressive decrease of cultural content in favour of commercial content. Overall, we have four distinct circles: the first two pertain to the cultural industries, while the others to the creative industries. We have, respectively, the 'non industrial' cultural core (visual arts, performing arts and heritage), the cultural industries (film and video, television and radio, video games, music, and books and press), the creative industries proper (design, architecture and advertising) and the related industries (those that produce technological interfaces for cultural and creative products: PC, MP3 players, mobile phones, etc.).

A more recent contribution to the understanding of this complex and rapidly evolving field is the Report on Creative Economy of the United Nations (UNCTAD 2008), which attempts to abridge different possible approaches to 'creativity', 'creative products', 'cultural industries', 'creative industries' and 'creative economy': A terminology that is being invoked increasingly often, but not necessarily in mutually coherent ways. Creativity as a measurable social process is, from the economic point of view, considered not only in terms of economic outcomes (transactions with a market value), but also of the whole cycle of creative activity, through the interplay of four forms of capital – social, cultural, human, and structural or institutional – that determines in turn the expansion of the 'stock' of creativity – the creative capital, namely, the ultimate outcome of the cycle (one could in fact adopt a 'circular flow' view of the creative process, where market transactions occurring in cultural and creative markets are not the end of the game, but are functional to the 'reproduction' of the stock of creative capital). UNCTAD defines creativity as 'the process by which ideas are generated, connected and transformed into things that are valued' (UNCTAD 2008: 10). The UNCTAD report attempts to provide a synthesis of the cornerstones of the literature on cultural and creative industries, mixing together the already cited DCMS classification; cultural studies' approaches, focusing attention on popular culture and on its channels of transmission and dissemination through the industrial media, such as film, broadcasting and press (Hesmondhalgh 2002); Throsby's concentric circles model; and the World Intellectual Property Organization (WIPO) framework developed for the industries involved directly or indirectly in the creation, manufacturing, production, broadcasting and distribution of copyrighted works (Musungu and Dutfield 2003). On this basis, the UNCTAD report arrives at stating that cultural and creative industries:

- work through cycles of creation, production and distribution of goods and services that use creativity and intellectual capital as primary inputs;
- rely upon a set of knowledge-based activities, focused on – but not limited to – the arts, potentially generating revenues from trade and intellectual property rights;
- deal with both tangible products and intangible intellectual or artistic services with creative content and economic value;
- are at the cross-roads among the artisan, services and industrial sectors; and
- constitute a new, dynamic sector in world trade (UNCTAD 2008: 13).

The UNCTAD Report gives an important contribution to the global acceptance and recognition of cultural and creative industries as a key option for local socio-economic development, and therefore paves the way to more generalized interest on the policy-makers' side. However, the rather particular mix of instrumental and non-instrumental motivations that are found across the spectrum of this rather heterogenous and diverse meta-sector is likely to make room for further discussion, both at the theoretical and at the policy level, and the early criticisms and perplexities of European Continental thinkers are not so easy to dismiss as obsolete (Hardt and Negri 2000, for a recent reprise).

Moreover, with the further unfolding of the post-industrial transition that is progressively unveiling a process of 'culturalization' of the whole economy, it becomes increasingly difficult to draw a line to establish which industries are creative and which are not; one can argue, for example, that 'all industries are cultural because they all produce products that besides having functional applications are also socio-symbolically significant' (Mato 2009: 73 and, for a critical rebuttal, Miller 2009). Certainly, there is ample room for believing that not only the local developmental impact of cultural and creative industries will increase through time, but also that their active role in driving the already cited, and ongoing 'culturalization' of the economy will be increasingly recognized and strategically accounted for by other economic players. In fact, one can argue that the next step of the process going through the previous pre-industrial and industrial phases will be, in terms of the positioning of culture and creativity along value chains, the further migration of the latter at the top of value chains, by producing content and meaning that will be deliberately appropriated (and, when necessary, paid for) by apparently non-cultural producers, to shape the perceptions of value of their customers by finding more and more sophisticated ways to associate such contents and meanings to their products: a trend that poses still further problems in terms of social sustainability and the instrumentality/non-instrumentality dilemma.

Cities as Creative Hubs

As we have seen in the previous discussion, there are various, distinct (but potentially interactive) channels through which culture-led local development may emerge. And, consequently, one is naturally led to think of the city as the natural location for these activities, provided that the city, from its very origins, has always been the place where new ideas were conceived and new models were experimented with (Hall 1998, Scott 1997, 2000). Then it seems natural to ask to what extent they relate to the issues of transformation of the post-industrial city, and how. On the basis of the above arguments, the question sounds nearly rhetorical, but it is a matter of fact that translating the previous discussion of some promising developmental factors and mechanisms into the actual practice of urban policy and planning is less obvious than one could expect.

The basic issue that comes to mind is that of localization. Cultural districts, meant as the spatial concentration of cultural facilities into one specific urban neighbourhood, is a typical presence in most North American cities, and not a recent one. In this context, agglomeration is a rather natural tendency, that responds to intuitive but economically compelling arguments: as cultural audiences have multiple and diverse interests, it is better to concentrate all the facilities in nearby locations, so that every single facility will have the opportunity to draw from whole basin of potential customers, and this concentration effect will most likely outweigh the competition effect deriving from the presence of so many other facilities that they might choose (Tym and Partners 1999). In the alternative scenario in which facilities are scattered through the city, there would be less competition but also significantly less potential customers, and especially so if a given facility is relatively small, far apart and less well known compared to major ones. Moreover, customers that choose one facility one night, having once learnt about the menu of possibilities, could be tempted to come back again another night and choose another one. At most, in larger cities where menus are very rich and diversified and audiences very big, there may be thematic clustering (the gallery district, the cinema district, the theatre district, and so on) (Stern and Seifert 2007).

This very reasonable spatial logic, however, reflects quite clearly a social attitude towards culture as entertainment (Clark 2003): In this view, cultural facilities are places whose aim is to attract final customers willing to pay to spend some time and enjoy. In a more mature developmental view of culture, however, as we have seen, there are many different functions that culture can and must perform, and this finds a reflection also in the spatial organization of cities. The attraction of potential audience is no longer the sole criterion: considering the dimension of cultural production and not just dissemination, as well as the link between culture and innovation and the attraction of talent and resources, new location criteria emerge. Among these: closeness to production facilities (recording, rehearsal, digital processing studios, etc.), to science and technology parks and to specific industrial clusters, to amenities, to colleges and universities, to downtown areas with particular symbolic value, and so on. If the emphasis moves from entertainment solely to entertainment *plus* production *plus* complementarity with other productive sectors *plus* etcetera, there are a number of different factors that act in a parallel fashion to determine location choices and thus how cultural activities fit into the spatial logic of the city (Alvarez and van Diggelen 2005). The spatial concentration of producers, which traditionally was a driving location factor only for certain fields of cultural production, most notably visual arts, is now taking over the full spectrum of cultural and creative production, also in view of the increasing complementarities between different disciplines.

Important differences exist in this respect between North American, European, and the new, emerging phenomenon of creative production in Far Eastern Asian countries (see for instance Hutton 2008, for a detailed comparative analysis). Here we do not have the space to discuss such differences in detail, but it can be useful to provide some very sketchy remarks. In Europe, where most cities are characterized by historical city cores rich with heritage, social life and fascinating ambience, industrial development has progressively pushed productive activities away to sub-central and peripheral areas, as such activities were generally noisy, polluting and caused congestion to the local traffic flows. But with the post-industrial transition where most 'new economy' productive activities have a very low environmental impact if any, there has been a massive 'flow-back' phenomenon that has led many new businesses, and in particular cultural and creative producers, to occupy central positions in order to exploit at best the position, image and prestige of the central areas (see e.g. Scott et al. 2008), thereby infusing new vitality to quarters that were for long mostly colonized by the offices of banking, insurance and financial services firms, as well as by other forms of 'tertiary' 9-to-5 occupations, and were inevitably subject to a late-afternoon desertification that made them grim, lifeless and even unsafe. The infusion of cultural and creative production to the city centre not only breaks down the 9-to-5 curse in that the time schedules of creative people are notoriously different from those of office clerks, but also causes a dramatic change in terms of nightlife vivacity, due again to very different work-time vs. leisure time dialectics between the two categories: rigidly separated for clerks, often intermingling for creatives.

In North America, contrary to Europe, most cities are lacking city cores characterized by a strong urban identity (in these cases, one should better speak of 'downtown' rather than of 'core'), and in addition most residential neighbourhoods are localized in the semi-central and suburban areas. This, of course, further intensifies the 9-to-5 effect, and at the same time weakens the need for a flow-back of cultural and creative activities during the new post-industrial phase. A notable exception are those cities, mostly on the East Coast (but also across the West Coast 'Cascadian' region with San Francisco, Portland, Seattle and Vancouver) where a proper city core can be found, maintaining more 'European' characteristics. In this more fluid context, there is some additional reason to keep cultural and creative activities compactly organized within a circumscribed space, to preserve their identifiability and urban centrality, but also in this case one witnesses the emergence of new districts characterized by richer mixes of functions and often different locations with respect to more established and traditional ones. There are also a few cases (Kansas City, for instance) where the relatively low prices of the urban downtown have caused the inflow of cultural and creative producers, thereby creating a new urban identity that is not linked to the idea of a (non-existent) city core, but rather to the spatial concentration of cultural facilities in the first place.

In the Asian Far East, the major difference is given by the fact that there is no particular interest or concern towards preserving and charging with symbolic meaning the parts of the city that are older or that host particularly ancient buildings, unless the latter carry a special functional or symbolic importance per se (temples, gardens, imperial residencies, and so on). Therefore, the city core vs. suburbs dialectics does not make particular sense in this context. The localization of cultural and creative activities follows pragmatic criteria of availability and adequacy of spaces and, at least momentarily, does not have a big impact on the overall urban identity, despite the fact that the interest and the private and public investments in this field are rapidly increasing.

In this vein, reconverting urban space for cultural use is becoming a trend with worldwide implications: in the recovered spaces, productive and entertainment facilities mix together in innovative (and sometimes surprising) ways, often blowing new life into former (and then abandoned) industrial spaces looking for new functional destinations. In the UK, for instance, a whole wave of new 'cultural quarters' is spreading across the country (Roodhouse 2006), and is giving a remarkable push to the social dynamics of places severely hit by de-industrialization and large-scale unemployment, whose benefits cannot be simply stripped down to the creation of a few new jobs, but also involve the opening of new meeting places that create opportunities for pro-social interaction, vocational training, social control of troublesome quadrants of the city, and so on (Keaney 2006). But also outside the UK – where the phenomenon has acquired a particular socio-economic relevance in view of the high social costs of the post-industrial transition in former industrial cities, and the consequent necessity to experiment with new, effective models of use of the urban space – in spite of the heterogeneity of the various experiences taking place in different economic, social and geographical contexts, there are also largely common factors that single out a global tendency for what we could call 'creative dismissions' (Sacco and Soru 2008). In its more general formulation, a creative dismission is the conversion of a facility previously used for non-cultural purposes into one for cultural and creative purposes. In many cases, these facilities are, as already suggested, factories or warehouses, but one can find examples of cases involving any possible kind of previous use, from hospitals or churches, to prisons, schools, military installments, and so on.

In some cases, the area under transformation is so huge as to generate a major impact even at the urban scale, such as in the case of the Zollverein area in Essen, or the Baumwollspinnerei area in Leipzig, both in Germany. The latter is a case of special interest (Sacco 2009), in that the project (which is a work in progress) is being entirely carried out by private capital, and consists of the re-conversion of a previous, quite large cotton mill factory (one of the hugest in continental Europe), that now hosts not only artist studios, art galleries, fashion design and crafts workshops, but also commercial outlets, only some of which are targeted to the area occupants (for instance, a large art supplies store). The project has been made possible through the coincidence of several factors: the low real-estate prices prevailing in Leipzig after the burst of the post-reunification price bubble; the presence in Leipzig of one of the most renowned (and sought after by collectors) panting 'schools' of the contemporary art scene; the availability of large and very apt spaces in a sub-central position in the city; the idea of creating not a mono-dimensional cultural quarter where only cultural and creative activities can be hosted, but a mixed-use quarter where the revenues from rents to classical commercial activities can be partly used to keep rental prices for creatives and galleries low enough to make them affordable – and, at the same time, maintaining the possibility to discriminate between potential tenants on the basis of quality. This latter aspect is particularly important, and interesting: once the place had become famous for the presence of prominent artists and galleries that were convinced to come to the Baumwollspinnerei, there has been, not surprisingly, a rush by artists and galleries to ask for spaces to exploit the notoriety and reputation effect. From the economic point of view, the landlords could have capitalized this rent by giving out spaces to best bidders, independently of quality; but the effect of this choice would have been that of destroying the reputation of the place by mixing up good artists and galleries with bad ones that could afford paying a lot to sit aside the former, and eventually this would have caused the flight of good quality and the eventual takeover of talentless affluence as the 'identity marker' of the place. For this reason, the landlords opted for a more sophisticated and long-term oriented perspective, namely, *subsidizing* good quality to ensure its sustainability, and exploiting quality to raise the rental prices for *non-cultural* tenants, whose willingness to pay would not feed back negatively on the place's cultural standing and reputation. In spite of its semi-peripheral position, Baumwollspinnerei has become a key player in Leipzig's cultural life (which, by the way, is quite lively and selective, because of its world-class orchestras and its long historic tradition in publishing), and is now attracting not only local visitors and creatives, but most of all international ones (the complex has been hosting galleries whose main headquarters are in North America, Mexico or India, as well as several international artists choosing to live and work in the premises), thereby affecting, although in slow and subtle ways, the 'central place geometry' of the city.

The example of Baumwollspinnerei is a clear illustration of how successful processes of creative dismissions are, once again, the result of site-specific alchemy rather than of straightforward applications of a book of recipes, and moreover that such alchemy calls for subtle exercises of non-conventional, intrinsically motivated forms of rationality, like the one that, in this particular case, prescribed giving up the extraction of a considerable amount of rent to preserve economically non-rewarding quality (and even subsidizing it). But in spite of these subtleties, the global geography of creative dismissions is astoundingly rich and constantly expanding. Not incidentally, most of the currently active projects have been organized into international, constantly expanding thematic networks – such as TEH-Trans Europe Halles or ENCC-European Network of Cultural Centres – that carefully monitor members' compliance with some fundamental aspects, such as absence of speculative motives, social inclusiveness, financial transparency, and so on. In other words, the vitality (and viability) of creative dismissions seems to rest basically upon the adherence to ethical codes that preserve the integrity of the cultural mission, irrespectively of their economic impact in narrow terms.

The creative dismissions theme provides a good illustration of the key issues that arise when thinking of the city as a creative hub. It is not merely a matter of dealing with talented people, or of providing the right facilities, and not even of investing enough resources to fuel the city's 'creative push'. It is basically a matter of finding a way to involve the whole social fabric of the city in the pursuit of cultural excellence and integrity, and in the construction of a collective identity in which culture becomes a cornerstone of the local approach to quality of life, social relationships and entrepreneurship. Localization choices are more a consequence than a cause of this. It is very well known that, in many cases, location choices of cultural and creative producers have been strategically exploited by real-estate developers to 'hip up' previously disregarded areas of the city, according to the classical mechanism: creatives go where space is cheap and abundant, they colonize the place and make it cool, then come the owners of trendy shops and concept stores, and finally the young urban professionals asking for luxury condos and the big fashion and design brands asking for archistar-signed megastores. At that point, creatives as well as previous inhabitants of the place are well on their way to somewhere else, in that they can no longer afford to pay the rent to stay where they once did (Kennedy and Leonard 2001).

In all of its cynicism and instrumentality, this mechanism is, at face value, highly compatible – if not highly representative – of the 'city as creative hub' phenomenon: After all, it perfectly illustrates the nomadic tendency of creativity, as well as the city's need for change and renewal. And certainly, real-estate developments transforming previously underrated areas into eagerly sought ones are spectacular examples of renewal. But this is only the appearance: The substance is that, in such mechanisms, cultural and creative activity is manipulated and instrumentalized in ways that undermine, rather than build up, the social cohesiveness of the city (Bianchini 1993, Pike et al. 2006), and that in the long run turn it into a cultural and creative *showroom* rather than in a cradle of new ideas and experimentations. This is what the creative dismissions movement has understood so well. Not incidentally, its ethical codes resist the assimilation to real-estate development mechanisms, and it maintains its social orientation and functions as far and as long as possible to preserve the spirit and practical implications of the cultural transformation of urban space. But clearly, the big stake is at another level: It is at the level of the governance of the urban space that the game is won or lost, depending on how sensitive urban policy-makers are with respect to the subtleties of post-industrial rationality. If the fundamental strategic choices are oriented by classical and familiar standards of instrumental rationality, all one can hope for is that the city witnesses a brief season of cultural and creative vitality, soon to be taken over by a massive gentrification that dries it up to the bone, thereby fostering the diffidence of inhabitants that learn how the seductive thrills of culture are essentially a Trojan horse used to conquer new urban spaces for real-estate speculation. And once this association is being made, it takes time and effort to dismantle it. In other words, the instrumental exploitation of culture paves the way to the immunization of the city towards further attempts at cultural stimulation: Once the real name of the game becomes apparent, people become much less willing to play, and the city loses relevant opportunities in a global scenario where the economics of knowledge rules. In a nutshell, then, the social (and economic) sustainability of culture-driven urban change boils down to the non-instrumentality of the cultural initiative upon which such change is built (Sacco and Tavano Blessi 2009).

Therefore, caution should be mandatory when conceiving a plan of cultural and creative development of the city. If a proper understanding of the underlying mechanisms and of their pitfalls is lacking, and if moreover there is the naïve belief that it is all about applying some recipe to bring creatives in and to serve their interests and expectations against those of the dumb, hostile-to-change, 'deadweight' non-creatives, the whole operation could backfire very badly, and quite soon, thereby delivering a deceitful message: that culture 'doesn't work' for cities. It isn't culture that doesn't work: What doesn't work is the idea that culture can be manoeuvred like a tool, with a sound and intentional ignorance of its complex codes of meaning (Hubbard 2006). Culture is not a productive factor in the traditional sense of the word: something that can be thrown into the productive process as an input with the certainty that, if the proper technical specifications are respected, will deliver something else, defined in advance, as output. Culture is a process with a very high dependence on initial and 'boundary' conditions. It cannot be engineered in the sense in which conventional production engineering does. It asks for new, and specific, methods of policy and social governance. Without this awareness, speaking of cities as creative hubs is mostly a source of misunderstanding and disappointment.

Towards a New Synthesis: System-wide Cultural Districts

Analyses of the role of culture as a catalyst for local development processes trace back to certain experiences in urban and regional planning, such as the urban regeneration plans carried out by the Greater London Council from the 1970s (DCMS 2004). These pioneering attempts were based on a somewhat 'conventional' (for macro-economists) strategic vision focused upon building cultural infrastructures and activities to impact upon the local economic dimension, and to activate thereby Keynesian multipliers to spur the whole economic system. Expanding on this first intuition, the reasoning on culture-led development processes has gradually extended to complementary fields such as, for instance, culture and society (Everingham 2003, Matarasso 1997), or culture and the built environment (Bianchini and Parkinson 1993, Graham 2002, Hutton 2006), to evolve into a fully fledged approach to local development.

Recent literature has put forward theoretical frameworks and classifications of cultural clusters (Scott 2000, Valentino 2003) or cultural districts (Evans 2001, Santagata 2006) as possible theoretical syntheses of the main aspects of culture-led local development, and at present there is extensive exploratory research on specific case studies, even if this growing amount of fieldwork also makes room for confusion and ambiguity. In this regard, Mommaas (2004) underlines that the cultural district label has often been attached to very different spatial ambits and scales, which have also followed different paths of development. He describes how, sometimes, the studies focus upon single buildings, more often upon urban quarters, but also upon entire cities and networks of small villages. Moreover, Hospers and Beugelsdijk (2002) have underlined that many policy-makers are tempted to apply somewhat mechanically cluster-based reasoning at different spatial scales, with little consideration of scale-specific characteristics, local idiosyncrasies, and so on.

Best practices research in culture-led development has shown that such mechanisms can effectively work only under certain conditions and spatial scales. Models that only count on one driver – like, say, innovation – do not explain convincingly the complexity of observed territorial dynamics. In addition, evidence in some regions demonstrates that productive agglomeration per se – even in the creative industries – does not necessarily contribute to the spreading or to the enhancement of innovative capacity (Simmie 2004), neither, a fortiori, to the improvement of social and human development dimensions. To make sense of culture-led economic development, we need more articulated and non-mono-causal models.

In general, one can reason in terms of two alternative approaches to culture-led development. The first is bottom-up, where the spatial agglomeration is neither designed nor driven by a specific (public) actor, but by a self-organized dynamics arising from complex spontaneous coordination of local actors, to be subsequently governed and regulated by the local public administrations. The second is, conversely, top-down, and is initiated by public or semi-public actors pursuing a specific developmental goal and incentivating/encouraging local private actors to take part and cooperate on the basis of the expectation of specific benefits. It is in principle more likely that top-down approaches tend to develop markedly instrumental approaches to culture with respect to bottom-up ones, and the reason is simple: as the involvement of cultural actors derives from the content of the developmental goal, it is the logic of the latter that determines why and how culture fits into it, how it may contribute, and to what extent. This is, for instance, a typical attitude of territorial marketing or touristic development plans, where culture concurs with other activities to define an optimal menu for would-be visitors/clients/investors; the presumed tastes and expectations of the latter are therefore the benchmark against which one should design the optimal supply package. Conversely, in bottom-up approaches it is the direct, intentional involvement of cultural actors that makes it at least plausible that the cultural dimension of the project emerges from a participative dialogue, and reflects an intrinsic sense-making motivation. As a matter of fact, in best practices one always finds, eventually, the coexistence of spontaneous and planned elements: whether the origin of the process is top-down or bottom-up, the system evolves towards a mixed configuration where both organizational principles are present and find an effective complementarity. But there are plenty of examples in which this synthesis is never attained, and the process eventually dries up either for insufficient coordination and inefficient allocation of the available resources (bottom-up lock-in), or for lack of intrinsic vitality and for the insufficient level of responsible involvement of the local players (top-down lock-in).

A *system-wide cultural district* (SWCD) is an organizational model that attempts to attain a fair balance between the two principles, thereby operating an effective synthesis of planned and self-organized components (Sacco and Tavano Blessi 2007, Sacco et al. 2008, Sacco, Tavano Blessi and Nuccio 2009 for a much more detailed presentation). On the basis of an extensive fieldwork on reference cases that can be considered best practices of post-industrial culture-led local development models, the SWCD approach isolates three main themes that are found to be recurring in the sample: attraction of resources and talents, competition-driven restructuring, and capability building. For each of these themes, it is possible to find widely recognized references that exemplify some aspects of the theme, even if they do not typically represent it exhaustively. For the attraction theme, the obvious reference is the already discussed Florida (2002a) creative class model, with all the caveats and limitations spelled out above. For the competitive restructuring theme, it is natural to think of Porter's (1989) view of the transition from an investment-based economy to an innovation-based one, as it is characteristic of well-adapted, mature capitalist systems. For the capability building theme, the inevitable reference is the Sen (2000) approach to economic development, properly rephrased in terms of the conditions for the development of a sustainable knowledge economy (and society). A properly articulated culture-led local development model must be able to modulate these three different dimensions in idiosyncratic ways that reflect the specificities and potential of the *genius loci*. Exactly like one-sided top-down or bottom-up approaches prove to be too unilateral to generate a sustainable culture-led development dynamics, conveying all of the policy efforts through just one channel typically brings about dysfunctional imbalances and distortions. Focusing solely on attraction leads, as already discussed, to disregarding local talent, instrumentalizing culture and exacerbating social dualism; the bias towards competitive restructuring leads to focusing upon the activities and assets that deliver ready-to-use sources of competitive advantage, and encourages an over-engineering of the cultural processes modelled on the prevailing standards of scientific-technological R&D; finally, exclusive emphasis on local capability building produces parochialism, that is, self-referential legitimization of indigenous talent, culture and codes of meaning, with the consequent marginalization for lack of adequate levels and quality of external social exchange. Overall, the three themes above can be spelled out into 12 characteristic factors; a well-developed local system should in principle perform quite well on all of them, as out-of-sample tests with real best practice case studies actually confirm (Sacco and Tavano Blessi 2007, Sacco, Ferilli and Tavano Blessi 2009).

The reason we speak of system-wide cultural districts is not straightforward and merits some illustration. Traditional districts (or clusters) are typically built around vertically integrated value chains, namely, a number of firms that efficiently coordinate by occupying each a specific segment of the whole productive process that transforms a certain amount of raw materials into a specific product, delivered on the final market. In this way, firms may exploit a considerable amount of economies of scale, scope, transportation, cognition, and so on, being able to compete even in a large global market in spite of their small size. In an investment-based or even price-based competitive scenario, where innovative leaps are mainly small and incremental (i.e. they consist in small tactical improvements of given products and processes), the classical vertically integrated district generally performs pretty well, and reacts quite quickly even to small changes in the competitive context it faces, while at the same time maintaining high levels of local cooperation and coordination.

In an innovation-based scenario, on the contrary, considerable and frequent innovation leaps are called upon to remain competitive, and this is hardly achieved in a local context where all firms focus on the same, restricted range of products and are in difficulty when asked to develop subtle ways of lateral thinking, establishing unprecedented connections with entirely different production chains, products and processes, that is, to perform the adaptive search tasks on multi-dimensional knowledge spaces that are typical of radical innovation. In this different setting, rewarding forms of integration are of the horizontal type, namely, they involve firms belonging to different and even very different production chains, but share a same inclination for, and experience with, innovativeness. The major difficulty with this alternative scheme, however, is that setting up a dialogue among diverse sectors is neither conceptually nor practically simple; some 'translation device' is necessary to establish a common platform where the various players can fruitfully engage in conversations that make sense. Actually, culture proves to be very well versed to serve as the basis of such a platform. The reason is relatively intuitive: the essence of cultural experience is a challenge for the expectations, the cognitive priming and the common sense of the audience, that through their involvement in the cultural experience, are forced to follow surprising and even unprecedented routes. Insofar as the experience would essentially match the expectations and cognitive presets of the audience, it would turn out to be basically uninteresting, and moreover quite boring. But once a culturally exposed audience get systematically trained to put into question their prejudices through regular attendance to stimulating events, shows, workshops, and so on, this readiness towards the unexpected also becomes applicable in their daily professional activity whenever they have to face non-standard problems calling for creative solutions; and all the more so the more their access to the cultural experience is proactive, intrinsically motivated, and continued. This is the channel through which culture becomes a translation device, turning different demands for market-specific innovative ideas into a joint search across abstract cognitive spaces. A search that, if properly conducted, delivers distinctive, context-specific implications for the different sectors involved in a common, culture-based human development strategy. This is why we speak of a *system-wide cultural* district: system-wide, in that it does not focus on a single, specific sector of the local economy but rather on a relatively large and diversified spectrum of sectors; cultural, in that culture is the main content of the platform that allows the translation, and an effective innovation-based communication, among the different sectors (Sacco, Ferilli and Tavano Blessi 2009). In this perspective, what makes an interesting case study of a *system-wide cultural district* is neither the presence of excellent but isolated and self-referential cultural institutions, nor the spatial concentration of a large number of cultural players, as it happens for traditional cultural districts; rather, it is the level of coordination among the various local cultural players, their strategic complementarity with the other (innovation-oriented) productive sectors, and the level and quality of culture- and knowledge-based sociability that, among other factors, may make a difference.

In a SWCD perspective, reasoning in terms of non-instrumental rationality is a necessity, in that the whole development dynamics stands upon the onset of a 'virtuous circle of competence', that is to say, an incremental diffusion of participative, knowledge-intensive modes of access to cultural opportunities and experiences that create the premises for a systematic improvement of the quality of cultural supply, for the consolidation of the social visibility and the identity-structuring potential of cultural platforms, and in turn for the further expansion of the competence pool of people and of the respective processes of social exchange and dissemination. This kind of dynamics would simply be unthinkable without a strong basis of intrinsic motivation. Also the appropriate modulation of attraction vs. restructuring vs. capability building aspects of the development dynamics is strongly guaranteed by a consistent intrinsic motivational base, which prevents the instrumental focusing on mono-thematic aspects as a form of rent-seeking behaviour that reduces the cultural development goals and practices to convenient excuses. Imagine, for instance, how an instrumental evocation of the competitive restructuring dimension could work for a group of private developers having a huge stock of dismissed factory buildings in their hands, or how the capability building dimension could provide room for lobbying of low-quality educational organizations in search of red tape forms of public procurement, not to speak of the already mentioned issues linked to the instrumental attraction of talents, and so on. Not surprisingly, one finds that when one reasons the other way round, that is, when one does not extract the intrinsic motivation component from best practices but looks for the motivational background in generic cases of culture-led local development, instrumental approaches and lack of social sustainability turn out to be strongly associated (Sacco and Tavano Blessi 2009).

Therefore, the relationship between viable culture-led local development processes and the prevalence of a new, non-instrumental paradigm of rationality is entirely bi-univocal. And the same holds a fortiori as a critical condition for the successful emergence and for the permanence of a SWCD: a pretty demanding condition, indeed – but, as a matter of fact, best practices always rest upon the capacity of attaining goals that for most of the others are way too ambitious.

Concluding Remarks

In this chapter, we have sketched a brief but relatively broad outline of the main issues underlying the expanding field of theoretical and policy studies upon culture-led local development models and processes. We have found that the hallmark of success in this context is the capacity to create the conditions for the emergence and diffusion of a strong basis of intrinsic motivation towards cultural experiences by the whole spectrum of players involved. This amounts to calling for a radical discontinuity in the standards of rationality that the prevailing orientations of current economic theory prescribes as the natural reference, namely, self-interested, instrumentally rational behaviour.

Failing to account for this subtlety in the design and implementation of urban policies oriented to the development of cities as cultural and creative hubs may cause serious harm in terms of social cohesiveness and, in the medium-long run, also of socio-economic sustainability. We hope that this new wave of research and policy thinking about appropriate approaches to the construction of knowledge economies and societies in a post-industrial setting will stimulate politicians and policy-makers to take such issues seriously, and to rethink the current policy agendas accordingly. Culture-led local development is a real opportunity for the well-being of the next future generations, but it deserves a fair chance, and in this case fairness necessarily means taking into account the premises within which culture really makes social (and, eventually, even economic) sense.

Bibliography

Aghion, P. and Howitt, P. 1997. *Endogenous Growth Theory*. Cambridge, MA: MIT Press.

Alvarez, M. and van Diggelen, L. 2005. *There's Nothing Informal About It: Participatory Arts Within the Cultural Ecology of Silicon Valley*. San Jose, CA: Cultural Initiatives Silicon Valley.

Bianchini, F. 1993. Culture, Conflict, and Cities: Issues and Prospects for the 1990s, in *Cultural Policy and Urban Regeneration: The West European Experience*, edited by F. Bianchini and M. Parkinson. Manchester: Manchester University Press, 199–213.

Bianchini, F. and Parkinson, M. 1993. *Cultural Policy and Urban Regeneration: The West European Experience*. Manchester: Manchester University Press.

Bontje, M. and Musterd, S. 2009. Creative Industries, Creative Class and Competitiveness: Expert Opinions Critically Appraised. *Geoforum*, 40(5), 843–52.

Bourdieu, P. 1984. *Distinction. A Social Critique of the Judgement of Taste*. London: Routledge.

Castello, A. and Domenech, R. 2002. Human Capital Inequality and Economic Growth: Some New Evidence. *The Economic Journal*, 112, 187–200.

Caves, R.E. 2000. *Creative Industries. Contracts Between Art and Commerce*. Cambridge, MA: Harvard University Press.

Clark, T.N. 2003. *The City as an Entertainment Machine*. New York: JAI Press.

Currid, E. 2007. *The Warhol Economy*. Princeton: Princeton University Press.

DCMS. 2004. *Sustainable Development Strategy 2004*. London: Department of Media, Culture, and Sports.

Debord, G. 1967. *La société du spectacle*. Paris: Buchet-Chastel.

Evans, G. 2001. *Cultural Planning: An Urban Renaissance?* London: Routledge.

—— 2003. Hard-branding The Cultural City – From Prado to Prada. *International Journal of Urban and Regional Research*, 27, 417–40.

Everingham, C. 2003. *Social Justice and the Politics of Community*. London: Ashgate.

Florida, R. 2002a. *The Rise of the Creative Class*. New York: Basic Books.

—— 2002b. The Learning Region, in *Innovation and Social Learning*, edited by M. Gentler and D. Wolfe. Basingstoke: Palgrave Macmillan, 159–76.

—— 2005. *The Flight of the Creative Class*. New York: Harper Collins.

—— 2008. *Who's Your City?* New York: Basic Books.

Garroni, E. 2010. *Creatività*, with a Preface by Paolo Virno. Macerata: Quodlibet.

Gimeno Martinez, J. 2007. Selling Avant-Garde: How Antwerp Became a Fashion Capital (1990–2002). *Urban Studies*, 44, 2449–64.

Glaeser, E.L. 2004. *Review of Richard Florida's* The Rise of the Creative Class. Available at: www.creativeclass.org.

Glaeser, E.L., Laibson, D. and Sacerdote, B. 2002. An Economic Approach to Social Capital. *The Economic Journal*, 112, 437–58.

Goldbard, A. 2006. *New Creative Community*. Oakland CA: New Village Press.

Graham, B. 2002. Heritage As Knowledge: Capital or Culture? *Urban Studies*, 39, 1003–17.

Hall, P. 1998. *Cities in Civilisation: Culture, Innovation and Urban Order*. London: Weidenfeld and Nicholson.

Hardt, M. and Negri, A. 2000. *Empire*. Cambridge, MA: Harvard University Press.

Harris, C. 1999. *Art and Innovation. The Xerox Parc Artist-in-Residence Program*. Cambridge, MA: Mit Press.

Hesmondhalgh, D. 2002. *The Cultural Industries*, London: Sage.

Horkheimer, M. and Adorno, T.W. 1972 [1944]. The Culture Industry: Enlightenment as Mass Deception, in *Dialectics of Enlightenment*, English edition. New York: Herder and Herder.

Hospers, G.J. and Beugelsdijk, S. 2002. Regional Cluster Policies: Learning by Comparing? *Kyklos*, 55, 381–402.

Howkins, J. 2001. *The Creative Economy. How People Make Money From Ideas.* London: Penguin.

Hubbard, P. 2006. *City*. London: Routledge.

Hutton, T.A. 2006. Spatiality, Built Form, and Creative Industry Development in the Inner City. *Environment and Planning A*, 38, 1819–41.

—— 2008. *The New Economy of the Inner City*. Abingdon: Routledge.

Judoson, R. 2002. Measuring Human Capital Like Physical Capital: What Does It Tell Us? *Bulletin of Economic Research*, 54, 209–231.

KEA European Affairs. 2006. *The Economy of Culture in Europe*. Study for the European Commission.

—— 2009. *The Impact of Culture on Creativity*. Study for the European Commission.

Keaney, E. 2006. *From Access to Participation: Cultural Policy and City Renewal.* London: Institute for Public Policy Research.

Kennedy, M. and Leonard, P. 2001. *Dealing with Neighborhood Change: A Primer on Gentrification and Policy Choices*. Washington, DC: The Brookings Institution Center on Urban and Metropolitan Policy.

Landry, C. 2000. *The Creative City: A Toolkit for Urban Innovators*. London: Earthscan.

—— 2006. *The Art of City Making*. London: Earthscan.

Lloyd, R. 2006. *Neo-Bohemia. Art and Commerce in the Postindustrial City.* London: Routledge.

Malanga, S. 2004. The Curse of the Creative Class. *City Journal*, winter: 36–45.

Marcuse, H. 1964. *One-Dimensional Man: Studies in the Ideology of Advanced Industrial Society*. Boston, MA: Beacon.

Markusen, A 2006. Urban Development and the Politics of Creative Class: Evidence from the Study of Artists. *Environment and Planning A*, 38, 1921–40.

Markusen, A. and Schrock, G. 2006. The Artistic Dividend: Urban Artistic Specialization and Economic Development Implications. *Urban Studies*, 43, 1661–86.

Matarasso, F. 1997. *Use or Ornament? The Social Impact of Participation in the Arts*. Stroud: Comedia.

Mato, D. 2009. All Industries are Cultural. *Cultural Studies*, 23, 70–87.

Miles, S. and Paddison, R. 2005. Introduction: The Rise And Rise Of Culture-led Urban Regeneration. *Urban Studies*, 42, 833–9.

Miller, T. 2009. From Creative to Cultural Industries: Not All Industries Are Cultural, and No Industries Are Creative. *Cultural Studies*, 23, 88–99.

Mommaas, H. 2004. Cultural Clusters and the Post-Industrial City: Towards the Remapping of Urban Cultural Policy. *Urban Studies*, 41, 507–32.

Musungu, S.F. and Dutfield, G. 2003. Multilateral agreements and a TRIPS-plus world: The World Intellectual Property Organisation (WIPO). Available at: www.geneva.quno.info/pdf/WIPO(A4)final0304.pdf.

Nye, J. 2004. *Soft Power: The Means to Success in World Politics*. New York: PublicAffairs.

Pike, A., Rodriguez-Pose, A. and Tomaney, J. 2006. *Local and Regional Development*. London: Routledge.

Peck, J. 2005. Struggling with the Creative Class. *International Journal of Urban and Regional Research*, 29, 740–70.

Peretto, P.F. and Connolly, M. 2007. The Manhattan Metaphor. *Journal of Economic Growth*, 12, 329–50.

Porter, M.E. 1989. *The Competitive Advantage of Nations*. New York: The Free Press.

Putnam, R.D. 2001. Social Capital, Measurement and Consequence. *Canadian Journal of Policy Research*, 1, 41–51.

Roodhouse, S. 2006. *Cultural Quarters: Principles and Practice*. Bristol: Intellect.

Sacco, P.L. 2009. Alchemies that Work: The Baumwollspinnerei Creative Quarter, in *Cultural Quarters*, edited by S. Roodhouse, 2nd edition. Bristol: Intellect, forthcoming.

Sacco, P.L. and Segre, G. 2009. Creativity, Cultural Investment and Local Development: A New Theoretical Framework for Endogenous Growth, in *Growth and Innovation of Competitive Regions*, edited by U. Fratesi and L. Senn. Berlin: Springer, 281–94.

Sacco, P.L. and Soru, A. 2008. *Creative Dismissions*. Venice: Mimeo, CUPR, IUAV University.

Sacco, P.L. and Tavano Blessi, G. 2007. European Culture Capitals and Local Development Strategies: Comparing the Genoa 2004 and Lille 2004 Cases. *Homo Oeconomicus*, 24, 111–41.

—— 2009. The Social Viability of Culture-led Urban Transformation Processes: Evidence from the Bicocca District, Milan. *Urban Studies*, 46, 1115–35.

Sacco, P.L., Ferilli, G. and Pedrini, S. 2008. System-wide Cultural Districts: An Introduction From the Italian Viewpoint, in *Sustainability: A New Frontier for the Arts and Cultures*, edited by S. Kagan and V. Kirchberg. Frankfurt: VAS Verlag, 400–60.

Sacco, P.L., Ferilli, G. and Tavano Blessi, G. 2009. *Culture as an Engine of Local Development Processes: System-wide Cultural Districts*. Venice: Mimeo, CUPR, IUAV University.

Sacco, P.L., Tavano Blessi, G. and Nuccio, M. 2009. Cultural Policies and Local Planning Strategies: What Is the Role of Culture in Local Sustainable Development? *Journal of Art Management, Law and Society*, 39, 45–64.

Santagata, W. 2006. Cultural Districts and Their Role in Economic Development, in *Handbook of the Economics of Art and Culture*, edited by V.A. Ginsburgh and D. Throsby. Amsterdam: Elsevier, 1101–19.

Sassoon, D. 2006. *The Culture of the Europeans: From 1800 to the Present*. London: Harper Collins.

Schumpeter, J.A. 1934. *The Theory of Economic Development*. Cambridge, MA: Harvard University Press. (English translation of Schumpeter, J.A. 1911. *Theorie der wirtschaftlichen Entwicklung*. Leipzig: Duncker and Humblot.)

Scott, A.J. 1997. The Cultural Economy of Cities. *International Journal of Urban and Regional Research*, 21, 323–39.

—— 2000. *The Cultural Economy of Cities*. London: Sage.

Scott, B., Yigitcanlar, T. and O'Connor, K. 2008. Creative Industries and the Urban Hierarchy: The Position of Lower Tier Cities in the Knowledge Economy, in *Knowledge-Based Urban Development: Planning and Applications in the Information Era*, edited by T. Yigitcanlar, K. Velibeyoglu and S. Baum. Hershey, PA.: IGI Global, Information Science Reference, 42–57.

Sen, A. 2000. *Development as Freedom*. New York: Anchor Books.

Simmie, J. 2004. Innovation and Clustering in the Globalised International Economy. *Urban Studies*, 41, 1095–112.

Stern, M.J. and Seifert, S.C. 2007. *Cultivating 'Natural' Cultural Districts*. Philadelphia: Reinvestment Fund.

Swart, J. and Kinnie, N. 2003. Sharing Knowledge in Knowledge-Intensive Firms. *Human Resource Management Journal*, 13, 60–75.

Throsby, D. 1999. Cultural Capital. *Journal of Cultural Economics*, 23, 3–12.

—— 2001. *Economics and Culture*. Cambridge: Cambridge University Press.

Tym and Partners. 1999. *Cultural Facilities: A Study on Their Requirements and the Formulation of New Planning Standard and Guidelines*. Hong Kong: Special Administrative Region, Hong Kong Legislative Council.

UNCTAD. 2008. *Creative Economy Report 2008: The Challenge of Assessing The Creative Economy: Towards Informed Policy Making*. Geneva: UNCTAD.

Valentino, P. 2003. *Le trame del territorio. Politiche di sviluppo dei sistemi territoriali e distretti culturali*. Milan: Sperling Kupfer.

Van Winden, W., van den Berg, L. and Pol, P. 2007. European Cities in the Knowledge Economy: Towards a Typology. *Urban Studies*, 44, 525–49.

Young, C., Diep, M. and Drabble, S. 2006. Living with Difference? The 'Cosmopolitan City' and Urban Reimaging in Manchester, UK. *Urban Studies*, 43, 1687–714.

Wu, C.-T. 2002. *Privatising Culture: Corporate Art Intervention since the 1980s*. London: Verso.

Chapter 12

Rural Areas as Creative Milieus: Evidence from Europe

Aliye Ahu Akgün, Tüzin Baycan and Peter Nijkamp

The Creative Milieu and Creative Capacity of an Area

Creativity studies have a long-standing history in the field of psychology, focusing on the innovative capacity of individuals, groups and organizations – particularly firms. In recent years, however, regional creativity has become the subject of many studies that focus in particular on urban and developed areas. Landry (2000), for instance, defines 'creative milieu' as a place – either a cluster of buildings, a part of a city, a city as a whole or an area – that contains necessary preconditions in terms of hard and soft infrastructures to generate a flow of ideas and invention. In addition, 'creative capacity' means the capability of any region to generate knowledge, and thus to achieve innovation and the diffusion of innovation activity, while ensuring the viability and sustainability of this process (Gülümser et al. 2010).

There are several reasons why such creativity studies focus mainly on urban areas. One is that in such areas the effectiveness of creative capacity can be measured over a long-term period. It is much easier to design creativity theories and to develop the creativity concept for urban areas that have already benefited from their creative capacity for a longer time because of the density of activities, actors and networks. Rural areas are far different from urban areas in terms of such densities; instead, in rural areas, activities and networks are more intense and actors are connected and acting through the intensity of their networks (Gülümser 2009). On the other hand, a creative milieu is a place where outsiders can enter freely, but also feel a state of ambiguity: they must neither be excluded from new opportunities, nor must they be so warmly embraced that the creative drive is lost (Hall 1998: 286). According to the previous description, rural areas can easily be turned into creative milieus, as they are usually open to visitors but defensive to the ones who want to be a part of their daily life.

Rural areas are usually evaluated as less developed than urban areas. Therefore, rural areas are often supposed to have poor living conditions. But, according to Sen (2000), in areas where poverty is obvious, the opportunities rest on the local activity potential, that is, on distinct local capabilities and functionings. Moreover, the establishment ways of creative milieus are nowadays increasingly not (exclusively) technology driven (Landry 2000). Besides poverty, rural areas used to be also largely different from urban areas in terms of technology infrastructures. But, in our modern age, they have increasingly enjoyed the benefits of the information and communications technology (ICT) era and can be distinguished less than in the past from urban areas and cities, apart from their demographic and natural characteristics (Akgün et al. 2010). In recent years, the u-turn of rural areas from abandoned to creative shows also that the achievement of sustainable urbanization as well as a reduction in the disadvantageous look of rural areas can be achieved.

Against this background, creativity theories tend to focus mainly on urban areas, although rural areas show an increasing trend and potential to offer creative milieus. Among examples of such creative rural areas, the members of the Associations of the Most Beautiful Villages in France, in Italy and in Belgium, can be found. Therefore, in our study we aim to pinpoint the factors that stimulate the progress of rural areas. In order to reach our research aim, first an index is created to identify the socio-economic progress of each of the rural villages investigated by means of a standard multi-dimensional analysis technique, viz. principal component analysis, while, next, the critical factors inducing this progress are identified by applying a recent artificial intelligence method viz. Rough Set Data Analysis. The data and information used in our study are retrieved from the results of a survey questionnaire conducted in 60 creative rural areas from Belgium, France and Italy in 2008.

The remainder of this chapter is organized as follows. The second section discusses rural creative capacity as a new approach, while the third section offers insights about the data and the methodology used in this study. The fourth and last section offers a retrospective and prospective view of the study, with some highlights about how to invert rural areas into creative milieus.

Rural Creative Capacity as a New Approach

Regional creative capacity is the starting point of an area's sustained competitive advantage, and its success route. Creative capacity is usually defined as the ability of an area to attract a creative class or creative talents. In recent years, rural areas have started to attract visible densities of such human resources. Clearly, the creative class attracted by rural areas can be different from the creative class referred to by Florida (2002): they are people in action that realize economic opportunities and create their own innovative activities in rural areas.

This attractiveness is not in the first place obtained by technology, availability of infrastructures or job opportunities but rather by the quality of life and locality characteristics of rural areas (Gülümser et al. 2009). The locality characteristics that are seen as the attractiveness are not only beautiful landscapes, local food or local handcrafts but also the diversity of economic activities, the availability of leisure activities and also the socio-economic structure of inhabitants. Therefore, rural areas no longer suffer from the deprivation of their capabilities (such as structural poverty), but may rather be seen as areas of opportunities. Rural areas have in recent years become more dynamic through the use of various intervening opportunities. These transformations will change the position of rural areas in the global market and the competitive arena (Gülümser et al. 2010). In other words, the heartland–hinterland paradigm loses its importance in our knowledge-based era as rural areas also enjoy the benefits from their localities (Brown and Grilliard 1981). In this respect, it is possible to define rural areas as settlements characterized by a unique cultural, economic and social fabric, an extraordinary patchwork of activities, and a great variety of landscapes (Cork Declaration 1996).

Rural areas, which can, more than urban areas, offer quality of life and beautiful landscapes with their diversified uniqueness and preserved resources, may play a crucial role in achieving sustainability and competitiveness in the complex world system. To sum up, the progress of rural areas in the global arena depends on their creative capacity, the mindset of the inhabitants and the use of technology in the area, locality, promotion, cultural heritage, and quality of life in the settlement (Gülümser et al. 2009). To obtain a rise in the socio-economic progress of rural areas, the economic activity – including services in the area – the access to the area and the distance of the area to the nearest centre and also available recreational areas are important, as these characteristics are the factors influencing people in visiting the rural areas.

It is noteworthy that, as a result of the diversified characteristics and the changing perceptions of rurality, including the reversal in mobility, rural areas have recently become some of the most attractive visiting and living places in Europe. Although they are sometimes seen superficially by many visitors as abandoned, underprivileged or poor places, we observe that creative and entrepreneurially oriented visitors have not only passed through but have even developed business ideas and invested in these rural areas. Nevertheless, many newcomers are unaware of the need to change closed cultural and social systems that sometimes tend to have a very defensive sense of community and locality. Landry (2000) claims that an area can be a creative milieu if it contains necessary infrastructural conditions, while Hall (1998) states that a creative milieu must be open for the outsiders to visit, although visitors cannot be a part of the daily life in the area.

Therefore, in the next section, we investigate which factors among the above-mentioned ones are the leading factors to increase progress of a rural area on the basis of the data collected from 60 members of the Associations of the Most Beautiful Villages in Europe.

Rural Areas as Creative Milieus

Prefatory Remarks

One of the successful attempts to make people aware of the opportunities for rural areas has come from France with the establishment of the Association 'Les Plus Beaux Villages de France', which later became the archetype for the Associations 'Les Plus Beaux Villages de Wallonie, Belgium', 'I Borghi più Belli d' Italia',

'Les Plus Beaux Villages de Quebec, Canada', and 'The Most Beautiful Villages of Japan'. Although the idea seems as simple as labeling particular villages as The Most Beautiful Villages in order to turn them into sustainable trademarks that can compete in a modern global economic arena, the process of membership is complicated, strict and very selective, so that not all candidates succeed in becoming members. Therefore, these attempts may be seen as a rural success story that protects the local cultural heritage and the existing membership, while attracting much attention from diverse groups as creative milieus.

The data and information used for our investigation are based on extensive survey questionnaires completed by relevant experts, that is, 60 members of the above-mentioned European Associations of the Most Beautiful Villages. Most data are qualitative in nature and call for a specific statistical treatment. Therefore, a recently developed artificial intelligence method, Rough Set Data Analysis, is applied to identify the most important factors that determine the factors leading to an increase in socio-economic progress of these rural areas.

The data and the information used for our assessment are based on extensive survey questionnaires filled out by relevant experts from 60 villages in Belgium (2), France (19) and Italy (39). The survey began in June 2008 by sending emails to the Associations' websites asking them: (1) to send the questionnaire for the village concerned together with the invitation letter directly to their members, or (2) to provide the contact details of the responsible person in their member villages. The questionnaires were translated into French and Italian in order to avoid any language problems.

The French Association immediately replied to our email, while the Italian Association replied late, as they were delayed by their General Assembly. Only the Italian Association helped efficiently by sending the questionnaire to its members. For the French case, via the website of the Association, we reached 81 French villages out of 152 members while 19 villages replied by posting the completed questionnaire, while in the Italian case we were able to reach 113 members of which 39 villages replied either by fax or by email. The hardest case concerned was the Belgian Association: the different organizational structure and limited profile of the Belgian Association caused difficulties in collecting data. We were able to reach only 10 members, and just two of them returned the questionnaire. Therefore, in the end we obtained a total of 60 returned questionnaires. Our questionnaire has four main parts:

1. general information;
2. environmental characteristics;
3. relations and connections with the outside;
4. membership.

These four parts were designed for specific purposes: Part 1 and Part 2 to reveal the similarities and the differences of the characteristics of the villages, viz. Part 3 to measure the progress of the villages, and Part 4 to evaluate the impacts of the Associations on the villages.

Rural areas are often seen as the leisure places of day-trippers or short-stay tourists, although the attractive image of an area depends not only on the leisure activities but also on its other dynamics and economic opportunities. Increasing socio-economic progress that changes the rural areas into a more developed area increases also the networks of rural areas. Therefore, rural areas can be potentially turned into creative milieus. In other words, rural areas can become creative milieus as their technologies, economies and labour markets are developing. In order to find out the leading factors of such a development, we used Rough Set Data Analysis (RSDA) which is a particular tool for analysing qualitative data on the basis of an artificial intelligence (AI) method. The next section explains the RSDA methodology and its application.

The Methodology and its Application

RSDA serves to pinpoint regularities in classified data, in order to identify the relative importance of some specific data attributes and to eliminate less relevant ones, and to discover possible cause–effect relationships by logical deterministic inference rules (van den Bergh et al. 1997). In principle, RSDA is a non-parametric classification technique (Nijkamp and Rietveld 1999) that has been developed as an AI method for the multi-dimensional classification of categorical data. It was introduced by Pawlak (1982) in the early 1980s and developed by Pawlak (1991) and Slowinski (1992). In recent years, RSDA has become popular in the social and economic sciences not only because of the advantage arising from its non-parametric character but also because of its ability to handle imprecise and qualitative data (Baaijens and Nijkamp 2000, Dalhuisen 2002, Nijkamp and Pepping 1998a and 1998b, Oltmer 2003, Vollet and Bousset 2002, Wu et al. 2004).

In order to deploy RSDA, we composed our dataset by defining 15 attributes under six groups and one decision attribute called 'progress' (Table 12.1). New insights have been gained during the last 10 years about the essential role of resilience for a prosperous development of society (Gunderson and Holling 2002). A growing number of case studies have revealed the connection between resilience, diversity and sustainability of social-ecological systems (Adger et al. 2001, Berkes and Folke 1998). Therefore, the variables used in this analysis seem to be contradictory, as the creativity in rural areas depends on contradictory elements. For instance, rural areas in order to become creative milieus need to achieve tolerance and openness inside their society as well as protecting their traditions, which can be seen as the capacity for resilience of the rural areas. In addition, the rural creative capacity not only depends on the creative sectors or creative economy as it is in urban areas but also on the traditions and values that already exist in rural areas (Akgün et al. 2010). Besides already existing capabilities of rural areas, there are also leisure activities that reflect the creativity of the area and thus, attract visitors to, for example, leisure facilities. Therefore, we used variables related to both economic activities and leisure settings of rural areas.

Table 12.1   Attributes used in the analysis

Code	Group name	Variable	Explanation	Type
A1	Economic activity	Agriculture	1 = yes; 0 = no	Dummy
A2		Manufacturing	1 = yes; 0 = no	Dummy
A3		Service	1 = yes; 0 = no	Dummy
A4	Distance	Distance	Distance to the nearest urban centre: 1 = 0–4; 2 = 5–11; 3 = 10–14; 4 = 15–19; 5 = 20–25; 6 = more than 25 km	Categorical
A5	Mode of Access	Car	1 = yes; 0 = no	Dummy
A6		Train	1 = yes; 0 = no	Dummy
A7		Sea	1 = yes; 0 = no	Dummy
A8		Bike	1 = yes; 0 = no	Dummy
A9	Inhabitants	Local	1 = yes; 0 = no	Dummy
A10		International migrants	1 = yes; 0 = no	Dummy
A11		Urban commuters	1 = yes; 0 = no	Dummy
A12		Seasonal	1 = yes; 0 = no	Dummy
A13	Recreational	Picnic	1 = yes; 0 = no	Dummy
A14		Other	1 = yes; 0 = no	Dummy
A15	Sports area	indoor	1 = yes; 0 = no	Dummy
D	Progress	Positive changes in the rural area	1 = not progressed; 2 = progressed; 3 = very progressed	Categorical

Although we obtained condition attributes coded by the letter 'A' from the results of the questionnaire, the decision attribute coded by the letter 'D' is obtained by the creation of a progress index. In order to identify different progress levels, we generated a progress index by applying a principal component analysis (PCA), which is used to transform the set of originally mutually correlated variables into a new set of independent variables. PCA is a non-stochastic approach and it mainly deals with the assessment of the common variance of the original variables. To calculate the progress index, we used five positive changes in rural areas:

Table 12.2   Commonalities to measure the progress in rural areas

Variable	Extraction
Inhabitants use more technology and their talent in their job	0.863
Number of leisure activities increased	0.844
Number of recreational facilities increased	0.812
Back-to-tradition increased	0.783
Number of job opportunities increased	0.726

the increase in the number of leisure activities, of recreational facilities, of job opportunities, in the use of technology and talents together by inhabitants, and in the willingness to return to tradition (Table 12.2). The use of such variables allowed us to use the word 'progress' to refer to the rural area as a creative milieu.

The results of the PCA show that the calculated index is composed of one component that is able to explain 80.6 per cent of the total variance. In addition, the results of this analysis also highlight that economic changes are relatively less important than the junction of technology and talent, which is the identifier of the creative capacity of the rural area (Table 12.2). Moreover, we classified these different progress indexes into three categorical levels, viz. not progressed, progressed, and very progressed. The progress index appeared to range between -2.37 and 1.48. Therefore, we classified the negative scores as not progressed (see Appendix).

Consequently, we were able to obtain an adjusted data table (see Appendix). After obtaining this new table, RSDA can be performed. In order to perform RSDA, a modular software system Rough Set Data Explorer (ROSE) was used in order to implement the basic elements of rough set theory and rule discovery techniques. This software was created at the Laboratory of Intelligent Decision Support Systems of the Institute of Computing Science in Poznan by Predki, Slowinski and Stefanowski in 1998 (Predki et al. 1998, Wu et al. 2004). There are also other attempts to create software for the application of RSDA, e.g. ROSETTA, but ROSE is the most user-friendly software to apply RSDA.

The basic idea in RSDA is to describe the data with rough sets (Rupp 2005). A rough set can be characterized as a set for which the classification of a group of certain objects is uncertain (Dalhuisen 2002). In our study, we defined important factors that are often associated with different progress levels of rural areas. In the application of RSDA, three main steps based on rough set theory must be carried out, viz. pre-processing, attribute reduction and rule induction.

The first step is pre-processing. This step enables the researcher to inspect the quality of classification and the accuracy of each of the categories of the decision attribute. This is done by dividing the lower approximation by the upper approximation of each category. In other words, if the quality and the accuracy of

Table 12.3   Approximations

	Approximations	Accuracy	Upper level	Lower level
1	Not progressed	1	23	23
2	Progressed	1	23	23
3	Very progressed	1	14	14
Accuracy of classification				1
Quality of classification				1

classification are lower than 1, then the chosen data and examples in the sample are not fully unambiguous concerning their allocation to the categories of the decision attribute. This information strengthens the conclusions made on the basis of the other steps of the RSDA. The results of the first step show that the rural areas in our sample are fully discernible regarding the three levels of progress (Table 12.3).

The second step of RSDA – the reduction – is used to form all combinations of condition attributes that can completely determine the variation in the decision attribute without needing another condition. In other words, in this step, minimal sets of attributes are found and these are called reducts. While finding reducts, RSDA can also find the frequency of appearance of all condition attributes in the

Table 12.4   Frequency of attributes included in the analysis

Attribute	Frequency	
	#	%
Agriculture	19	100.00
Distance	19	100.00
Seasonal	19	100.00
Picnic	11	57.89
Service	10	52.63
Train	10	52.63
Other recreational	10	52.63
Indoor	10	52.63
Manufacturing	9	47.37
International migrants	8	42.11
Urban commuters	6	31.58

Table 12.5 Rules of the analysis

Rule	Progress	Cases #	Cases %	Economic activity	Distance	Inhabitants	Recreational facilities	Sport facilities	Access mode
R1	Not progressed	5	21.74	No manufacturing No services			Other	No indoor	No train
R2	Not progressed	7	30.43	No manufacturing		No commuter	No picnic No other		Car
R3	Progressed	7	30.43	Agriculture	20–25 km				
R4	Progressed	4	17.39	Agriculture Manufacturing		No commuter Seasonal			
R5	Progressed	9	39.13	Agriculture				Indoor	
R6	Very progressed	5	35.72	Agriculture	0–4 km				

reducts. If, among them, one or more attributes has a frequency of 100 per cent, this is called the core. The result of the second step in our case is that there are four combinations of condition attributes that determine the variation in the different progress levels (Table 12.4). In addition, three condition attributes, viz. agriculture as the main economic activity; distance to the nearest urban centre, and seasonal inhabitants are the core elements which are included in each of the 19 reducts. This step shows that the cores are the most important attributes to determine the different progress levels, while the other attributes are relatively less important.

The third and last step is rule induction. This provides rules that explain both the exact and the approximate relations between the decision and the condition attributes. An exact rule guarantees that the values of the decision attributes correspond to the same values of the condition attributes. Therefore, only in that case it is always possible to state with certainty if an object belongs to a certain class of the decision attribute. In addition, if a rule is supported by more objects, then it is more important, for instance, in classifying the different rural areas. In our RSDA application, we excluded rules which are supported by less than four cases. Therefore, we were able to formulate six exact rules that are supported by more than three cases (Table 12.5).

Therefore, the two rules that explain why there is no progress in the area are as follows: *Rule 1*: If there is no manufacturing and services as economic activity but there is recreational facilities without indoor sport facilities, then there is no progress in the area; *Rule 2*: If there is no manufacturing, no commuter, no recreational facilities but access by car, then there is no progress in the area.

Among six rules, there are four rules that explain that somehow there is progress in the area. Therefore, these rules are: *Rule 3*: If the area is 20 to 25 km away from the nearest urban centre, then there is progress in the area; *Rule 4*: If the area has manufacturing as an economic activity without any commuters, then there is progress in the area; *Rule 5*: If there is indoor sport facilities in the areas, then there is progress in the area; *Rule 6*: If the area is located between 0 to 4 km away from the nearest urban centre, the area is very progressed. In the next subsection, we will discuss the results of the analysis and also the above-mentioned six rules.

Results

Rural areas are becoming more creative every day and they could become the future creative milieus as they are already the dreamed living environment of many people. In order to investigate the important factors for being a creative milieu, we applied RSDA analysis. The results of the analysis showed that the existence of an agriculture sector and the distance of the rural area to the nearest urban centre are the most important factors for explaining the progress of rural areas.

Furthermore, the results of the RSDA analysis provided the rules for identifying progress in rural areas. According to the rules, the main economic activity is very important for differentiating a progressed rural area from one that has not progressed, while having indoor sport facilities for inhabitants and visitors is also as important as the main economic activity. According to the rules, we can also state that the more the rural area has progressed also holds for the existence of indoor sport facilities. In addition, the accessibility of the rural area has a contradictory effect on the progress of rural areas. Therefore, we cannot state a specific indicator for the distance or for the types of inhabitants.

We can state that the economic development and leisure activities for outsiders are very important for a rural area to convert itself into a creative milieu. Therefore, rural areas can definitely have a place in the global arena. However, we focused mainly on already creative rural areas and our sample is very limited, so to generalize our findings is not appropriate, as there are still an immense number of rural areas suffering from a lack of economic development. But our results offered important highlights on what to do to convert rural areas into creative milieus. In the last section, we will conclude our analysis by discussing highlights obtained from the results.

Retrospect and Prospect

Creativity has become one of the main identifiers of the development of an area. Although urban areas are often defined as creative milieus, recent trends in mobility have proven that rural areas can also be named as creative milieus. Therefore, in this study we aimed to investigate leading factors that stimulate the progress of rural areas to become creative milieus.

Rural areas that offer the necessary preconditions with a high level of hospitality and defensive localism at the same time can become the ideal creative milieus with a relatively small effect on the global market. From this point onwards, we conducted our research in 60 creative rural areas from Europe that are members of the Associations of the Most Beautiful Villages. To investigate these creative milieus, we focused on the changes in creativity and attractiveness that occurred in these areas first by applying principal component analysis (PCA) to obtain a progress index with the aim of classifying rural areas and second by applying RSDA to investigate the role of leading factors for the progress of such areas. The results of the PCA showed that the combination of technology and the local capacity of the rural areas are very important in classifying a rural area as a creative milieu and, thus, classify it as progressed or not. In addition, the results of the RSDA showed that the economic development of the rural areas is the most prominent factor for being a creative milieu.

Clearly, focusing mainly on already attractive and known villages prevents us from developing an overall picture for all rural areas. Nevertheless, this study provides insights at least into why certain villages have become 'creative milieus' in Europe. On the other hand, the study also highlights the importance of economic development for rural areas in order to obtain a diverse creative and sustainable system. But, the economic development in rural areas cannot be achieved independently from urban areas, but rather the city and the village should be seen as a unique, complex and dynamic system. In other words, urban and rural regions cannot be considered separately and so policies cannot create a sharp distinction between urban and rural regions. Therefore, subsequent research may focus on modelling a system that can indicate how to increase the progress and the continuity of rural areas, while developing strategies to turn rural areas into creative milieus with their urban surroundings.

Bibliography

Adger, W., Neil, P., Kelly, M. and Huu Ninh, N. 2001. *Living with Environmental Change: Social Vulnerability, Adaption, and Resilience in Vietnam.* London: Routledge.

Akgün A.A., Baycan-Levent T., Nijkamp P. ve Poot J. 2010. Roles of Local and Newcomer Entrepreneurs in Rural Development: A Comparative Meta-Analytic Study. *Regional Studies.* DOI: 10.1080/00343401003792500.

Baaijens, S. and Nijkamp, P. 2000. Meta-Analytic Methods for Comparative and Exploratory Policy Research: An Application to the Assessment of Regional Tourist Multipliers. *Journal of Policy Modelling*, 22(7), 821–58.

Berkes, F. and Folke, C. (eds) 1998. *Linking Social and Ecological Systems. Management Practices and Social Mechanisms for Building Resilience.* Cambridge: Cambridge University Press.

Brown, L. and Grilliard, R.S. 1981. On the Interrelationship between Development and Migration Processes, in *Geographic Research on Latin America: Benchmark 1980*, edited by T.L. Martinson and G.S. Elbow. US: CLAG, 357–73.

Cork Declaration. 1996. The Cork Declaration – A living countryside, paper presented at The European Conference on Rural Development. Cork, Ireland. 7–9 November.

Dalhuisen, J. 2002. *The Economics of Sustainable Water Use: Comparisons and Lessons From Urban Areas.* PhD Thesis, Amsterdam: Vrije Universiteit.

Florida, R. 2002. *The Rise of the Creative Class.* New York: Basic Books.

Gülümser, A.A. 2009. Rural Areas as Promising Hot Spots: Sustainable Rural Development Scenarios. PhD Thesis. Istanbul Technical University, Istanbul, Turkey.

Gülümser, A.A., Baycan-Levent, T. and Nijkamp, P. 2009. Beauty is in the Eyes of Beholder: A Logistic Regression Analysis of Sustainability and Locality as Competitive Vehicles for Human Settlements. *International Journal of Sustainable Development*, 12(1), 95–110.

—— 2010. Measuring rural creative capacity, *European Planning Studies*, 18(4), 545–64.

Gunderson. L.H. and Holling, C.S. (eds) 2002. *Panarchy: Understanding Transformation in Human and Natural Systems.* Washington, DC: Island Press.

Hall, P. 1998. *Cities in Civilization*. New York: Pantheon.

Landry, C. 2000. *The Creative City*. UK: Earthscan.

Nijkamp, P. and Pepping, G. 1998a. A Meta-analytic Exploration of the Effectiveness of Pesticide Price Policies in Agriculture. *Journal of Environmental Systems*, 26(1), 1–25.

—— 1998b. A Meta-analytical Evaluation of Sustainable City Initiatives. *Urban Studies*, 35(9), 1481–500.

Nijkamp, P. and Rietveld, P. 1999. *Classification Techniques in Quantitative Comparative Research: A Meta-comparison*. Research Memorandum, Vrije Universiteit, Amsterdam.

Oltmer, K. 2003. *Agriculture Policy Land Use and Environment Effects: Studies in Quantitative Research Synthesis*. PhD Thesis, Amsterdam: Vrije Universiteit.

Pawlak, Z. 1982. Rough Sets [J]. *International Journal of Computer and Information Sciences*, 11(5), 341–56.

—— 1991. *Rough Sets: Theoretical Aspects of Reasoning about Data*. Dordrecht: Kluwer Academic Publishers.

Predki, B., Slowinski, R., Stefanowski, J., Susmaga, R. and Wilk, S. 1998. ROSE: Software Implementation of the Rough Set Theory, Rough Sets and Current Trends in Computing. *Lecture Notes in Artificial Intelligence*, 1424, 605–08.

Rupp, T. 2005. *Rough Set Methodology In Meta-analysis – A Comparative and Exploratory Analysis*. Discussion Papers in Economics, Instituts fur Volkswirthschaftslehre, Technische Universitat Dramstadt.

Sen, A. 2000. *Social Exclusion: Concept, Application, and Scrutiny*. Social Development Papers No.1, Office of Environment and Social Development Asian Development Bank, Manila, Philippines.

Slowinski, R. 1992. *Intelligent Decision Support: Handbook of Applications and Advances of the Rough Sets Theory*. Dordrecht: Kluwer Academic Publishers.

van den Bergh, J.C.J.M., Button, K., Nijkamp, P. and Pepping, G. 1997. *Meta-analysis in Environmental Economics*. Dordrecht: Kluwer Academic Publishers.

Vollet, D. and Bousset, J.P. 2002. Use of Meta-analysis for the Comparison and Transfer of Economic Base Multipliers. *Regional Studies*, 36(5), 481–94.

Wu, C., Yue, Y., Li, M. and Adjei, O. 2004. The Rough Set Theory and Applications. *Engineering Computations*, 21(5), 488–511.

Appendix   The data table

ID	Village	Country	Progress	D	A1	A2	A3	A4	A5	A6	A7	A8	A9	A10	A11	A12	A13	A14	A15
1	Lagrasse	France	0.39	2	1	1	0	20	1	0	0	0	1	1	1	0	1	1	0
2	San Donato V di C	Italy	0.71	3	1	1	0	9	1	0	0	0	1	1	1	1	1	0	1
3	Gordes	France	-2.37	1	1	0	0	20	1	0	0	0	1	1	0	1	0	0	0
4	Morano Calabro	Italy	0.26	2	1	0	0	0	1	0	0	0	0	0	0	1	1	0	1
5	Saint Lizier	France	-0.83	1	0	0	0	2	1	0	0	0	1	0	1	0	0	1	0
6	Fources	France	-0.35	1	1	0	0	12	1	0	0	0	1	1	0	1	1	0	0
7	Novara di Sicilia	Italy	-0.07	1	0	0	0	0	1	0	0	0	1	0	0	0	0	0	0
8	La Flotte-en-Re	France	1.00	3	1	0	0	4	1	1	0	0	1	1	0	0	0	1	1
9	Gourdon	France	0.10	2	0	0	0	9	1	0	0	0	1	0	0	1	1	1	1
10	Bova	Italy	0.40	2	1	0	0	7	1	0	0	0	1	1	0	1	1	0	0
11	Bienno	Italy	0.54	2	0	0	1	0	1	0	0	0	1	0	0	0	1	0	1
12	Volpedo	Italy	0.54	2	1	0	0	10	1	0	0	0	1	1	1	0	1	0	1
13	Neive	Italy	1.02	3	1	0	0	2	1	0	0	0	1	0	1	0	0	0	0
14	Zavattarello	Italy	-0.99	1	0	0	0	42	1	0	0	0	1	0	0	1	1	0	1
15	Fagagna	Italy	0.40	2	1	1	0	15	1	0	0	0	1	0	0	0	1	0	0
16	Castel di Tora	Italy	1.48	3	1	0	0	31	1	0	0	0	1	1	1	1	1	0	0
17	Geraci Siculo	Italy	0.87	3	1	0	0	15	1	0	0	0	1	0	0	0	1	0	1
18	Civ. di Bagnoregio	Italy	-0.99	1	0	0	0	1	1	0	0	0	1	0	0	1	0	0	0
19	Gradara	Italy	0.27	2	0	1	0	4	1	0	0	0	1	1	0	1	0	0	0
20	Cusano Mutri	Italy	0.56	2	1	0	1	40	1	0	0	0	1	0	0	1	1	0	1
21	Mombaldone	Italy	-0.52	1	1	0	0	22	1	1	0	0	1	0	0	1	0	1	0
22	Borgio Verezzi	Italy	-2.37	1	0	0	0	2	1	0	0	0	1	0	1	0	0	1	0
23	Castel del Monte	Italy	1.16	3	1	0	0	45	1	0	0	0	1	1	1	1	1	1	1
24	Furore	Italy	1.16	3	0	0	0	7	1	0	1	0	1	0	0	0	1	0	0

ID	Village	Country	Progress	D	A1	A2	A3	A4	A5	A6	A7	A8	A9	A10	A11	A12	A13	A14	A15
25	Saint Quirin	France	-1.80	1	0	0	1	18	1	0	0	0	1	0	1	1	1	0	1
26	Orvinio	Italy	0.10	2	1	0	0	45	1	0	0	0	1	1	1	0	1	1	1
27	Tourtour	France	-0.08	1	0	0	0	25	1	0	0	0	1	1	0	1	0	0	1
28	Giglio Castello	Italy	-0.22	1	0	0	0	0	1	0	0	0	1	0	0	0	0	0	0
29	Stilo	Italy	-1.12	1	0	0	0	0	1	0	0	0	1	0	0	1	0	0	0
30	S.terre de R.	France	1.48	3	1	1	0	35	1	1	0	0	1	1	0	1	1	0	1
31	Oramala	Italy	-2.37	1	1	0	0	4	1	0	0	0	1	0	0	0	0	0	0
32	St. Agnes	France	0.90	3	0	0	0	8	1	0	0	1	1	0	1	1	1	0	0
33	Vernazza	Italy	0.56	2	0	0	0	14	0	1	0	0	1	0	0	0	0	0	0
34	Cutigliano	Italy	-0.66	1	1	0	0	40	1	0	0	0	1	1	1	1	1	0	1
35	La Roche-Guyon	France	-1.60	1	1	0	0	20	1	0	0	0	1	0	1	1	0	0	0
36	La Bas. Clairence	France	0.43	2	1	0	0	7	1	0	0	0	1	0	1	0	0	0	1
37	Asolo	Italy	0.08	2	0	1	0	18	1	0	0	0	1	1	0	1	0	1	0
38	Moresco	Italy	1.02	3	1	0	0	10	1	0	0	0	1	1	1	1	1	0	0
39	Montsoreau	France	-0.06	1	0	0	0	14	1	0	0	0	1	0	1	1	1	0	0
40	Pet. s Gizio L'Aqu.	Italy	0.71	3	0	0	0	7	1	0	0	0	1	1	0	1	1	0	0
41	Bettona	Italy	-2.37	1	1	0	0	15	1	0	0	0	1	1	1	1	1	0	0
42	Le Bec Hellouin	France	-0.21	1	0	0	0	6	1	0	0	0	1	1	1	1	1	1	0
43	St. Benoit du Sault	France	0.09	2	0	1	0	20	1	0	0	0	1	1	0	1	1	0	1
44	Ars en Re	France	0.39	2	1	0	0	30	1	0	0	1	1	0	0	1	1	0	1
45	Crupet	Belgium	0.23	2	1	0	1	20	1	0	0	0	1	0	1	1	0	1	0
46	Chardeneux	Belgium	0.37	2	1	0	0	16	1	0	0	0	1	0	1	1	1	0	0
47	Pietracamela	Italy	0.57	2	0	0	0	20	1	1	0	0	1	0	0	1	1	0	0
48	Campo Ligure	Italy	-2.37	1	0	0	1	35	1	0	0	0	1	1	1	1	1	1	1
49	Navelli	Italy	-0.06	1	1	0	0	35	1	0	0	0	1	1	1	0	1	1	0
50	Mirmande	France	0.55	2	1	0	0	20	1	1	0	0	1	0	0	1	1	1	0

Appendix continued   The data table

ID	Village	Country	Progress	D	A1	A2	A3	A4	A5	A6	A7	A8	A9	A10	A11	A12	A13	A14	A15
51	Belves	France	1.01	3	1	1	0	35	1	1	0	0	1	1	1	1	1	0	1
52	Montefioralle	Italy	-0.34	1	0	0	0	2	1	0	0	0	0	0	1	0	0	0	0
53	Canale	Italy	-0.83	1	1	0	0	7	1	0	0	0	1	0	0	1	0	1	0
54	Chiusa	Italy	0.55	2	1	0	0	0	1	1	0	0	1	1	1	0	0	0	1
55	Roussillon	France	0.72	3	0	0	0	10	1	0	0	0	1	1	1	1	1	0	1
56	Brisighella	Italy	0.09	2	1	0	0	0	1	1	0	0	1	1	1	1	0	0	1
57	Massa Martana	Italy	-0.22	1	1	0	0	10	1	1	0	0	1	0	0	0	0	0	1
58	Ricetto di Candelo	Italy	1.16	3	1	0	0	0	1	1	0	0	0	0	0	0	1	1	0
59	Buonconvento	Italy	0.40	2	1	0	0	25	1	1	0	0	1	1	1	1	0	1	1
60	Offida	Italy	0.55	2	1	0	0	0	1	0	0	0	1	0	0	0	1	0	1

PART IV
Creative Cities:
New Methodological Approaches and Planning Instruments

Chapter 13
City Design, Creativity, Sustainability

Francesco Forte

Continuity in Urban Design and Planning

The 1950s and 1960s

By the mid-1960s the social processes characterizing the modern metropolitan industrial city were well known. They had been studied by the ecological school of Chicago in the first half of the century, emphasizing the correlations between spatial factors and social behaviour, connected to income, race, and the industrial sector (Bogue 1950, Dickinson 1947, Geddes 1915, Hatt and Reiss 1957, Le Corbusier 1967), and they had been specified further as a result of the body of knowledge gained in trying to decode the connotations of the modern city (Chapin and Weiss 1962, Dobriner 1958, Friedman and Alonso 1964, Mumford 1961, Nairn 1965, Reissman 1964, The Pittsburgh Regional Planning Association 1963, Webber et al 1964).

Dominance and sub-dominance, influence and sub-influence, distribution by gradients and circular rings, optimizing logic applied to the choices of the users of land, the invasion of the pre-existing and social succession – all these multiple aspects investigated by researchers, aimed at decoding what was occurring, were thought to characterize the social dynamic of the contemporary territorial articulation of the industrial city. The sustained urban population growth caused by immigration and shifting from the rural world to the city formed the scenario in which the processes were manifest. The desire for privacy drove the behaviour of the social groups therein, who were already beneficiaries of the *affluent society*, and this fuelled the suburban expansion of the settlements, and giving land use as an anthropological connotation, and the simultaneous change in the social and architectural equilibrium typical of the historic settlement. Therefore, the low density 'suburb' experienced by the affluent population and the high-density construction blocks aimed at promoting social housing to satisfy the needs of the *poor society* were the bipolar modalities that have characterized the landscape of life in the modern metropolis and megalopolis.

In the 1960s the face of cities such as Boston (Boston Redevelopment Authority 1966), showed these characteristics, which were secondary relative to the 'signic' (Lynch 1962) attraction of the great monuments of the tertiary sector or the public buildings present on the skyline of the cities, but which characterized the form and the settlement structure of the metropolis, by spatial extension of the processes, the role of installing infrastructure, the underlying problems, and the amenities available. Significant research has described its character, for example, the work by Hoover and Vernon on New York in 1959 (Hoover and Vernon 1959), or *Megalopolis* by Jean Gootman in the 1960s (Gootman 1961). In volumes published towards the end of the 1960s, this intellectual elaboration is frequently reported, citing authors – such as L. Rodwin, K. Lynch, B. Frieden, J. Dyckman, M. Webber, C. Bauer Wurster – of permanent relevance in comprehending the form of the structure that had developed and that of the structure of the form of both metropolitan and urban settlements (Aquarone 1961). The social role of planning was quite clear as expressed by Melvin Webber (see Figure 13.1[1]).

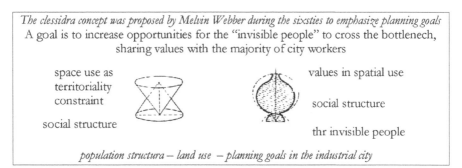

The clessidra concept was proposed by Melvin Webber during the sixsties to emphasize planning goals
A goal is to increase opportunities for the "invisible people" to cross the bottlenech, sharing values with the majority of city workers

space use as territoriality constraint

social structure

values in spatial use

social structure

thr invisible people

population structura – land use – planning goals in the industrial city

Figure 13.1   Melvin Webber's theories on the role of planning

The process was not neutral in relation to public policies. This item had to become a research field, as decisions were being made in government.[2] The succession of parliamentary reports over time aimed at exploring the correlation between policies and phenomena.

The all-embracing nature of possible public action was emphasized, in order to limit the inefficiency that had characterized the administration by sector and by separation (Rodwin 1961). So 'city managers' were established to whom appropriate powers and the necessary urban decisions were assigned. And in some cities the first civic movements arose to affirm that 'citizen's rights' should be taken into account in making the decisions that the public function of governance had to make. The first advocacy planning movement dates back to that very period, in the United States, that recorded the first change in the modalities of

---

1   The bottleneck that separates the two bulbs of the hourglass, whereby the intrinsic imbalance of urban society has been interpreted, was represented in Forte *Metodologia urbanistica, ricerca operativa, modellistica urbana* (1972), following theories on the role of planning processes elaborated in the 1960s by Melvin Webber.

2   In the United States the former Housing and Home Finance Agency was transformed into the Department of Housing and Urban Development in the mid-1960s (public law 89–174, 79 Atat 667, 5 U.S.C, 624).

the relationship between governors and citizens, shifting towards the practice of sharing 'deciding together' (Forte, 1971) (see Figure 13.2). And this acquired multiple awareness had implications for the image of the territory informed by the body of knowledge involved in the project, because of the implicit bipolar form of structure, as manifested in the strategic plans for the metropolitan development of Greater London in the 1950s, and of Boston, Washington and San Francisco in the United States, in the 1960s.

Columbia Distrct Regional Plan, Washington D.C., 1961

a new town within the Disctrict. Reston, Virginia. Creativity shaping natural environment

Boston, Mass., Urban Renewal The town center model, 1966

Baltimore in the sixties, looking to the future.
The renewal center city plan, reusing the historical port area.

Figure 13.2   Citizens' participation and strategic plans for metropolitan development: some examples

When these two phenomena occurred in the urban areas of that period, there were no doubts about the future of the city of historical industrialization. In the advanced Western democracies the objective set is the containment, on the one hand, of expansion, through new city planning policies – the Town Planning Act in Great Britain dates back to 1947 – or, on the other hand, of degradation – the Renewal Act in the United States dates back to 1949. In some metropolitan contexts success was pursued through specific modalities of the intersection between innovative research, productive development, spatial order, as in Silicon Valley, California, or along the metropolitan circular Route 128 of the city of Boston in Massachusetts in the USA.

In the Mediterranean city the proposed objective was reconstruction, the strengthening of housing provision, the adjustment of the historic city to the demand for basic infrastructure following the increase in the use of the private vehicle, a more marked industrialization. Little attention was reserved for the cultural heritage of the Italian Mezzogiorno, thought to be 'dependent' on other economic sectors typical of modernity (Compagna 1967), according to the logic of the fragility of the tourist economy.

The 1970s and 1980s

In the 1970s and 1980s, there occurred a momentous revolution in the relations between production, consumption, settlement and community culture, supported by technological innovation in informatics and biomedical science, as the characteristics of what would later be defined as post-modernity became consolidated. The structure of flows, based on worldwide commercial transactions and organized by global networks, supported by technological innovation in informatics and telecommunications, replaced the structure of places and their general function. A vast territory has interposed itself between flows and places, as vast as the man-made environment is global, and globalization has interjected itself (Salamon 1990). And in replacing categories, the differential territorial dynamics that were found in nations have gone out of focus. In Europe the 'blue Banana' emerged, like the 'sun belt regions' in the United States. The affirmation of the 'Pacific Tigers' imposed innovative goals for the traditional industrial nations. In researching motivations for processes and reasons for public policies, categories have become established that innovate normal considerations of the structure. The sociological dimension has joined the economic-productive dimension; the environmental-ecological dimension has supported the landscape dimension; the plurality of subjects and stakeholders has joined single decisional structures; the efficacy of governance and the vertical and horizontal subsidiarity has contributed to the efficacy of the government. It has been understood that efficiency may not be realized if it is pursued without these supportive contributions (see Figure 13.3).

In some privileged areas, where geopolitical considerations have accompanied city planning considerations, the originality of the conditions that were maturing was realized; but in the majority of places this realization came slowly. In the 1970s in Baltimore in the United States there was consistent attention to urban policies that were necessary because of the effect of the emergence of the Pacific nations on the global market, and of the innovative structure that the Panama Canal assumed at the onset of an approaching new century. In Great Britain the dismantling of the Port of London in order to make room for the urban re-conversion of the area of Greater London was being decided as an alternative to the new city policy for open space proposed by Peter Hall in 1963 (Hall 2003); the

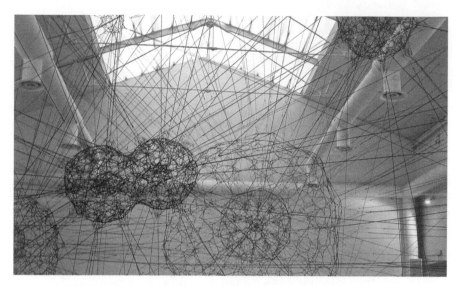

Figure 13.3 Thomas Saraceno, Argentine artist working in Frankfurt am Main in Germany, International Art Exhibition sponsored by the Venice Biennale in 2009. The staged installation art well represents the sociological condition resulting from globalization

proposed objective was densification by developing a new city inside the existing city. In France in those years, new types of urban policy were being formulated, replacing the practice of the metropolitan districts, consolidating in this context the policies of regional decentralization and of the new municipalities and cities that formed the new Paris in the following years. In Barcelona, Pasqual Maragal, having become Mayor, founded the Institute for Metropolitan Studies, aimed at supporting the innovation of the urban governance, contributing pragmatically to what had been learnt from the pre-existing similar centre founded at Johns Hopkins in 1974. John Dyckman, Professor Emeritus, left Berkeley, to establish the World Industrialization Observatory at Lille. The observation of the urban project searched for certainties to confirm the role of history, ideas and models (Choay 1980). In Italy in those years there was too much preoccupation with research into the conditions for governance, the necessary 'parallel convergences', in the connotations to be assigned to the new regionalism, and in 'governing events', to be able to pay attention to the pressing global processes. And in some regional contexts, such as Campania in the Italian Mezzogiorno (Forte 1979), the lack of attention to these processes was motivated by the commitment necessary to confront dramatic 'emergencies', such as the cholera epidemic in the 1970s and, in the early 1980s the reconstruction following earthquake damage (Forte 2006).

Comments on Mutation

The Role of Local Atmosphere to Promote Change

Those who first perceived the fresh eastern wind, blowing around the world, anticipated the race to acquire a role in carrying out the changes, so giving a *creative* response to the necessity for innovation determined by the new geopolitical condition. Freedoms and that condition of legality that protects its proceedings have created a favourable habitat for the consolidation of the process.

Flow economics, taking place in a context of competition, rivalry, with locational flexibility, have without our realizing it supplanted the self-sufficiency of locations. The economic activities have reflected the restructured role of manufactured products, by the effect of re-conversion of process or product, externalization of functions, and territorial delocalization. Immaterial production has expanded following the necessity to conceive innovation of process or product through research and by launching new ways of organizing the processes of consumption and production. And, likewise, there is the correlation between rights, legality, decision-making, foresight, structure and strategy, structure and workability, bad governance or good governance. This implies not just the conformity to the rules but above all their appropriateness to the historical period (Sternleib 1990).

As a result, there has been an impressive growth in the service sector, for families, for industries, for governance, as units that influence the socio-political identity, in their turn interacting with the modifications in demand and the modalities of consumption. The impressive growth in the need for *communicative interaction* has stressed the role of producers of immaterial services, connected to the expansion of the demand for education, research, entertainment, information, free time, individual and communal consumption, and also commerce and financial services to support the decision-makers. Finance has come centre stage with the establishment of new ways of organizing savings and accumulation. In a global catchment area, a role has emerged for specialized niche production typical of small and medium sized firms that in Italy have used creativity to adapt to changing expectations and lifestyles by searching for new markets.

Considerable social transformations have accompanied productive innovation with consequences for the socio-political and sociological dimension, whose nature re-emerges in the contemporary city through the nostalgic dimension of what used to be (see Figure 13.4).

The primary necessities have widened, by investing in the right to work, to safety, to exercise liberties, to competition and rivalry, together with the need for urban hygiene, clean air, unpolluted water, for a cooperative and supportive city, for participation, for the quality of the environment, and for a sharing and co-deciding geared to decision-making on action plans.

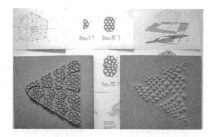

Figure 13.4   **A molecular social structure, based on the freedom to be, appears to be a character of the contemporary urban region's inhabitants. The shapeless metropolis acquires a structure by the edges of cells, settlement components without coherence. Plan goals should avoid fragmentation, give identity to edges, increase an interchange between cells**

At the same time, the break-up of the Soviet Union, exemplified by the demolition of the Berlin Wall (1989) and the beginning of the second phase of European integration, generated significant legislative innovations from the Parliament of the Italian Republic regarding the restriction on development, protecting nature and the landscape and environment: in other words, concerning new citizen's rights.[3]

There have been some positive effects on the economy of some cities consequent to these changes. Accompanied by marked internationalization processes, the hierarchical processes already operating in the urban structures of nations have been accentuated. As pointed out by Richard Florida (2004), world, national and regional cities have become important, with a role that goes beyond their population size in the urban hierarchy. Instead, they are connected to the type of activity, and their 'logo' has tended to synthetically characterize the predefined action guidelines of 'metropolitan cities' in order to attract activity, investments and population, with the objective of competing in a world of cities (ANCE 2005).

The role attributed to the 'event' has been acquired as an expression of creativity in governing (Barcelona,1991: the Olympics and the regeneration of the city).

---

3   Legge 18 Maggio 1989, n. 183 'Norme per il riassetto organizzativo e funzionale della difesa del suolo'; legge 8 Giugno 1990, n. 142 'Nuovo ordinamento delle autonomie locali'; legge 7 Agosto 1990, n. 241 'Nuove norme in materia di procedimento amministrativo, e di diritto di accesso ai documenti amministrativi'; legge 6 Dicembre 1991, n. 394 'Legge quadro sulle aree naturali protette'.

An Effect on the Structure of Urban Social Change: Revitalization of Historic Urban Centres

An urban population of service providers has acquired space in the city scene in that period. And in establishing themselves as major actors of the structural process, they have come up with ways of being, feeling consuming and communicating. The role of 'creative' has been assigned to this social group, reductively proposed as a 'creative class', interpreters of discontinuous and innovating values that were previously the domain of the enlightened bourgeoisie that guided the second Industrial Revolution. It was a case of generations of innovative service producers, capable of transferring all their creative feeling into the productive process, together with the needs that emerged from the production revolution that was occurring. Among the ways of expressing oneself originally, the originality in the choice of localization in the urban space was recognized. The people who are the 'creators' of innovation have attributed a role to the usual locational advantages of central urban sites typical of historical cities, and consequently of historical settlements. These advantages ensue from proximity and accessibility, restoring meaning to symbolic values connected to memory and to its concrete manifestation. The process of the revitalization of historic cities has been facilitated by the marginalization of significant parts of the urban settlement after the exodus of residents and activities. And gentrification has become consolidated, in the definition given by Ruth Glass in 1964 of what was occurring at the time in London, whereby a 'gentleman' would reuse the buildings of the lively 'flea market' of Notting Hill, because this use attracted visitors to London. The already-known processes of social substitution made possible by social mobility tended to increase, as a result of the increment of differential revenue of selected areas of the city, and the connected locational cost. New services of habitation have accompanied the social re-utilization, strengthening the attractiveness of sites. Many of the *gentrification* processes in the historical cities are the result of this innovative individuality of styles and ways of life impacting on locative choices. In many contexts that characterize historic cities, the social use of space has been disturbed by the imposition of locational choices that affect the fruition of goods and the demand for qualifications. As a result of such processes, these findings do not always receive positive comments: for example, the critical reflections on the historic centres of Prague or Florence.

The necessities imposed on governments by the re-conversion of cities' historic role has inspired public policies aimed at supporting and encouraging *gentrification*. Many of the successful practices in the revitalization of historic centres have gone hand in hand with the process of social dynamics based on the establishment of *creative* social groups. The public policies of urban renewal of the central areas of cities; the relocation of the functional and productive role of the port areas of historical cities, as has occurred in London, Marseille, Barcelona, Valencia, Bilbao, Baltimore and San Diego; the programmed re-utilization of disused industrial sites resulting from the first urban industrialization, both through the production of lofts, that is houses re-designed in their typology, and connected services using private capital, which through mixed innovative use have in some cases been configured as cities within the pre-existing city, for users able to appreciate the semi-central location in the building form structure, as has occurred in New York and Milan; the inclusion within the historic settlement of mass transport networks, a public policy chosen in Naples; the reformulation of laws of the pre-existing urban legislation; all these lines of action have characterized the public policies that have accompanied and supported the configuration of the historic urban space, revitalized by the *creative people* through new creative decisional processes.

Complex processes have accompanied the social re-utilization of the areas impacted by the gentrification that has accompanied productive modernization. Among these processes, an important role has been assumed by the tourist use of historic cities and the related economics, with the resulting dichotomy between the restoration of monuments, and the quality of services to residents, a dichotomy that has been reinforced in recent decades as an effect of normative partial definitions of state–regional duties, and communal duties. The resulting social fragilities have been foreseen, as in the considerations on urban renewal proposed by W. Alonso (1964, 1965). Where they have been incorporated in public decision-making, they have had the effect of giving rise to two-pronged public actions aimed at both human capital and cultural capital, widening the possibilities for conservation and renewal.

At that same time, the creative spontaneity typical of gentrification was accompanied by innovation in the government of the urban space, pursued through policies of intense public residential building, organized through districts located in the peri-urban space or in new cities contained in the metropolitan space. In commenting on that time, the decision-making wisdom of some of the policies promoted by local authorities demands attention, as is evident when the innovation has related places to access networks, habitable areas to productive areas, and the lifestyle of the user and the public decision-making both in the peri-urban districts and in the new cities (Quaroni, 1963).

Acting *creatively* has permeated every aspect of contemporary society. Creative action implies the ability to transform the conception consequent to the necessity, to the intuition, or to the dream of concrete practicability of making a success, of businesses, families, government institutions, in a dynamic context of: expansion, market transactions and improvement of the welfare conditions, and even of changes in political expectations (Osborn 1967). Creative action is fuelled by urban inter-ethnicity and by the multiculturalism consequent to the transmigration and nomadism implicit in globalization. Creative action is shrewd: it studies the horizon, calibrates the effort, and interprets the actors and the decisional processes in order to be efficient, and therefore incisive and proactive. It is set against the simplicity of the conceptual process, which is self-referencing and explicitly inconclusive. Industriousness can be inconclusive, but may also promote processes that effectively influence the living conditions of selected actors, through the development of the creative process stimulated by presumably meaningful ends. An intimate correlation holds true between the growth of a productive economic system and the verifiable diversification in the organization and technology of processes and products, the social implications and the urban multi-ethnicity, and the ability of governments of areas or nations to promote favourable conditions with the establishment of innovative behaviour and creative action. And, moreover, a correlation may exist between the evasion of communitarian rules, or practices founded on legality in their enactment, and the exercise of *amoral* creativity, as is the case of the organized criminality 'system', which established itself in the 1990s in the regions of the 'Mezzogiorno Italiano', characterized by violence and arrogance, and which is a source of insecurity for urban life. And it has been necessary to ascertain the connection between the establishment of this amoral creativity and the irresponsibility in the exercise of public functions, which is the effect of the shabbiness and decisional division, as transmitted by the failure by the government in Campania with the rubbish cycle, or the dissipation of public financial resources. The amoral creativity has in some cases cast the state as its actor, hurrying through its deliberating organs' decisional models in which the exercise of the public function has tended to reward specific interests.

Environmental Sustainable Development and Design

The perception of 'Limits to Growth' has emerged since the 1970s through the exploration undertaken by the Club of Rome, but the inspiration for the *sustainable city* emerged in the 1990s. In the first decade of the new century the necessary instrumentation has reflected on public policy aimed at encouraging the energetic self-sustainability of cities, neighbourhoods, districts and housing units, basing itself on the separation of consumers–producers and network administrators. *Energy certification* has become mandatory for all new buildings, through innovation in national laws. Wind farms have become elements of landscapes, with positive reverberations in the harmony of natural landforms. Photovoltaic power plants are economically feasible as a result of national and regional policies giving incentives to investors.

The aspiration for a sustainable city (Fusco Girard and Forte 2000) has introduced an original critical condition in interpreting the needs, impacting on the symbolic formalization that typifies the creative city, and its creative people. Creativity as an expression of human energy, aimed at solving contradictions of context, has established itself as a process and product for tackling uncertainty. And this observation has led to a perception of the global flow network as a manifestation of a new process of energy consumption, with associated waste impacting on the human habitat, or rather on the condition of the soil, water and air that we use, which all presents opportunities for the exercise of creativity. Innovative methods for satisfying the needs and rights of the post-industrial community have been outlined, and they have characterized successful situations in the government of urban change following the conception of the sustainable city (Fusco Girard and You 2006). 'Strong thought and action' about urban development has characterized some of the government modalities practised. Clarity of the medium and long-term goals to be pursued in local government through growth has promoted comfort and reduced uncertainty, with a positive influence on success. Both architecture and city planning have been shaken by this experiment in strong thought.

The convergence between public intentions and symbolic formal aspirations has characterized the experience of the relations that have taken place between people and power. The industriousness that was made explicit during the first and second Industrial Revolutions was full of creativity, and reaches its apex when it addresses the primary human necessities, codified by the *rights* sanctioned through the conquest of civil conditions. The cities of the feudal kingdom conceived by Luigi Vanvitelli in Caserta on behalf of his customer, King Charles III of Bourbon, has intrinsic creativity in the vision that it transmits, where the configuration of the form of the structure associates the administration of the kingdom with the symbolism of the royal palace.[4] And, moreover, this creativity is present in the 'Città Borghese', which was later built in Caserta, as a result of the enterprise of a productive bourgeoisie, operating under a legitimating scheme of the public rules sanctioned by the liberal monarchic state, which through the law protects the activities of individual liberties. These are the histories of the negative conditions that promote innovation and rules, and creative discontinuity in the rules already in use, as testified by the flow of theory and action that pervades urban architecture, on this basis creative innovation promoting 'the rational city' of modern plans. It can be interpreted as an intense de-contextual episode, inspired by amorous practices, aimed at solving the perception of hardship through cognitive processes, of consciousness acquisition, of consequent effective action, public or private (see Figure 13.5).

Ebenez Howard in his *Garden city of To-morrow* (1902) efficiently interpreted the needs in the theoretical new city, and in the productive decentralization (see Figure 13.6).

The safeguarding of the rights of the neighbour promotes the urban 'rule', which with the designed representation interprets the 'form' of the built or constructible environment, as in the project for the modern city of the twentieth century, which the events of Amsterdam well represent (see Figure 13.7).

As has already occurred, the establishment of new ways of articulating the relationships between social aspirations and production has imposed answers even on the symbolism typical of the urban and therefore on the manifestations in the sensitive form of the modalities of space production (public, communitarian or private).

---

4   Creativity in city design during the eighteenth century was tested in Caserta in Italy, St. Petersburg in Russia, and Washington, DC in the USA.

Figures 13.5a and 13.5b
>   In Caserta, in Luigi Vanvitelli's project, matrix and symbol
>   are both present in the conception, but with weakening of
>   the vision in carrying out the creative process, or rather the
>   administration, and substituting the symbol of the 'city of
>   equality' represented by the factory city. The photograph
>   shows the old factory of San Leucio

Figure 13.6   E. Howard. Welwyn Garden City was created on the basis of the theory of the 'garden city of tomorrow', exploring the advantages that the urban profits can generate if the city is governed by the association of interests. The figure shows the form of the structure of Welwyn Garden City (Howard 1898, 1902)

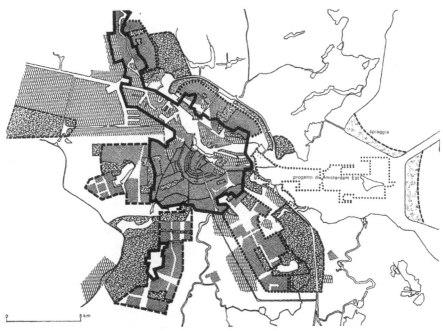

Il piano regolatore generale di Amsterdam è stato ap-
provato nel 1935, ed è stato eseguito nei trent'anni suc-
cessivi; quindi oggi lo possiamo giudicare come una rea-
lizzazione concreta.

Figures 13.7a and 13.7b
         Van Esteren's plan for Amsterdam, 1936, revisited in the 1950s
         (Morbelli 1977)

The anthropomorphic aspirations proposed in the 1960s by E. Saarinen must
be brought back to these spatial influences (e.g. in the TWA Building, New
York, in the gymnasium of the University of Yale, New Haven, Connecticut)
(see Figure 13.8).

Figures 13.8a and 13.8b
The inspiration in the works of Eero Saarinen

This inspirational motif has been picked up by Z. Hadid (the serpent of the HST (High Speed Train) shaking through the desert of the planned *technological park* of Afragola, on the periphery of Naples), and of Renzo Piano (the tortoise shells of the Rome auditorium). The search for the role of the fourth dimension in configuring the structure of form characterizes the elaboration of Frank O. Gehry qualifying and focusing on the way of production of the public space in a context pervaded by the symbolism underlying the sensitive form; and characterizes, moreover, the inspiration for the metropolitan acropolis that Richard Meier transmitted through the Getty Centre Museum of Los Angeles (see Figures 13.9 and 13.10).

The theory of the religious fence applied in Los Angeles by Josè Rafael Moneo in configuring the public area provides a picturesque reinterpretation of the original neo-gothic building (see Figure 13.11).

In the production undertaken by public investors, the role attributed to the vernacular is reflected in the renewal of the historic settlement of Ponticelli in Naples, a project that began in the second half of the 1980s. But the inspirational motif of the pre-selected modalities of action must be seen in the light of the theory of 'innovative conservation' of the historic settlement, in turn an expression of the decoding of sociological aspirations, not connected to the transactions of the market and therefore to the preferences of the consumer (see Figure 13.12).

Figure 13.9   Richard Meier, Getty Center Museum – Los Angeles, 2004

In the production carried out by private investors, the spatial style connected to its use by emerging 'creative people' has stimulated vernacular architectural production, or rather of architectural signs and forms resulting from transformation of parts of the city already in urban use, guided by the usual culture of structure and form of the buildings (see Figure 13.13), expression of habits and cultural interpretations common to ordinary social subjects placed in the consumer society as a result of an acquired knowledge and sometimes applied with creativity (Herzog 2006).

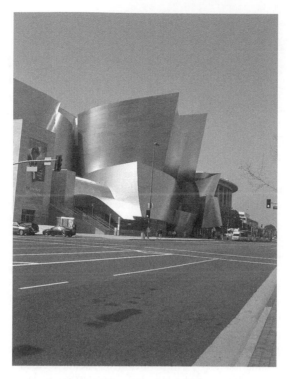

Figure 13.10 Frank O. Gehry, Walt Disney Concert Hall, 2005

Figure 13.11 Josè Rafael Moneo, the cathedral of Our Lady of the Angels, Los Angeles, 2002

Figure 13.12 The historic settlement of Ponticelli in Naples as it appears after renewal

And a stratification of signs of popular vernacular origin characterizes the identity of the 'new city' that inspires Leon Krier's search for the historical picturesque, or the movement for the 'new city planning' as theorized in the United States by Peter Calthorpe (Calthorpe and Fulton 2001) (see Figure 13.13).

Policy planning and city architecture had a meaningful sense in Berlin, as pointed out by Hans Stimman (2009). 'Architecture functionalism' has oriented its reconstruction during the 1960s. But a city order does not create values (counterpart of the Naples shape, lacking in order, but rich in values, in spite of the tensions of modern life).

Berlin became a city with no memory at its centre. The reconstruction of Berlin put into practice the break with history, defined by a famous critic in 1964 as 'assassination city'. The solution of social need in Berlin has been identified with the 'solution to the housing needs' (Engels 1872), characterized

Figure 13.13 'Vernacular' architecture in San Diego, California, USA

by ribbon development along the arterial roads leading from the city centre, and setting aside the theory of neighbourhood designed in the modern movement. The spatial conception of the pre-modern European city, and how to build pre-modern, represents a culture that traced explicit choices for the future. In Berlin the cultural break with the functionalism approach has emerged in the 1960s, with the logo 'a future for the past'. The logo 'live in the centre' adopted by Institute of Berlin Architects (IBA) initiated the confrontation, and represents an intensive creative cultural approach. The logo requires critical thinking about by the reciprocity of the contemporary values with the values of the past, assumed as a theoretical foundation for city and architectural design It is not sufficient to implement this through the architecture for the city, but it is necessary to design the architecture of the city. On this basis, despite the split in interpreting the how to do, a large part of the Baroque town was rebuilt. The process has affected the city centre. With the aim of constructive policy from 1996 to 1999 the city plan was drawn up for an area of 30 square kilometres and 300,000 inhabitants. The Senate in May 1999 approved the plan of the 'inner city' (Mitte, Berlin's historic City Centre), for the development and construction of the city centre. The plan did not pursue a break with tradition, but instead continued the ongoing dialogue between tradition and innovation to develop Berlin as a European City. It urged a meeting between property developers and real estate. It was necessary to have a thorough elaboration of the project, and the Senate worked hard on this. Under the plan the city inside has established the paradigm of modern building inherited from the Congressi Internazionali di Architettura Moderna (CIAM), involving the separation of functions, network traffic, the size of urban green spaces, the removal of private property for the benefit of social housing, with the goal of building a new post-industrial city.

Urban Plan Rules

In the care demanded by building regulations, the modalities of consumption chosen by the creative people have been important in legitimating innovative *rules*, operating in the recovery, even at times the conservation, of the existing architectural heritage, such as in the portioning and fusing of pre-existing properties, in the reuse of attics, in the technological adjustment aimed at economizing on energy, operations made admissible in the city planning and building guidelines drawn up by the Italian councils in the 1990s and, recently, aimed at maintaining the existing architectural heritage and improving its performance thanks to sensitive reuse.

By carefully observing the historic settlement of our Mediterranean cities, previously used strategies and new opportunities can be inferred. And in relating these strategies to the rigidity of the local planning regulations one can verify original modalities for reusing space, for revisiting the historical heritage, or rather an adaptation of land use antithetical to the planning regulations, put in place to defend subjective survival conditions, which are frequently the origin of contradiction with other significant values, of history and memory. The art of getting by these very actions explains the phenomenology typical of the operations of planning architecture, which always returns a conceptual activity based on real needs, which a more marked correlation between form and substance should allow comprehension giving rise to innovation concerning the rules sanctioned by the 'public function'. The occupation of public land for the protection of houses at street level should be interpreted as a request for innovation in the public responsibilities of urban conservation practices, acquiring the form of a request for denied citizenship rights and not only as the repression of illegality. And, analogously, illegal occupation of residences should be related to the inadequacy of the public function in pursuing locational aspirations, demanding adjustments in the rules governing the use of urban space. Only critical judgement is able to discriminate between the reasons for operating, and is solicited in its specificity by the context conditions, both civil and spatial. The regulations sanctioned by public institutions have characterized the planning of the modern city. The regulations have interpreted the legitimate limitation of the *enjoyment* of property, by regimenting aspirations, and behaviour. Addressing the unfolding of creative action through the organization of the administrative function and connected assessments demanded by the procedure and the bureaucracy which takes care of its progress, the regulation has revealed itself as further support for conformism, homogenization, stasis and even illegality, and so it is an obstacle in the unfolding of creative potential. In antithesis to this *controlling state*, some writers have drawn attention to the objective of making light and easy the form–norm–decision correlation. Simplifying rules to invigorate forms has become an area of attention for research in creating the positive conditions for the affirmation *of a city for citizens* launched in the process of conscious *liberalization*, with clear impacts for city planning. It is a direction strongly conditioned by local culture, immersed in the locally manifest social cultural and political contradictions, but is still only outlined, and often denied by decisions aimed at strengthening the controlling ability, based on the perception of civil incapacity for self-regulation in configuring positive reciprocal relationships, among interests, values and individuals.

The contradiction between controlling state and free markets has been resolved through the affirmation of programmed fragmented legality inserted through the policy of sectors, and the tightening of control on activities not contemplated by the sectors. The institution of an *agreed programme* stipulated in the Italian legislation by law n. 142 of 1990, and reinterpreted in regional Italian law extending to the private individual, the co-participation in the agreement, with a majority ratification, interprets the spirit of the fragmentation programme very well. The city of fragments and of sectors relegates to a marginal role the need for form and structure, encouraging action by sectors and functions, the concomitant detachment of parts, giving rise to agglomerations devoid of vertebrae and structure. Even this consideration must be brought back to the impact of transformations induced by creative people, giving rise to *a creative administrative practice* manifestation of, and intention to, mediate instances. But we must be conscious that the sacrificial lamb in this process is the conception of the 'indivisible city', which in the culture of city planning has been explained in the 'organism city', and then in the unitary project for territory and city, of the 'organic everything' as conceptualized by Luigi Piccinato (1948).

A causal relation exists between the legally binding general city plan and the style of the local communal government. In other words, there is an opposition between the legally binding and the dynamic process of decisions. The city that intentionally promotes forms of structure codified by the conformative plan is mediated, thought out, designed, and provides rules, basing on these the hope for quality of life, peace of mind, of action, and variety of choice. The city that directs processes based on industrious hope needs coherences that follow from 'vision-scenarios', to be reformulated progressively with correlated decisional processes verified through contextual 'evaluations' (Fusco Girard and Nijkamp 1997). And in this perspective it is reasonable to pay attention to action guidelines that outline desirable futures of settlements typifying the 'organism city' founded on the innovative porosity of their structure. Traces of innovative directions to explore can be seen in the role attributed to the 'scenery' or vision in the long period of planning cities. The planned horizon is the base of this decision methodology, and the territory as a 'horizon' broadens the notion of form and structure; it widens the scope of the reciprocal relationship with the territorial morphology. In the new ensemble, the parts assume the role of the group structure, as strategic choice or complement, and therefore are never autonomous, as in the organic analogy with cells and membranes. One perceives once again the territory of the differences articulated by the relationships of density, mass, flow intensity, and not by elements or fragments devoid of structure. In this context, the new role to be attributed to regional landscape planning makes sense, as justifiably announced in the Italian Republic by the 'Cultural Heritage and Landscape Code' (Codice dei Beni Culturali e del Paesaggio, 2004), based on principles in conformity with the European Landscape Act (Convenzione Europea sul Paesaggio, Florence, 2001), making it mandatory to have a regional landscaping plan, with regulation based on compatibility judgement, verified by 'landscape relationship'. And it makes sense to re-propose the meaning of 'conservation' as already proposed by Roberto Pane, as aimed at combining 'permanence' and vitality, tying urban policies regarding the historical heritage to this perception.[5]

Traces of innovative directions to be investigated can be seen in the new role attributed to national, regional and local, fiscal and associated monetary policies, wherever in conformity with the dynamism and tensions that characterize the condition of markets and individuals in a global society (Burdett 2006).

In this perspective, the 'urban and territorial perequation' assumes a role as theorized by us (Forte 2000, 2004). We have proposed the 'perequational approach' as a method for evaluating the quality of the planning and structural propositions regarding the future city, and therefore as a principle of city planning, or the 'planning criterion' as defined by the Italian Constitutional Court. The method

---

5   See Forte 2008b.

involves smoothing the foundations of the modern state by seeking tax equity. Equity in the use of urban land has led us to assume the equalization criteria in public decisions on land use, with implementation devolved to private investors.

This approach underpins structural convergences with legal regulations on land rents, land development costs and development rights, as set out by American researchers such as Harvey M. Jacobs (2008, 2009). And we therefore consider it erroneous to treat the 'city planning perequation' as an instrument, to be used to specify the way to implement the transformations. This thin distinction is little understood, but the regions of the Italian Republic that have comprehended it, embodying it in legislation, have definitely accelerated their entry into the network of transnational flows, and therefore increased their chances of success.

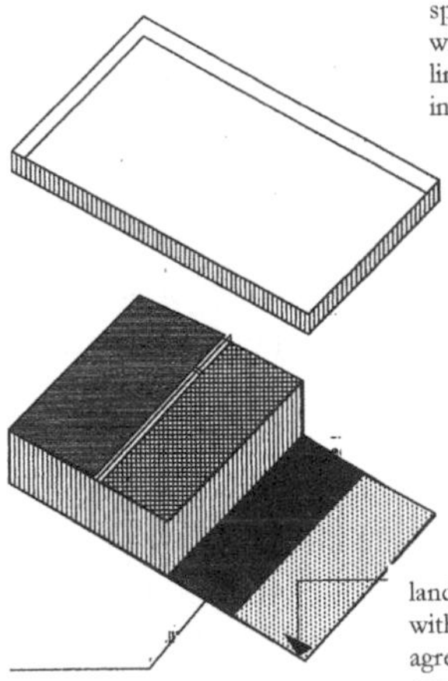

An ordinary development right is made admittable within city plan rules designed by a local government, effective on a land surface of a spatial unit called "distric". All properties within a district have a building capacity non linked to activities functional definition, for instance X mqsus/mqst

The building capacity is cumulated through translation from the total surface to a selected land surface defined by a land zone. Within a functional zone are designed public spaces defined by surface standard, roads, squares and parking areas, and the building land area and its surface. A specific Index measures the building capacity admitted on the functional zone, and on the building land area.

$Jt = X\ mqsus/mqSt$

The urban development right is measured by an Index: Jt, square mt. of floor space for meters of land surface

land to be transferred to public properties without payment and on a private-public agreement, to be put in use for activities of general public interest, like public parks, or facilities of general community interest

land surface regulated in its consistency by law standard, to be transferred to public properties without payment on the base of a private-public agreement, to be put in use for gardens, sport, parking, local streets, integrated with functional uses (residential, services, productive activities like hotels, manufacturing, commercial or amenities centres).

A maximum development right Index may be defined. Differential between maximum and ordinary development Index is aimed to give incentives for private action to produce private social homes, or to transfer land to public properties; as to regulate home consistency production under the public administration control.

CRITIRIA HAVE BEEN FORMALLY DEFINED WITHIN THE LAND GOVERNMENT LEGISLATION ENACTED BY ITALIAN REGIONS IN THE LAST DECADE

Figure 13.14 Perequation criteria: Zoning in urban planning

Perequation Principles

The whole settlement is structured by components, characterized by districts of change. The implementation of an *urban district* by transforming previous land uses, undertaken by private entities, must be accomplished without charge to the municipality, or other public entities, as regards land property. Private real-estate developers working on a district have to implement facilities, equipment, parks, social housing and other public works or facilities of public interest, as defined in the zoning enshrined in an urban development plan of the city (see Figures 13.14, 13.15 and 13.16).

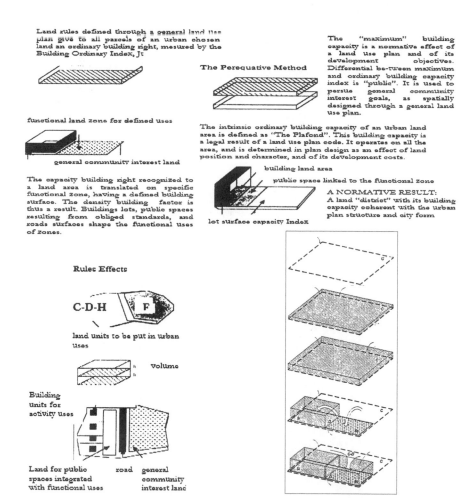

Figure 13.15a   **Perequation criteria applied to urban land rights**

Figure 13.15b   **The transfer of development rights aimed to the foundation of a development district (bottom right)**

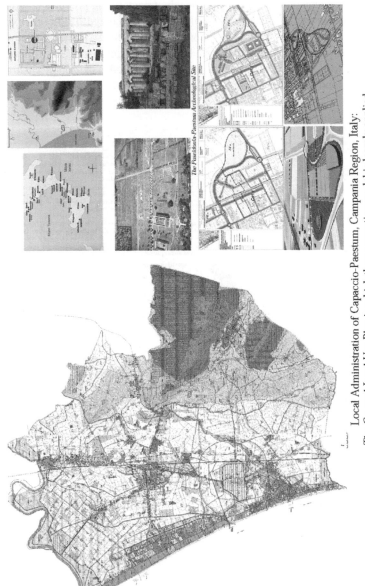

The Poseidonia-Paestum Archeological Site

Local Administration of Capaccio-Paestum, Campania Region, Italy:
The General Land Use Plan in which the perequative model is largely applied
Prof. F. Forte Planning Office, 2008-2010

Figure 13.16a

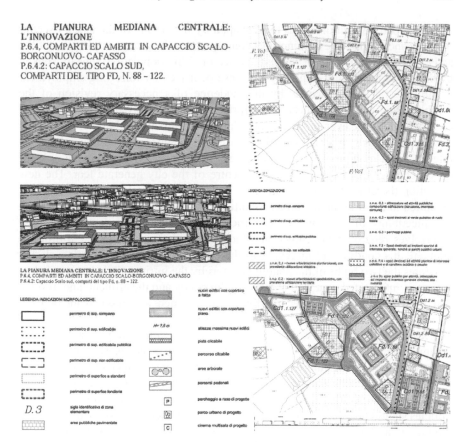

Figure 13.16b **The perequative model applied to the General Land Use Plan of Capaccio-Paestum (Campania region, Italy, 2008–10)**

The Contemporary Age: The Deconstruction

A Future with Hope, Threats and Opportunities for City Design

The alterations to what we have tested following the damage to the tested processes of accumulation, production, consumption and saving that has taken place in recent months have particular impact for the urban population. The crisis has hit production activity mainly located in cities, and, because of causal chain reactions, it has had an impact on families, on consumption, instilling fear or hope. With good reason Giulio Tremonti recently entitled one of his books 'Fear and Hope' (*La paura e la speranza*, 2008). Sergio Rizzo, commenting on the research carried out by Cittalia, the Associazione Nazionale dei Comuni Italiani (ANCI) foundation for research, has stressed that 'Cities are afraid. Of the crisis' (2009).

The urban population seems to be shaped by the culture of the tertiary sector, within a social structure of minorities consequent to the great complexities of the activities that participate in the tertiary sector. This population appears shaken by the fall in the quality of life, by the sense of insecurity following employment instability and actual income, by the evidence of a necessary revision of the symbolism of the great industrial enterprise, by the fragility of housing policies and social policies, even more than by criminality and multi-ethnicity.

The emphasis placed on urban insecurity appears instrumental to an intention that manipulates the effectiveness of change. The consequent risks of de-structuring[6] that overshadow the structure of the city generate fear. The new condition therefore focuses attention on the creative modalities that need to be exploited in confronting the urban condition that currently seems to be spreading the uncertainty that follows the de-structuring process. Considerations impose themselves regarding the choice of which public policies to enact, in order to alleviate this condition of insecurity of the city, and to widen the bottleneck that separates the two bulbs of the hourglass whereby the intrinsic imbalance of urban society is interpreted, that is, that between the needy and the affluent, and through which a role has been attributed to the urban government and to social and housing policies.

De-structuring impacts on the poorest urban community restricting the communication between the two bulbs, suffocating tensions typical of social improvement, denying future prospects to small businesses that still today give a face to the city. Public policies should, therefore, try to control and attenuate the effects of de-structuring, persevering in the search for sustainability of change. Policies should be effective in the short term to influence the de-structuring of the job market and of enterprise, but should also create favourable conditions for the establishment of innovative and original structural conformations, resulting from a new creativity centred on the needs highlighted by the de-structuring process through the selective concentration of investments and fiscal equity. De-structuring focuses attention on alternative values to processes, products and consumption inspiring the many planning *arrangements* resulting from the ineffective local government, or the utopian approach of many land use plans, or the very 'strong experiences' practised in directing the evolution of the frequently unsustainable post-industrial city. The decisions that will be taken in the city government will have an effect on the *city*, in its communitarian values, and in its structure and form, influencing directives that will shape the characteristics of the following five-year periods of the new century.

---

6 By 'de-structuring' is meant forces that destroy a successful equilibrium pursued within an enterprise or a family.

Design within the Crisis, a Path for a Creative Approach

The modalities through which to promote the satisfaction of needs, by improving the structure of the settlements, therefore propose themselves as a significant reflection of the new human condition resulting from a de-structuring of post-modernity. And a new source of creativity must be unfolded in order to put into practice in this new project the theory that recognizes the necessary conditions of knowledge, hope and optimism of intention. The attention to the *modalities* and to the rules has made it possible to prevent the principle of freedom from transforming into arrogance, arbitrary judgement and conflict, with clear manifestations of this in the body of knowledge committed to the project, and therefore on the form of the structure and the structure of the form of the metropolitan city. In the attempt to limit the uncertainty that overshadows the road to the future, two contrasting fundamental guiding principles are outlined, to which decisions should be directed, and which follows the principle of freedom by influencing policy.

First, there is the principle of *rigidity*, which refers to checks typical of the autocratic state, of the bureaucratic conformation of the administration, of the fear of innovation and so of the slowing down of creative action promoted by creative people. It is the inauspicious direction of 'cascade' planning, of the region–province–council model that denies institutional autonomy, legitimated by public institutions that are not very reliable in their scientific sagacity, to majorities of hybrid coalitions regarding their aspirational principles, whose exponents, although elected, are known to have lost – because of processes that have taken place – the popular recognition that related them to democracy.

As an alternative, there is the second principle, which is explained by self-regulation, based on civil consciousness, economic–financial, legal, and technological and organizational, and on the organization of the management of the public institution of government aimed at acquiring light and speedy processes, also taking advantage of the technological apparatus available at the current time. The consequent synthesis of these two principles should pursue the equilibrium of expectations typical of industrious dialectic democracies, which fuses principles of efficiency and effectiveness, placing limits, through replacement powers, to the inefficiency of the government of autonomy. Something that tends to uphold the hypothesis has become established in the laws of the Italian Republic, though the reformulation of the Fifth Title of the Republican Constitution.

These two alternative pathways play a substantial role in highlighting the connotations of the form of the territorial structure that the activities will have; of the concrete expression of a social structure with an accentuated diversification of values; and of the structure of form that will explain the landscape of the future city, by conditioning the contradictions that are consequent to the process of de-structuring. The clear statement of the non-negotiable and, vice versa, of the negotiable appears to be the essence of this necessary convergence.

The existing built city structures promise a large resource, which puts local authorities in a position to seek private investment in works of adaptation. This policy choice may support the construction economy, and the dissemination of products resulting from the uptake of solar technology, water recycling and energy consumption. In this perspective, we understand the success that has been obtained during 2009 by means of the exceptions to existing rules introduced through the urban land laws enacted by all the Italian regions, with brief proposals for what are called *House Plans*. The limits that can be glimpsed in making these experiences were revealed. Fragility and weaknesses still exist due to lack of awareness of differences between values, the heterogeneity of spatial structures, the insufficient evaluation of correlations between transportation and urban land use. A more acute awareness of this heterogeneity could promote greater consistency in the way of action made permissible by public laws.

These findings highlight the role that we should attribute to knowledge of the origin of the diversity of spatial order. The role of the *land survey* within a site information system acquires new meaning as a support, for effective decisions. Public investment in adapting the network to the mobility of persons and goods should be re-examined within the crisis condition. Public investment in urban rail networks testify temporal non-matching programmes under condition of crisis, calling attention to the more effective use of existing networks of road and rail mobility, their proper maintenance, renewal of incentives for private means of transport, parking facilities.

The notion of non-negotiable and, vice versa, of the negotiable requires the consciousness of the differences of values, of the heterogeneity that permeates the spatial values, of negotiation of the enhancing practices, of the necessary relationship between public transport and urban use of the land, adjusting the recognized and catalogued heterogeneous ways of acting, always legitimated by public function.

All this should be pursued both in the conservation policies of the historic settlement and in the planning policies of transformation in open spaces, which assumes more marked aesthetic intention in goals that underlie the implementation of architecture-city planning. The conscious relationship between means and ends, that has characterized the rationality of the industrial city should continue to be a value of an efficient local government, even in directing the adaptation of the structure of the urban form and its management in confronting the accentuation of uncertainty.

Setting the target beyond the current de-structuring should involve the improvement of the management of local public services, in terms of efficiency and efficacy, and of the network of multiple services that characterize cities, refocusing attention on the criteria for housing developments consequent to the nature of the services, uses, cost of production, access rules (Forte 2008a).

There could be a reinterpretation of the theory of human settlement nuclei that was proposed in the *garden city*, assigning the role of production of biomasses or the production of compost resulting from the rubbish cycle to the greenbelt, already previously contemplated.

Proximity to footpaths and canals may compensate for the high densities required by some buildings. This is already seen in the use of thermal currents generated by channelling by turbines to achieve self-sufficiency for cooling and heating (see Figure 13.17a, 13.17b, 13.17c).

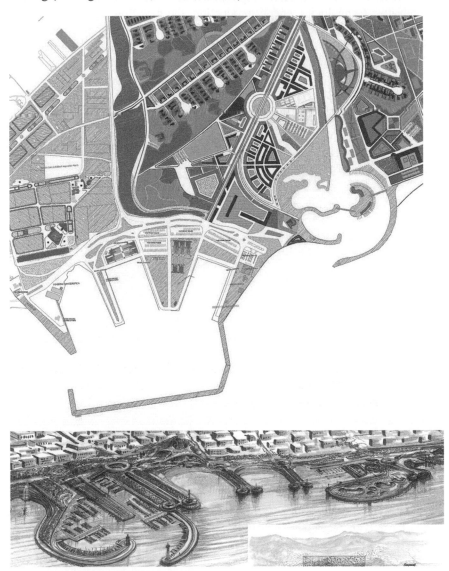

Figures 13.17a and 13.17b
Design approaches explored, consistent with the definition of problems

Source: Prof. Forte's Planning Office Archives.

Dams protecting ports from marine kinetic energy could be reinterpreted as containers for the production of energy, using solar and wind power as a wind and photovoltaic field, as well as the energy from the tides. New port cities along the Mediterranean coast could be re-proposed as testament to the positive meeting of innovative technologies, energy production, the new flow economy, and the green economy of the new settlement. In the pluralism sought out by actors we should all be potential self-producers of our energetic demands, self-regulators of our rubbish and waste cycle, and self-generators of our water demand (see Figure 13.17b above).

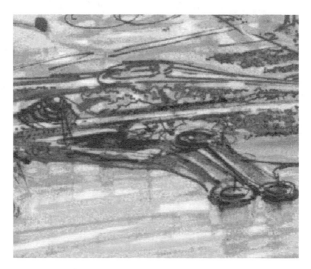

Figure 13.17c   Design approaches explored, consistent with the definition of problems

Source: Prof. Forte's Planning Office Archives.

Conclusion

The conditions we are currently experiencing demand a reflection on the role of the correlation between sustainability and creativity in the planning of the city of the twenty-first century, both in the areas characterized by demographic stability and in those with accentuated dynamic. The originality that can be foreseen is consequent to the current de-structuring and the necessity to re-establish aims, objectives and instruments. As has occurred in the past, the planned perspective is always the habitat for the survival of humankind, and of the structures of society that direct it, realizing its aspirations through the form and structure that these cities will assume. Technological innovations can help in this direction, but do not solve the problematic nodes. Government institutions should take on the role of *politics*, which is not that of *governing events*, but of anticipation of, and foresight concerning, a community's aspirations, directing events in such a perspective through improved *government* and efficient *governance*. Therefore, it is not simply the role of surviving, or of simply sitting out, but the role of conforming and configuring the arrival at the intended goals. Urban planning has its own specific role within public policy, through conceptual and creative processes derived from the research into the processes of sustainability that anticipate and aim for desired and shared goals in the evolution of life.

Bibliography

Alonso, W. 1964. The Historical and Structural Theories of Urban Form: Their Implication *for* Urban Renewal. *Land Economics*, 40, 2.

—— 1965. *Location and Land Use*. Cambridge, MA: Harvard University Press.

Aquarone, A. 1961 *Grandi città e aree metropolitane in Italia*. Bologna: Zanichelli.

ANCE (Associazione Nazionale Costruttori Edili, Studio Ambrosetti). 2005. *La città dei creativi, visioni & progetti*. Roma: Edilstampa srl.

Bogue, D.J. 1950. *The Structure of the Metropolitan Community, a Study of Dominance and Subdominance*. Ann Arbor, MI: University of Michigan Press.

Boston Redevelopment Authority. 1966. *General Plan for the city of Boston and the Regional Core, 1965–1975*. Boston: Boston Redevelopment Authority.

Burdett, R. 2006. *La Biennale di Venezia 10. Mostra Internazionale di Architettura. Città, architettura e società*. Venezia: Marsilio Editori.

Calthorpe, P. and Fulton, W. 2001. *Planning for the End of Sprawl: The Regional City*. Washington: Island Press.

Chapin, F.S. and Weiss, S.F. 1962. *Urban Growth Dynamics*. New York: John Wiley and Sons.

Choay, F. 1980. *Le regle et le modele, Sur la thèorie de l'architecture et de l'urbanisme*. Paris: Edition du Seuil.

Compagna, F. 1967. *La politica delle città*. Bari: Laterza.

Dickinson, R.E. 1947. *City, Region and Regionalism, A Geographical Contribution to Human Ecology*. London: Kegan Paul.

Dobriner, W. (ed.) 1958. *The Suburban Community*. New York: C.P. Putnam's Sons.

Engels, F. 1872. *La questione delle abitazioni*. Roma: Edizioni Rinascita.

Florida, R. 2004. *The Flight of the Creative Class*. New York: Harper Business.

Forte, F. 1971. *International Association for Metropolitan Research and Development, Toronto, Canada, Improving Citizens Participation*. Report produced for the Naples Study Committee, mimeo.

—— 1972. *Metodologia urbanistica, ricerca operativa, modellistica urbana*. Napoli: Guida Editore.

—— 1979. *Stato e regioni nella politica regionale per il Mezzogiorno italiano*. Napoli: Guida Editori.

—— *Il progetto urbanistico e la disciplina perequativa*. Napoli: Edizioni Scientifiche Italiane.

—— *Struttura e forma del piano urbanistico comunale perequativo*. Napoli: Edizioni Scientifiche Italiane.

—— 2006. *Politiche urbane, Napoli: storie, problemi, opportunità*. Roma: Inu Edizioni.

—— 2008a. *Architettura-Città, un programma, un progetto: nuove città, città nella città*. Roma: Gangemi Editore.

—— 2008b. *Roberto Pane, la sfida urbanistica*, paper presented to the conference Roberto Pane tra storia e restauro. Architettura, città, paesaggio, Naples, 27–28 October.

Friedman, J. and Alonso, W. 1964. *Regional Development and Planning*. Cambridge, MA: MIT Press.

Fusco Girard, L. and Forte, B. 2000. *Città sostenibile e sviluppo umano*. Milano: Franco Angeli.

Fusco Girard, L. and Nijkamp, P. 1997. *Valutazioni per lo sviluppo sostenibile della città e del territorio*. Milano: Franco Angeli.

Fusco Girard, L. and You, N. 2006. *Città attrattori di speranza*. Milano: Franco Angeli.

Geddes, P. 1915. *Città in evoluzione*. Milano: Il Saggiatore.

Glass, R. 1964. *London, Aspects of Change*. London: Macgibbon and Kee.

Gootman, J. 1961. *Megalopolis*. New York: The Twentieth Century Found.

Hall, P. 2003. *London 2000*. London: Faber and Faber.

Hatt, P.K. and Reiss, A.J. 1957. *Cities and Society*. Glencoe: The Free Press.

Herzog, L.A. 2006. *Return to the Center*. Austin: University of Texas Press.

Hoover, E.M. and Vernon, R. 1959. *Anatomy of a Metropolis, a New York Metropolitan Region Study*. Cambridge, MA: Harvard University Press.

Howard, E. 1898. *Tomorrow: A Peaceful Path to Reform*. London: S. Sonnenschein.

—— 1902. *Garden City of To-morrow*. London: S. Sonnenschein.

Jacobs, H.M. 2008. The Future of the Regulatory Taking Issue in the US and Europe: Divergence or Convergence? *Urban Lawyer*, 40(1), 51–72.

—— 2009. An Alternative Perspective on US–European Property Rights and Land Use Planning: Differences Without any Substance. *Planning and Environment Law*, 61(3), 3–12.

Le Corbusier. 1967. *The Radiant City, Elements of a Doctrine of Urbanism to be Used as the Basis of our Machine-age Civilization*. New York: The Orion Press (first edition *La ville radieuse*, 1933).

Lynch, K. 1962. *Site Planning*. Cambridge, MA: MIT Press.

Morbelli, G. 1977. *Città e piani d'Europa*. Bari: Edizioni Dedalo.

Mumford, L. 1961. *The City in the History*. New York: Harcourt, Brace and World.

Nairn, J. 1965. *The American Landscape, a Critical View*. New York: Random House.

Osborn, A.F. 1967. *L'immaginazione creative, sviluppo ed utilizzazioni*. Milano: Franco Angeli Editore.

Piccinato, L. 1948. *La progettazione urbanistica*. Ristampa 1988. Padova: Marsilio Editore.

Quaroni, L. 1963. *Città e quartiere nell'attuale fase critica di cultura*. Quaderni La Casa, 3. Roma: Ed. De Luca.

Reissman, L. 1964. *The Urban Process, Cities in the Industrial Societies*. New York: The Free Press.

Rizzo, S. 2009. La vita nelle aree urbane. '*Corriere della Sera*', *Focus*, 28 February, p. 8.

Rodwin, L. 1961. *Housing and Economic Progress*. Cambridge, MA: Harvard University Press and The Technology Press.

Salamon, L.M. (ed.) 1990. *The Future of the Industrial City: The Challenge of Economic Change in America and Europe*. Baltimore: The Johns Hopkins University, Institute for Policy Studies.

Sternleib, G. 1990. The Global Economy and Industrial Cities, in *The Future of the Industrial City: The Challenge of Economic Change in America and Europe*, edited by L.M. Salamon. Baltimore: The Johns Hopkins University, Institute for Policy Studies.

Stimman, H. 2009. *Berliner Altsstadt, Von der DDR-Staatsmitte zur Stadtmitte*. Berlin: DOM Publishers.

The Pittsburgh Regional Planning Association. 1963. *Region with a Future*. Pittsburgh: University of Pittsburgh Press.

Tremonti, G. 2008. *La paura e la speranza, Europa: la crisi globale che si avvicina e la via per superarla*. Milano: Mondatori.

Webber, M. et al. 1964. *Explorations into Urban Structure*. Philadelphia: University of Pennsylvania Press.

Wright, F.L. 1958. *The Living City*. Dublin: Edizione Mentor Book.

Chapter 14

The Urban Metabolic Pattern: Dynamics and Sustainability

Mario Giampietro, Gonzalo Gamboa and Agustin Lobo

Introduction

The first part of this article describes the discontinuity in the trajectory of human development represented by the transition from pre-industrial rural society to modern urban life in the nineteenth and twentieth centuries. As we will see, this transition was directly related to changes in energy metabolism, namely the switch from the use of renewable energy inputs to fossil fuels as primary energy source. The total dependence of modern urban society on fossil energy as primary energy source brought about an important change in the *role* of the city in society. Whereas in pre-industrial rural society cities could exist only by the mere grace of the surplus produced in rural areas, in post-industrial urban society cities are in charge of directing the flows of fossil energy in the entire society inclusive of rural areas. This provides to cities the required degree of freedom to become drivers of creativity and innovation.

In the second part we present an approach situating the energy metabolism of the modern city in relation to the metabolic patterns expressed by the various household types composing the city. Such a multi-scale integrated analysis of the urban metabolism makes it possible to study the mechanisms and the consequences of changes across different hierarchical levels (the whole, the environment and the parts inside the black box). When carrying out this analysis we see the progressive increase of the share of human activity allocated, within society, to adaptability (away from the repetitive activities associated with working in the primary and secondary sectors). In this chapter, we present only preliminary findings. However, we believe that it is important to develop this typology of analysis. In fact, to study the sustainability of cities in relation to their ability to change, it is essential to develop analytical tools capable of linking the overall metabolic pattern of the city to demographic change (ageing and immigration) and also to the interface that urban metabolism has with the metabolism of the ecosystems embedding the city.

From Pre-industrial Rural Settlements to Modern Urban Society

According to the latest report on the State of the World Population more than 50 per cent of the world population was living in cities in 2007 (UNFPA 2008). This figure represents a dramatic break with the past; in the pre-industrial era the percentage of the population living in cities rarely exceeded 10 per cent. The unprecedented phenomenon of urbanization at the global level indicates a clear discontinuity in human development. But what is more relevant is that this phenomenon is tightly linked to structural and functional changes that have taken place in society's energy metabolism.

It is well known that the structure of pre-industrial societies was strongly constrained by the ability of rural areas to produce food and fuel (wood) surpluses for the ruling classes living in the 'cities' (Cottrel 1955, Debeir et al. 1991, Tainter 1988) (Figure 14.1). The limited potential of rural activities to generate a surplus of resources for a distinct class of urban dwellers represented a *biophysical* constraint to the expansion of cities. This biophysical constraint, or bottleneck, was governed by two distinct urban/rural ratios, one relating to human time and the other to colonized land:

1. The *ratio* of hours of activity of the urban dwellers, the *rulers* (lords with their courts and soldiers – the only members of society having a relative degree of freedom in their daily activities) requiring a given set of inputs for their relative metabolism, *divided by* the hours of activity of the farmers, the *ruled*, whose activity provides the inputs for both the rulers *and* the ruled (the members of society experiencing a forced pattern of behaviour).
2. The *ratio* of hectares of colonized land in the cities *divided by* the hectares of colonized land in the countryside.

Knowing: (i) the profile of energy expenditures per hour of activity of a king and his court (the rulers) on the one hand and the energy expenditures per hour of activity of the farmers and their dependents (the ruled) on the other hand; and (ii) the productivity of the working hours associated with the hectares in food production in the ruled compartment, one can calculate a range of values for the possible ratios 'rulers/ruled' that will generate a feasible configurations for society (Giampietro and Mayumi 2009).

In fact, the forced relations between farmers (generating surplus), colonized land and lords (consuming surplus) have long been known to be key factors in determining the ratio between the rulers and the ruled in pre-industrial societies. 'Before the end of the first millennium AD ... the European kings and other big landlords were forced to move from one of their estates to another in the course of the year. They and their followers would remain for some time in a castle or manor, hunting the game and consuming the local wine, cereals, meat, fruits and so forth. When the surplus was consumed, they moved to another of their properties.' In this passage from Boserup (1981: 97) we can recognize a more general pattern followed by traditional societies.

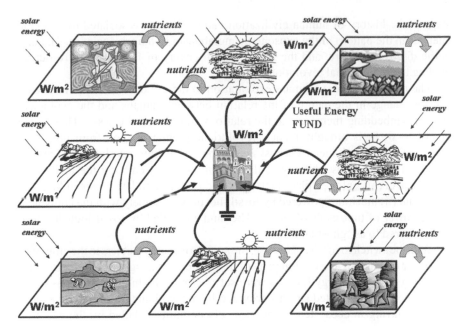

Figure 14.1   Concentrating the energy surplus produced by the countryside

Source: Giampietro and Mayumi, 2009, *The Biofuel Delusion*, Earthscan, reproduced with permission.

Human density over the territory depended on the availability of an adequate supply of resources, i.e. food and energy (e.g. wood). As soon as the requirement of food and/or energy in a particular place and over a particular time period surpassed the density of the supply in that place and period, humans were forced to move on to another spot where resources were available (nomads or shifting cultivation). So we can say that even the kings and the lords had to cope with the constraints of a limited surplus provided by the rural context of the city.

The early utilization of exosomatic[1] energy brought about some degree of freedom in the relation between urban settlements and rural areas. Early forms of utilization of exosomatic energy are mainly accounted for by fire, animal power, wind power and other occasional power sources such as waterfalls, river streams and geothermal events (Giampietro et al. 1997). For this reason, techniques improving the harnessing of fire, wind, waterfalls and animal power have been crucial in providing an edge among competing pre-industrial civilizations (Cipolla 1965, Cottrell 1955, Debeir et al. 1991, White 1943, 1959). Indeed, two crucial

---

1   *Exosomatic* energy metabolism refers to technical conversions of energy inputs into end-uses that take place *outside* the human body, but under direct human control. *Endosomatic* metabolism refers to physiological conversions of energy inputs (food) into end-uses that take place *inside* the human body.

steps in the history of human civilization can be directly associated to dramatic changes in the pattern of the exosomatic metabolism of human societies. The first is the discovery of fire, and the second the discovery of fossil energy resources leading to the industrial revolution.

The movement from biomass exploitation to the use of fossil fuels implied a historic change in the nature of the relation between humans and the ecological systems embedding them, and in the relations among social classes. That is, the use of fossil fuels brought about a revolutionary emancipation from both land (Mayumi 1991) and social class system. The key to this revolutionary change is that: (i) fossil fuels (initially coal, later oil and gas) embody the activity of ecosystems (concentrated plant matter) of the past accumulated over millions of years, and (ii) machines powered by fossil fuel provide society with the equivalent of the power of thousands of slaves. The land embodied by fossil fuels has also been referred to as 'ghost land'.

With the switch to fossil energy use, the supply of energy carriers to the cities was no longer dictated by the natural pace of nutrient cycles in agro-ecosystems. After switching to fossil energy humans were set free from the biophysical constraints represented by the limited surplus provided by the exploitation of natural energy flows occurring in ecosystems. The clear discontinuity entailed by the adoption

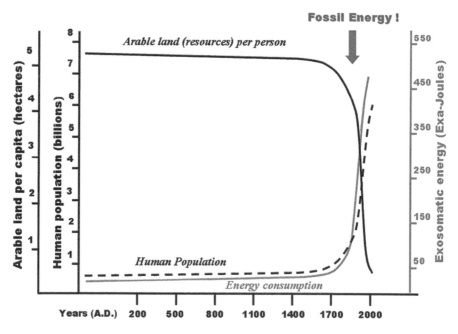

Figure 14.2   Discontinuity in human development in relation to fossil energy use

Source: Giampietro and Mayumi, 2009, *The Biofuel Delusion*, Earthscan, reproduced with permission.

of fossil energy is illustrated in Figure 14.2. It shows a sudden and simultaneous change in human population size, fossil energy consumption per capita, and colonized land per capita (the source previously supporting the expenditure) over the last 2000 years. The simultaneous increase in population size and per capita energy expenditure resulted in a spectacular increase in overall energy expenditure of society. Using an analogue, humankind underwent a dramatic change like a blue-collar family abruptly changing from a humble lifestyle constrained by a meagre hourly wage, to a Hollywood lifestyle supported by a fat bank account resulting from winning the lottery.

As a result of the industrial revolution, the flow of energy carriers required by modern society to generate useful work is no longer going from the countryside to the cities. On the contrary, the industrial sector of developed societies, centred near the cities, provides the flow of energy carriers (fuel for machinery and production of fertilizers and pesticides) consumed in the rural areas (Figure 14.3).

The metabolism of modern socio-economic systems is based on flows with an extremely high energy intensity (per hour) and energy density (per hectare). Vaclav Smil (2003) has provided a comparison of the power density ranges of different typologies of supply of exosomatic energy as related to different primary energy sources (left graph in Figure 14.4) and a comparison of the power density ranges

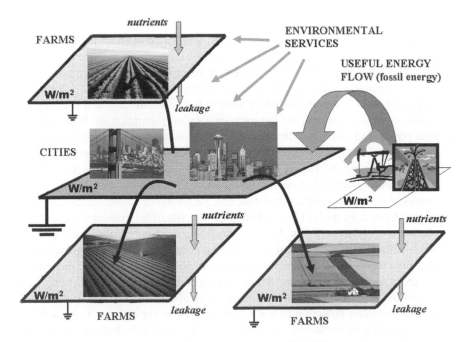

Figure 14.3 The flow of fossil energy from the cities to the countryside

Source: Giampietro and Mayumi, 2009, *The Biofuel Delusion*, Earthscan, reproduced with permission.

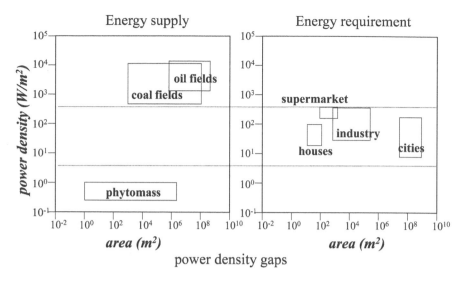

Figure 14.4 The density of typologies of supply and requirement of exosomatic energy

Source: Giampietro and Mayumi, 2009, *The Biofuel Delusion*, Earthscan, reproduced with permission.

of different typologies of land use associated with the pattern of metabolism of developed societies (right graph in Figure 14.4). Differences in power densities among typologies of supply and consumption of exosomatic energy are so big – orders of magnitude – that a logarithmic scale must be used.

Thus, whereas in the pre-industrial society both endosomatic and exosomatic metabolism were dependent on the pace of surplus generation by land, labour, animal power and technical infrastructures such as water mills or sailing ships, in post-industrial urban society both types of metabolism are dependent on the pace of fossil energy flows (providing power and technical inputs) generated by stock depletion. These fossil energy flows are regulated by corresponding flows of money, and therefore reflect the economic performance of the socio-economic systems. This is at the basis of the explosion of diversity of behaviours that the urban environment guarantees to its dwellers. To better understand the extraordinary achievements of human technical progress two points need to be recognized. The human population has increased by 3 billion in the last four decades, more than in the previous 40,000 years, and the per capita consumption of fossil energy has been continuously increasing in that same period! At the same time, there has been a concomitant reduction of natural resources per capita (ours being a finite planet!), notably of colonized land per capita (Figure 14.2). Indeed, due to the special characteristics of fossil energy, in particular the emancipation from land, it has been possible to dramatically increase both population size and per capita energy consumption in spite of the reduction in available resources per capita.

Urban Metabolism in Relation to Patterns of Household Metabolism

The Methodological Approach

When considering the metabolic pattern of developed societies, more than 95 per cent of GDP and of total energy consumption is associated with land uses taking place in less than 2 per cent of available land (which can be included within the category 'Urban Land', which in turn is defined inside the category 'colonized land' – more about the definition of these categories below).

The overall consumption of resources (energy, food, water, market products) and the overall generation of waste (garbage, CO_2 emissions, pollutants) of a city are closely related to the physiological concept of *metabolism*. Like any other living system, socio-economic systems have to continuously metabolize a flow of energy and matter to maintain and reproduce the structures and the functions associated with their identity (Giampietro and Mayumi 2009, Prigogine and Stengers 1984). Indeed, the very existence of a city entails a continuous flow of material and energy inputs and an associated flow of material and waste outputs (Dyke 1988, 1989).

At ICTA−UAB, we are developing an integrated analysis of the urban metabolic pattern. In particular we apply the methodology of multi-scale integrated analysis of societal and ecosystem metabolism (MuSIASEM) in order to characterize and analyse the viability and desirability of patterns of production and consumption of socio-economic systems across different dimensions of analysis and different hierarchical levels of organization (Giampietro 2000, 2001, Giampietro and Mayumi 2009, Giampietro et al. 2009, Ramos-Martin et al. 2009).

This approach makes it possible to establish a bridge between the analysis of the metabolic patterns of relevant household typologies and the resulting metabolic pattern of the whole city. In this way, this analysis makes it possible to analyse the effect of demographic trends, immigration/emigration, and various economic variables on the overall metabolism of the city. It also allows us to represent scenarios and simulations in spatial terms using geographic information systems (GIS).

The approach builds on the general concept of multi-scale integrated analysis of *societal* metabolism that has been presented in several earlier publications (for an overview see Giampietro and Mayumi 2009). The quantification is based on the fund-flow model proposed by Georgescu-Roegen to represent the metabolism of socio-economic systems (Georgescu-Roegen 1971). Two concepts are key to this model, *fund-elements* and *flow-elements*. Funds are the elements that remain present over the duration of the analysis ('structural elements' that have to be reproduced), such as human activity, while *flow-elements* are those that disappear during the window of analysis – for example, the flow of energy, food, and other products consumed ('consumables').

Linking Demographic Variables to Patterns of Human Activity

The demographic structure of a society or a city is intricately linked to human activity, in terms of hours available to perform various activities. In our approach, we express the human activity fund in terms of hours of activity available per year (e.g. 8,760 hours of total activity per year per capita) as illustrated in Figure 14.5. The analysis in Figure 14.5 refers to a hypothetical society of 100 persons. These 100 individuals translate into a fund of total human activity (THA) of 876,000 hours/year (100 persons × 8760 hours in a year).

At the hierarchical level of individual beings we can define a set of *structural* types determining what the system *is*. Put another way, we characterize the population as being made up of a set of structural types (categories) of individuals, each category having a determined size depending on the demographic structure. In Figure 14.5 we distinguish six structural types based on age (3 groups: $x_1 < 16$; $16 < x_2 < 65$; $x_3 > 65$) and gender (y_1 = males; y_2 = females), thus generating a 3 × 2 matrix. The population of 100 individuals (or 876,000 hours per year) is distributed over these six types according to the existing demographic structure.

At the same time, we have to define what the system *does* at the level of individual human beings. In other words, we have to define categories, or

Grammar and dictionaries for a developed society (100 people)
LEVEL OF INDIVIDUALS – level *n-3*

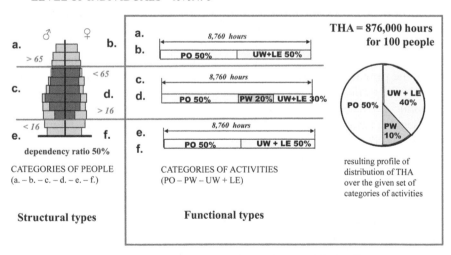

Instances of Structural Types: **a.** = 13; **b.** = 13; **c.** = 25; **d.** = 25; **e.** = 12; **f.** = 12 ➔ **100 people**

Figure 14.5 A grammar establishing a relation between demographic structure and profile of human activities

Source: Giampietro and Mayumi, 2009, *The Biofuel Delusion*, Earthscan, reproduced with permission.

functional types of human activity. In Figure 14.5 we distinguish three functional types of human activity (HA) to which time is dedicated: physiological overhead (HA$_{PO}$), comprising activities essential to human survival such as sleeping, eating and personal care; paid work (HA$_{PW}$); and other activities such as household chores, leisure and education (HA$_{HC+LE}$).

The simple grammar in which each structural type maps onto a known pattern of human activities (functional types) is shown in the middle of Figure 14.5. In this way, it is possible to map the structural profile of the population (the given pattern of structural types) onto an overall profile of distribution of human activities (a distribution of functional types at the level of the whole society). In Figure 14.6 we apply this method to Catalonia, a modern society. It clearly shows that different types of individuals (e.g. elderly, children, adults) do express different patterns of activities, and therefore they do have different preferences and requirements for goods and services. This confirms that socio-demographic variables do matter for studying the patterns of production and consumption of a society.

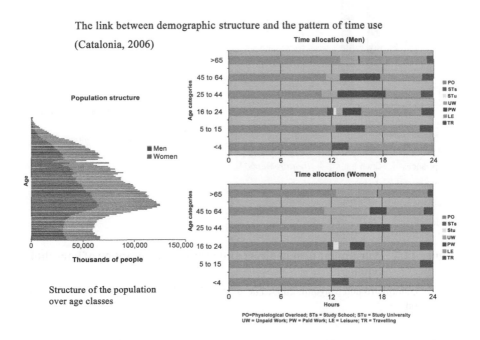

Figure 14.6   Demographic structure and time use with reference to Catalonia, 2006

Source: Gamboa 2009.

The Metabolic Pattern of Households

At a higher hierarchical level we can link structural household types to their specific metabolic pattern. For example, we can characterize household types in relation to *direct* exosomatic energy consumption (energy that is directly consumed by the household typology) and *indirect* exosomatic energy consumption (energy that is consumed at the level of the economy to produce the goods and services consumed by the household typology). Having selected a set of categories for such a representation, we can establish a relation between the specific structures and functions of different typologies of households, as illustrated in Figure 14.7.

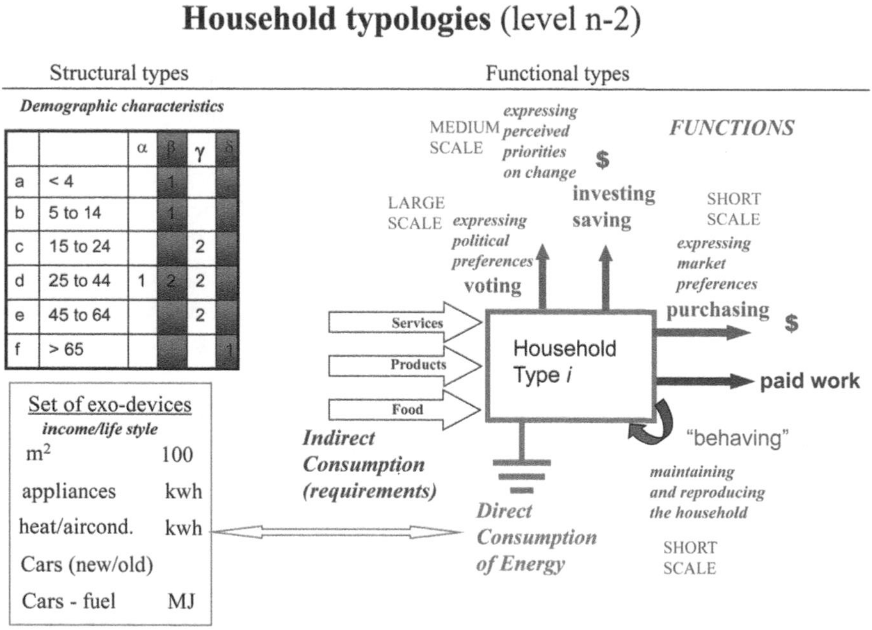

Household typologies (level n-2)

Figure 14.7   A grammar to map direct and indirect consumption of household typologies

Source: Giampietro and Mayumi, 2009, *The Biofuel Delusion*, Earthscan, reproduced with permission.

At this point, having this information, we can study how a different profile of distribution of the population over defined typologies of households will change the overall characteristics of the household sector as a whole. An example of this relation is shown in Figure 14.8 for the metropolitan area of Barcelona (Catalonia), based on a simplified grammar with only four household types relevant to that society. A more elaborated grammar (based on nine household types) can be found in Gamboa, 2009.

Household Sector: HH – level n-1

Urban HH – level n-2

Urban household types – level n-3

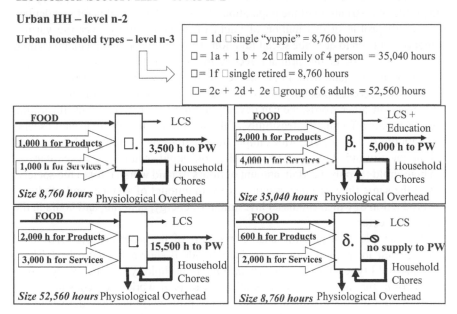

□ = 1d □single "yuppie" = 8,760 hours
□ = 1a + 1 b + 2d □family of 4 person = 35,040 hours
□ = 1f □single retired = 8,760 hours
□ = 2c + 2d + 2e □group of 6 adults = 52,560 hours

FOOD → LCS

1,000 h for Products ▷ □. 3,500 h to PW

1,000 h for Services ▷ Household Chores

Size 8,760 hours Physiological Overhead

FOOD → LCS + Education

2,000 h for Products ▷ β. 5,000 h to PW

4,000 h for Services ▷ Household Chores

Size 35,040 hours Physiological Overhead

FOOD → LCS

2,000 h for Products ▷ □ 15,500 h to PW

3,000 h for Services ▷ Household Chores

Size 52,560 hours Physiological Overhead

FOOD → LCS

600 h for Products ▷ δ. no supply to PW

2,000 h for Services ▷ Household Chores

Size 8,760 hours Physiological Overhead

Figure 14.8   Characterizing the urban metabolism of household types for the metropolitan area of Barcelona

Source: Data from Gamboa 2009.

At the hierarchical level of households, we can define again a set of *structural* types determining what the system *is*. Again we use the fund element human activity as the basis for the definition of household types. We thus characterize the population of a city as made up of household types, the sizes of which are expressed in terms of hours of human activity. According to the previous analysis we can also define what these structural types *do*, the functions that they express. In the example based on a simple grammar of only four types, presented in Figure 14.8, we distinguish the following household types:

- Type α: *single adult household* – only one person, economically active; size: 8760 hours/year (1 × 8,760).
- Type β: *classic family household* – consisting of an adult couple and two children; size: 35,040 hours/year (4 × 8760).
- Type γ: *immigrant flat household* – consisting of six adults all working; size: 52,560 hours/year (6 × 8760).
- Type δ: *single retired household* – only one person, retired; size: 8760 hours/year.

Given the chosen set of types used to describe the population of a city, we can use the profile of distribution of the population of households over this set to calculate the overall profile of distribution of hours of human activity over the various categories chosen to describe social functions (physiological overhead, household chores, paid work in the primary sector, in the secondary sector, leisure and cultural activities, paid work in the service sector, paid work in the government).

Next, it becomes possible to define what the system *does* at the level of the whole city. In other words, we can develop a quantitative assessment illustrating the constraints affecting the feasibility of the interaction that the household sector has with the rest of society. An example of this analysis is shown in Figure 14.8. Within the simplified representation of the household sector, we can see that: (i) the households supply a certain amount of hours of paid work (PW) to the society; and at the same time: (ii) they require a certain amount of services and products, the production of which entails an investment of energy, material, and hours in the paid work sector.

An example of a possible characterization of the selected four household types in their interaction with the production side is shown in Figure 14.8. There are household types, the so-called *dissipative* household types, such as type δ in Figure 14.8, that are net consumers of hours of paid work; they consume more hours of paid work (embodied in services and products) than they deliver to the paid work sector. On the other hand, there are also household types, the so-called *hypercyclic* household types, such as type α in Figure 14.8, that are net providers of paid work hours to society.

Having established this relation, we see that certain demographic changes such as population greying – a growing proportion of elderly in the population – will translate into a change in the profile of distribution of the population over the household types (relative increase in type δ in Figure 14.8). It will also change the mix of products and services required by society. In this case, the demographic change will translate into a sharp increase in bio-economic pressure; that is, a reduced supply of work hours and a concomitant increase in demand for goods and services. These economic effects of greying of the population can in principle be counterbalanced by a relative increase in strongly *hypercyclic* household types, that is, households supplying much more work hours than they demand, such as households made up of several adults (type γ in Figure 14.8). We have indeed found this to be the emergent type of household in a preliminary analysis of the province of Barcelona; households composed of several adult immigrants living together have been the fastest growing typology of household over the past 10 years, together with the single elderly households.

This is an example in which a phenomenon of emergence – the system becoming something else by adding new typologies – can be predicted by examining the critical situation caused by actual demographic trends. The possibility of foreseeing the appearance of new categories required for describing typologies – something impossible for semantically closed systems of inference such as dynamical systems – is a major plus of our methodology.

As mentioned, this analysis does require a larger set of household types than the one exemplified in Figure 14.8. Preliminary data referring to an analysis based on a set of nine household types have been presented by Gamboa, (2009) and Ramos-Martin et al. (2009). In this way it becomes possible to study how the distribution of human activity and the relative requirement of products, services and metabolic rate are affected by demographic changes and by the 'movement' of hours of human activity across the various typologies of households (according to the type of analysis illustrated in Figure 14.6). That is, a given number of people – say, 10,000 – will require a completely different pattern of services, products and energy if they are organized in 1667 households of type g, 2500 households of type β, or 10,000 households of type α (single workers) or type δ (single elderly).

Analysing the Metabolic Pattern of Households using GIS

After estimating the typical metabolic rate of energy and/or material flows per hour of human activity in the different categories of human activity included in the analysis (e.g. MJ of energy per hour of activity in the household, or MJ of energy per hour of working in the service sector, litres of water consumed per hour of activity in manufacturing; kg of garbage per hour of activity spent at home) it also becomes possible to represent these flows in spatial terms.

To do so, one has to assess the density of human activity of defined categories in space. For example, the hours allocated to the category 'physiological overhead' (sleeping and personal care of individuals) are mostly spent in the house, and depending on the location of residential areas this can be associated with a defined distribution of density of human activity per hectare. In the same way, the hours of human activity allocated to the category 'paid work in the service sector' are linked to specific areas of land use associated with this category (either in shops and office for services or in industrial plants for manufacturing).

The research group led by Dr Agustin Lobo at ICTA–UAB has developed spatial maps referring to the allocation of human time in the metropolitan area of Barcelona. Examples from the work of Lobo and Baez (2009) are presented in Figure 14.9. Using the data of census it is possible to track the location of hours of human activity spent out of work in residential land uses (Figure 14.9a). Using data on commercial activities, it is possible to locate the time spent in specific subsectors of the paid work sector, such as the industrial sector, service sector and commerce (Figure 14.9b). In this way, one can check the gradients between the spatial location of time allocated to work and non-work activities, which may be used for studying transportation flows (Figure 14.9c). An example of this integrated analysis overlaying a view of Barcelona is illustrated in Figure 14.9d.

It is noteworthy that knowing: (i) the spatial distribution of the whole set of categories of human activities for a city; (ii) the density of hours of human activity per hectare of the various categories; and (iii) the metabolic rate (per hour of human activity) relative to various relevant flows – e.g. food, water, energy, urban waste and key pollutants – it becomes possible to have a much better analysis of the metabolism of the city. In other words, the various flows determining the overall metabolism of the city can be explained, simulated and, in case of structural changes in the city, predicted by looking at demographic and economic changes, determining the profile of distribution of the population over the set of household types and changes in the characteristics of the household types.

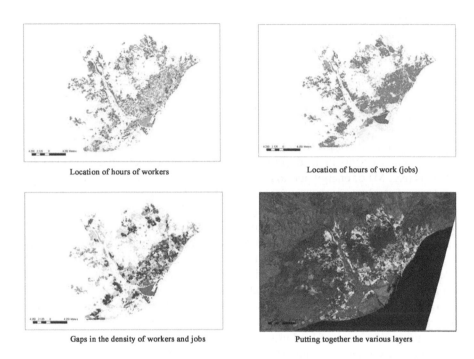

| Location of hours of workers | Location of hours of work (jobs) |
| Gaps in the density of workers and jobs | Putting together the various layers |

Figure 14.9a–d   Spatial representation of the allocation of human time for the metropolitan area of Barcelona

Source: Lobo and Baez 2009.

Conclusion

The incredible set of transformations experienced by human civilization in the last century is associated with the phenomenon of urbanization. Cities are the nervous system regulating the entire set of activities expressed by modern societies. For this reason, many tend to blame cities for the lack of sustainability of the current pattern of metabolism. However, this criticism is not justified. Our civilization is an urban civilization. Therefore the challenge associated with sustainability is not about getting rid of cities, but rather is about learning how to use the creativity made possible by the existence of the cities to explore alternative patterns of development. In this struggle, cities will have to lead the transition towards a more sustainable metabolic pattern. Because of this fact, however, it is essential to understand the key factors determining the feasibility of urban metabolic patterns and to develop methodological approaches capable of studying trends and generating scenarios of changes. The applications of the MuSIASEM approach to the analysis of urban metabolic patterns are still in the exploratory phase. However, the preliminary results presented in this chapter suggest that this approach does have a great potential for this type of study.

Bibliography

Boserup, E. 1981. *Population and Technological Change*. Chicago, IL: University of Chicago Press.

Cipolla, C.M. 1965. *Guns, Sails, and Empires: Technological Innovation and the Early Phases of European Expansion, 1400–1700*. New York: Minerva Press.

Cottrell, W.F. 1955. *Energy and Society: The Relation between Energy, Social Change, and Economic Development*. New York: McGraw-Hill.

Debeir, J.-C., Deleage, J.-P. and Hemery, D. 1991. *In the Servitude of Power: Energy and Civilization through the Ages*. Atlantic Highlands, NJ: Zed Books.

Dyke, C. 1988. Cities as Dissipative Systems, in *Entropy, Information and Evolution*, edited by B.H. Weber, D.J. Depew and J.D. Smith. Cambridge, MA: MIT, 355–67.

—— 1989. *The Evolutionary Dynamics of Complex Systems: A Study in Biosocial Complexity*. Oxford: Oxford University Press.

Gamboa, G. 2009. *Social Metabolism of the Household Sector in Catalonia: Time Allocation and 'Embodied' Time Consumption from the Service Sector*. Deliverable EU-Project DECOIN, Catalonia Case Study Document C. Available at: www.smile-fp7.eu/deliverables/CATcasestudyDocumentC.pdf.

Georgescu-Roegen, N. 1971. *The Entropy Law and the Economic Process*. Cambridge, MA: Harvard University Press.

Giampietro, M. 2000. Societal Metabolism. Part 1 of 2: Introduction of the Analytical Tool in Theory, Examples, and Validation of Basic Assumptions. Special issue of *Population and Environment*, 22(2), 97–254.

—— 2001. Societal Metabolism. Part 2 of 2: Specific Applications to Case Studies. Special issue of *Population and Environment*, 22(3), 257–352.

Giampietro, M. and Mayumi, K. 2009. *The Biofuel Delusion: The Fallacy behind Large-scale Agro-biofuels Production*. London: Earthscan Research Edition.

Giampietro, M., Bukkens, S.G.F. and Pimentel, D. 1997. The link between resources, technology and standard of living: Examples and applications, in *Advances in Human Ecology*, volume 6, edited by L. Freese. Greenwich (CT): JAI Press, 129–99.

Giampietro, M., Mayumi, K. and Ramos-Martin, J. 2009. Multi-scale Integrated Analysis of Societal and Ecosystem Metabolism (MuSIASEM): Theoretical Concepts and Basic Rationale. *Energy*, 34, 313–22.

Lobo, A. and Baez, M.A. 2009. *Geographic Distribution of Human Time and Energy Throughput in the Barcelona Metropolitan Area, Report on Environmental Science #2*, ICTA, Universitat Autònoma Barcelona. Available at: www.recercat.net/handle/2072/40697.

Mayumi, K. 1991. Temporary Emancipation from Land: From the Industrial Revolution to the Present Time. *Ecological Economics*, 4, 35–56.

Prigogine, I. and Stengers, I. 1984. *Order Out of Chaos*. New York: Bantam Books.

Ramos-Martin, J., Cañella-Bolta, S., Giampietro, M. and Gamboa, G. 2009. Catalonia's Energy Metabolism: Using the MuSIASEM Approach at Different Scales. *Energy Policy*, 37, 4658–71.

Smil, V. 2003. *Energy at the Crossroads: Global Perspectives and Uncertainties*. Cambridge, MA: MIT Press.

Tainter, J.A. 1988. *The Collapse of Complex Societies*. Cambridge: Cambridge University Press.

UNFPA. 2008. *UNFPA Annual Report 2007*. New York: United Nations Population Fund. Available at: www.unfpa.org/upload/lib_pub_file/777_filename_unfpa_ar_2007_web.pdf.

White, L.A. 1943. Energy and Evolution of Culture. *American Anthropologist*, 14, 335–56.

—— 1959. *The Evolution of Culture: The Development of Civilization to the Fall of Rome*. New York: McGraw-Hill.

Chapter 15

Exploring Creative Cities for Sustainability: Towards Applications of Relational Visualization

Joe Ravetz

Introduction

The context and laboratory for this discussion is the global urban situation – urgent and potentially catastrophic. Three billion new urban residents are expected by 2050: nearly all will be in the developing world, and most are likely to live in slums (Neuwirth 2005, UNDESA 2007). Many such cities have high levels of unemployment, poverty, corruption, drug trafficking, ethnic tension and fundamentalism.

Overarching the urban problem are the global life-support challenges, recently summed up as 'planetary boundaries': climate, acidification, ozone, nitrogen/phosphorus cycles, aerosols, water, chemicals, biodiversity and land use (Rockström et al. 2009). Each of these global challenges are interdependent on the global system of cities – cities are the sites of most environmental impacts and resource consumption, and of most environment and resource-related markets and technologies. But from experience of current urban conditions and trends in both North and South, achievement of the global life-support goals appears to have very little chance of success – there are too many interlocking interests, systemic barriers and inbuilt power structures (Roberts et al. 2009).

We can observe that the global urban agenda is not so much a linear problem, as a 'systemic' and 'wicked' type of problem – complex, problematic, emergent; on the intersection of power, wealth and ideology; with multiple perceptions and values, and – significantly for this chapter – on the edge of chaos and creativity (Rittel and Webber 1973). It is also an 'ecological' kind of problem, as a complex adaptive system that calls for complex adaptive types of responses (Waltner-Toews et al. 2009).

And it is clear that to make real progress on the global life-support agenda, cities will need not only new technology, but innovation in markets, governance structures and social patterns. It is also clear that in order to promote social economic and political innovation, cities will need wider and deeper forms of cultural creativity and social enterprise.

So the agenda for 'creative cities for sustainability' becomes crucial on a global scale – not only a local agenda for cultural industries in tourism and gentrification, but a wider global agenda for innovation and creative action, on social, economic, cultural and political fronts. This is also a multi-level agenda, which links the top-down and bottom-up approach, focusing on creative everyday practices 'by the people and for the people' (Figure 15.1).

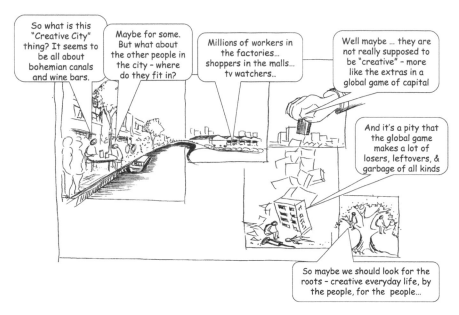

Figure 15.1   An agenda for creative sustainable cities

At the same time, it is also clear that these are more than mechanistic processes that can be studied by reductive analysis alone. We need to look for other more holistic and communicative channels, to explore and appreciate the multiplicity of the interconnections and the opportunities.

So, this chapter does not aim at a normal research structure of hypothesis, evidence and results. Rather, it is a kind of exploration of challenges and opportunities emerging from different parts of society, and a demonstration of new approaches in research and social learning. To do this we focus on some innovative examples to represent a very wide field. We look at cities from the point of view of the peri-urban areas, rather than conventional urbanist centres. We look at the sustainability agenda from the point of view of the Low Carbon Economy (LCE), which is very topical in UK and EU policy. We look at the concept of creativity in cities on several levels, both personal and political.

To bring this together we demonstrate an emerging body of theory and practice: an application of 'relational thinking', which has been developed to respond to such complex and multiple challenges (Ravetz 2010c). And to demonstrate how this works in both analytic and experiential modes, we show visual examples, in a range of possibilities from concept mapping to figurative drawing, and from static images to an interactive process of 'relational visualization' (Figure 15.2).

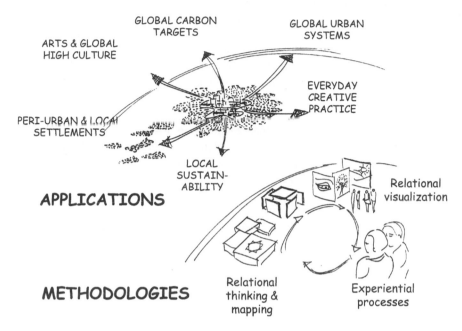

Figure 15.2 Scope of the chapter

So the structure of the chapter includes first an exploration and unpacking of the main agenda:

- 'city' or 'city system' – here focusing on the peri-urban geography;
- 'sustainability' – focusing on the 'low carbon economy';
- 'creativity' – linking between personal and political creative process.

Following that are some methods and techniques that are suited for this kind of agenda:

- 'relational thinking', a new approach to systems mapping and transition/ evolutionary analysis;
- 'relational visualization': a new approach to multiple channels of dialogue and experiential process.

Finally we show some examples from work in progress – current activities in the UK city-region of Greater Manchester.

New Thinking: 'Cities' and the Peri-urban

As we explore the creativity of a city, or cities, urbanism, metropolitan regions, or global urban systems, there are few simple definitions. The very idea of a city is open to question – cities are increasingly seen not as freestanding objects, but more as components within an 'urban system' – the question is, what is the nature and logic of that system? Such a question is increasingly challenging, as the material economic and spatial functions of cities are shifting to other dimensions, such as cosmopolitan global networks and 'spaces of flows' (Borja and Castells 1997, Sandercock 2003, Soja 2000). It could be argued, for instance, that the executive lounges of airports around the world are more connected to each other, both functionally and culturally, than each airport with its nearest urban slum. It can also be argued that the massive problems of urban poverty and alienation can be re-constructed as opportunities. However, we can draw some lines around different levels of urban system:

- first, there is an 'world urban system', external to any one city – describing the interactions between different city sizes and types, as a kind of central spine of the global economy (Sassen 1994);
- second, in larger countries there is a national urban system: this is both a geographic community of industrial, service and residential functions, and also the national policies and markets that provide the parameters for change;
- somewhere between is the 'city-region system'; this is a city together with its functional hinterland or bio-region, containing a labour market, water supplies, food supplies, and so on (Whitaker 2006);
- there is an 'internal urban system' – a set of components and dynamics of cities and settlements within their built-up areas, however these are defined. This could divide further into smaller and smaller districts, neighbourhoods, blocks and streets.

Within these layers, the scope of what is the city or urban is likewise open to many alternative views. There is a popular saying that 'the future is urban', which focuses on the mega-cities of over 10 million population, which are set to double in number in 20 years. In reality the bulk of the urban population will continue to live in settlements of between 25,000–500,000 people, although many of these are components of larger metropolitan areas, or 'polynucleated urban regions' (Hall and Pain 2006). There are different ways to define such urban areas, in physical boundaries, economic activity or political units.

An innovative approach is to focus on the peri-urban, rather than the urban, as the archetypal territory of the twenty-first century. Recent work in the EU FP6 project PLUREL has contrasted the intensive urban hubs, with the much larger and more extensive peri-urban areas (www.plurel.net). These are surrounding areas, often without clear definitions between rural and urban, and generally a complex

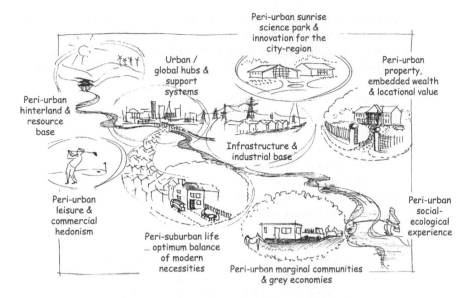

Figure 15.3 Twenty-first century peri-urban geographies and imaginations

mix of drivers and pressures in many different sectors, at different spatial scales (Nilsson et al. 2008, Ravetz 2010a). The peri-urban role is often as a hinterland and support system for the more concentrated and dominant urban activity, and provides a very relevant demonstration of the dynamic of cultural capitalism (Figure 15.3):

- One effect is the dynamic transition of 'metropol-ization' of former rural communities – a transition of economic activities, social types and spatial patterns of work and lifestyle, towards global metropolitan culture and networked flows.
- Alongside economic and social change there is a dynamic transition of 'cognitive cultural capitalism' for new and globalizing social and cultural lifestyles, attitudes and perceptions (Scott 2000 and 2006).
- One implication is the restructuring of the 'spatial ecology' of land uses and land use activity relationships. This looks at the interdependency of multiple land uses – for example, airports/water plants/high-value housing/ heritage landscapes – which each provide a range of functions and services to others, as part of a larger city-region system (Ravetz 2000). This points to a dynamic transition of 'localization' where in an age of mobility and networking, the city-region Green Infrastructure is one of the main channels for communities to re-invent their role and identity in a common landscape.
- We focus on this peri-urban agenda, partly as an under-researched and overlooked dimension of the city system, and the urban Low Carbon agenda. But particularly it offers new angles on the creative city agenda,

by focusing on the tensions, contradictions and dilemmas at the level of sites, land uses and landscape dynamics (Clay 1994). This can be explored in a simple functional way, or in a wider and deeper structural perspective:

– peri-urban zones of chaotic and creative experiments, urban residues and so on;
– peri-urban class competition for territory and control;
– peri-urban zone of capital accumulation in the circuit of urban property;
– peri-urban competition of city dominance over rural interests;
– peri-urban development as re-invention of new socio-economic-cultural roles and agendas in obsolescent areas;
– peri-urban economies as creative destruction of obsolescent activities;
– peri-urban community initiatives as sites for new social movements, socio-cultural enterprise and so on.

This conjunction or conflict of economic, social and cultural agendas can provide the sites for peri-urban creative responses and processes, as much as any inner city site. These might focus on the restructuring roles of peri-urban areas, or in reconfiguration of the city-region and regional structure. Putting all this together is a challenge for linear rational thinking. In the search for responses to a complexity agenda, a stakeholder workshop in 2008 piloted a technique of 'relational visualization' as a multi-channel dialogue. This helped to link local with regional perspectives, and to explore conflicts and conjunctions between different actors (Figure 15.4).

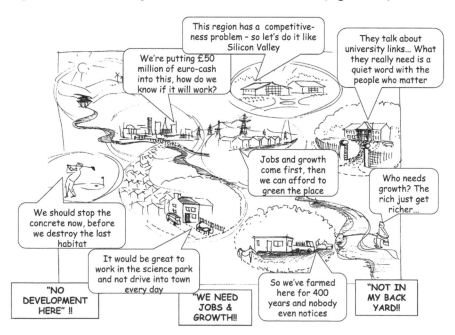

Figure 15.4   Peri-urban contradictions and dilemmas

This visualization shows a series of dialogues or exchanges between the key actors in a complex set of relationships, in this typical 'system of innovation' in the peri-urban situation. Focusing this down to the actor level offers a way into understanding the 'actor system', that is, the set of typical pressures, context factors, goals and responses that surround any typical actor or stakeholder. Scaling up from the actor/agency level to a structural and geographical level, we can frame the dynamics of urban and peri-urban development, in terms of a causal layers framework (Ravetz et al. 2011). This goes beyond the conventional DPSIR (driving forces/pressures/states/impacts/responses) to look at different levels of self-organizing systemic change (Figure 15.5).

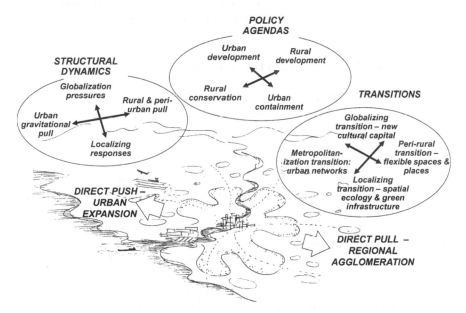

Figure 15.5   Peri-urban dynamics and transitions

First, there are the direct 'push' factors of urban expansion: population and economic growth; jobs and services; housing and infrastructure. Then there are direct 'pull' factors of regional agglomeration, where the key force is the consolidation at the regional level of a modernized and competitive economy and infrastructure.

Behind this the structural dynamics can be seen as tension between globalization – new forms of expropriation and interdependency – and localization – new forms of culture and place identity. There is also a spatial dimension which can be seen as an outward push (sub-urbanization and counter-urbanization), countered by an inward pull (urbanization and re-urbanization). Policy agendas and responses are spatially or socially specific (i.e. certain places, communities or social interests are involved):

- urban development agenda, as with science/business parks;
- urban containment agenda, as with green belt and similar policies;
- rural development agenda, as with modernization of the rural economy;
- rural conservation agenda, as with zoning of landscapes and habitats.

Over-arching these are the transitions identified above – globalizing or localizing: and metropolitan-izing or rural-izing. Each type of transition is changing the combined systemic structures, patterns, norms and values at various levels of complexity (de Roo and Silva 2010, Rotmans et al. 2001). The point here is that such transitions are not neutral mechanical processes – rather they are driven by creative processes, and themselves open up new spaces for creative processes, as in the section 'New Methods: Multi-Channel Dialogue' below).

New Thinking: The Low Carbon Urban Economy

The technical and policy consensus is now for an 80 per cent cut in carbon emissions by 2050: even this is doubted by many scientists, who call for a 90 per cent+ approach (Hansen 2009: 242). This is a long way beyond marginal change, and more like a transformation of socio-technical systems; this is an agenda for the 'Low Carbon Economy' (LCE) which is both radical and essential for survival. And as cities are the sites for carbon emissions, and also for new markets and technologies, it appears that the fate of the global climate lies in its global city system.

Underlying the LCE agenda, and all the policies on urban energy supply and demand, are some radically new ways of thinking – that is, paradigms – about resources, economy and society (Ravetz 2009a):

- First, that carbon accounting and trading is intended to become a parallel strand to the mainstream economy. This is could be not so much a marginal adjustment, as a new paradigm for economic activity and resource use – living on a small and fragile planet with a complex set of balances, which are easily destabilized.
- Underlying this is the 'carbon cycle' concept – the flow of carbon through the physical and economic metabolism, with sinks and storage, efficiency and emissions, at various points along the cycle.
- In parallel is the 'embedded carbon' concept, where imported goods or technologies can be allocated their share of carbon emissions that are generated along their supply chains.
- The implication is then to look along the supply chain with a preventative angle, looking for emissions avoided, or activities avoided, which involve disturbing the carbon cycle, or offsets in many shapes and forms.

- This then implies a whole new generation of performance benchmarking and labelling, for production processes, consumer products, consumer services and financial services. In turn there will be an industry for monitoring, accreditation, evaluation, technology assessment, carbon credit banking, carbon insurance, and so on.
- There are also many levels, from the household to the neighbourhood, local authority, multi-area or regional scale. There are difficult choices concerning global effects between local actions, or vice versa. But the emergence of 'transition towns' and 'zero-waste neighbourhoods' shows the scope of public engagement and awareness raising.

Above all there is the challenge of 'policy integration'. It is likely that carbon policy will be vastly more effective when linked to other benefits – in construction and urban development; transport and accessibility; industrial competitiveness and product design; agriculture and landscape management; and so on. This stretches the conventional model of 'silo thinking' and departmental policy towards a more holistic model, which we are only just beginning to grasp (Figure 15.6).

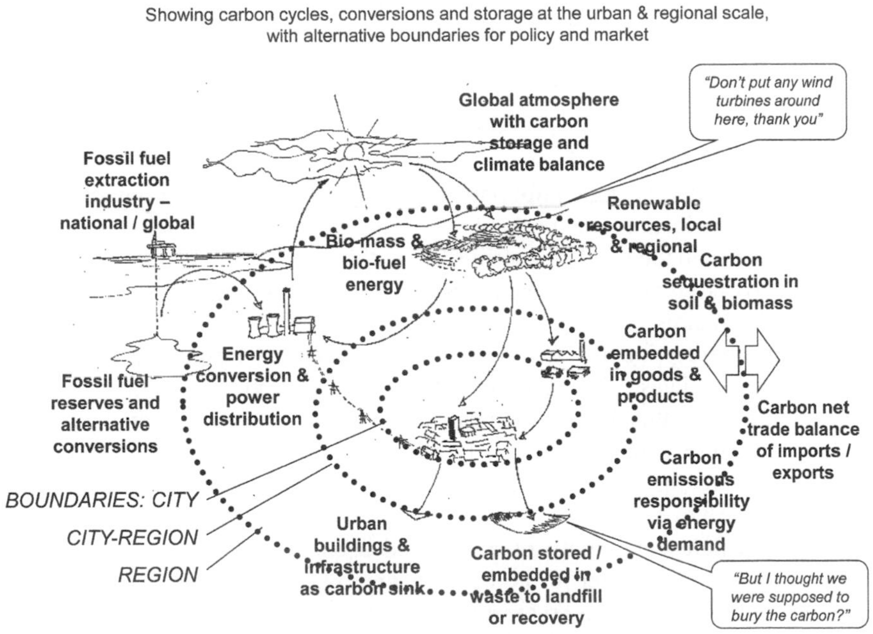

Showing carbon cycles, conversions and storage at the urban & regional scale, with alternative boundaries for policy and market

Figure 15.6 Low carbon economy in the city-region

Institutional Perspective

One way to bring together this complexity is via transition theory, and the 'transition management' agenda towards LCE development paths (DECC 2009). Such transitions are likely to involve combinations of technology, economics, regulation, infrastructure development and lifestyle/consumption choices – all these combined in new and synergistic ways.

The climate agenda is often framed as a 'market failure' agenda with solutions to be found in taxation and trading systems. It is also framed as a 'citizenship' agenda that reflects the changing sense of public or corporate responsibility, from local to global, and so on. In terms of a 'technology agenda' there are major issues on infrastructure development, innovation and supply chain constraints, and technology as an instrument of exploitation. In environmental science terms there are multiple layers of offsets, joint implementation, emissions trading – each of which raises challenges on ethical values, political values, economic distribution and socio-cultural commitment.

Again the underlying paradigm is that of a *relational system* – again, not a simple collective model, but rather an emerging set of relationships and partnerships between many types of organization, in a multi-lateral world of interdependency and mutual responsibility. This can work on many levels – in fact it needs to work on many levels, which combined are more robust than any simple economic or governance system.

For instance, a city-region emissions trading market could be a first step towards an interactive system that links all energy/carbon traders, suppliers, producers and consumers into a single information sharing system. This then involves new cultural–ethical–institutional concepts of responsibility, stewardship, equity and collateral, participation and citizenship. Then, this needs a much more responsive and reflexive system of governance, which can set the market conditions and regulatory standards.

In turn, these principles translate into new modes of business organization, investment vehicles, decision-making structures, virtual networks and communities. And this, in turn, demands new ways of integrated spatial planning and infrastructure, coupled with new forms of citizen/consumer empowerment.

The over-arching concept here is that of a 'relational' system, where the LCE transition is achieved through enhanced multi-level relationships and exchanges, between a whole community of actors and agencies. Figure 15.7 shows the application of this concept to the housing improvement agenda: each pathway involves multiple actors in co-creating a chain of added value, which includes economic, social, cultural and political interactions (Ravetz 2010b). In turn, the development and evolution of such combined pathways requires a high level of social, economic and cultural creativity and innovation.

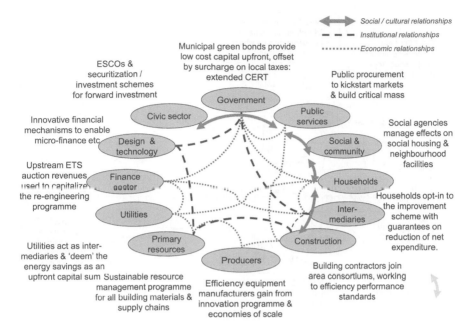

Figure 15.7 Relational low carbon economy

New Thinking: Urban Creativity and Innovation

The agenda for urban creativity and innovation has many dimensions: here we outline three themes. One is the creative cultural perspective as expressed by Florida (2002), and as 'cultural creatives by Ray and Anderson (2000). Another is more regional level thinking on innovation as a non-linear process, and 'systems of innovation' as a relational agenda (Braczyck et al. 1998). This points to an underlying creative process that links the personal to the political.

Creative Industries as Cognitive Systems

The nature of the twenty-first-century urban creative system was explored in a stakeholder consultation programme centred on Manchester and Salford, UK, with a series of stakeholder discussions centred on the digital development project SUREGEN (www.suregen.co.uk). The context was an exploration of the virtual interfaces and digital opportunities between planning, community development, urban creative industries, and so on. This is one application of a broader concept of 'cyberinfrastructure' that relates social and economic emergence to digitally enabled governance, science and education (Ravetz 2009b).

The example in Figure 15.8 shows the exploration of a virtual city in relation to a real city. Here the policy-makers and planners are struggling to understand and manage the complexity of the real city, and so revert to virtual city models in

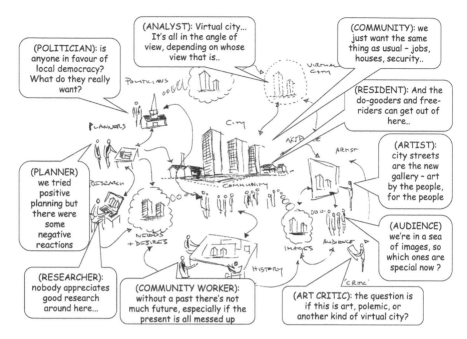

Figure 15.8   Virtual city, creative dialogue

two or three dimensions. In the centre is a community that has been displaced and fragmented through the process of urban and economic restructuring; but through a process of reclamation and visualization of their collective memories and history (perhaps with digital assistance), they are able to mobilize resources and innovate. Some of this involves the urban creative industries, here shown as the commercial fine art sector; this brings its own system of critics, curators, artists, intermediaries, collectors; together with some kind of contested but growing involvement of the community at the back of the room.

The output of this is, first, a set of commentaries and dialogues, in caricature form, between the different groups involved. What emerges is a realization that the very different worlds and worldviews of planners, IT developers, local communities and art dealers are interconnected to some extent, through shared goals and relationships. At the same time there are tensions, contradictions and misunderstandings about these goals, the nature of the reciprocity, and the underlying structures of power and ideology.

This can be explored as a cognitive 'dissonance', a tension or conflict in material and ideological terms, which can be exposed by such a visualization. The policy and governance regime for such interconnected problems should ideally be responsive to such nuances. But policy and governance itself is often constructed in simple linear terms, and driven by political game strategy, and so there is a strong probability of policy failure. This appears to be the case in many urban regeneration situations, where 'creative industries' have a multiple agenda, with gentrification as a subtext of urban cultural renewal.

Urban Cognitive Capitalism

To follow through the story above, we can explore the more 'structural' issues in the peri-urban from a cognitive–cultural–capitalist perspective, as described by Allen Scott (2000).

The creativity agenda brought together a theoretical approach to urban studies, with a very practical menu for action by policy-makers keen to get 'their' city moving upwards in the global urban hierarchy. The context is a global urban creative system, held together by a complex set of relations of competition and cooperation. There may be positive effects in local employment, reclaiming of dereliction and rediscovery of lost cultures. At the same time, there can be negative effects in the expropriation of cultural and physical spaces, with new patterns of exclusion and marginalization.

The implications of this approach are many. For instance the peri-urban territories can be zones that are contested between global and local dynamics, very different from the 'respectable suburbia' of conventional urban studies. The LCE agenda can be seen as a form of contemporary 'niche' capitalism, that buys in the cultural creative classes following in the trail of the eco-bohemians, and offers tokens to maintain the appearance of progress. The creative agenda and cultural industry sector itself exists in a tension between commercializing forces and the creative leading edge, just beyond the mainstream capitalist frontier. Each of these effects can be accelerated by digital social technologies as informational catalysts, and by their nature can be highly uneven in distribution.

Personal and Collective Creativity

The act of creation, or the attainment of a creative state of mind can be characterized as an immersion in the unconscious, as in 'U-Theory' (Scharmer 2007); a dynamic connection, almost as a Zen-like realization (Watts 1957); as a process of 'deductive tinkering' with a random mutation type testing of alternatives (Beinhocker 2006); or as an holistic experiential response to whole systems (Nachmanovitch 1990 and 2007). In visual terms we can imagine a set of linear problems and opportunities, combining to form an emergent realization of a higher order, and the process of attainment in a deep immersion with no guarantee of outcome.

At the same time the individual level of 'cultural creativity' is a small section of a much wider field. We need to re-examine many common activities and processes in the light of the creativity perspective. For instance, for the urban LCE policy agenda above, the social innovation in ecological-institutional economics is crucial – and high levels of interpersonal and collective creativity are needed to enable and encourage that. Even such a simple act as buying food from a shop can be either reductive and alienating; or it can be culturally creative and

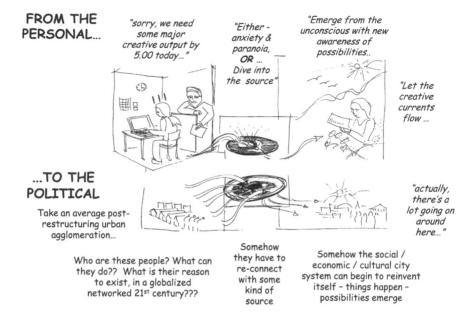

FROM THE PERSONAL...

"sorry, we need some major creative output by 5.00 today..."

"Either - anxiety & paranoia, OR ... Dive into the source"

"Emerge from the unconscious with new awareness of possibilities..

"Let the creative currents flow ..."

...TO THE POLITICAL

Take an average post-restructuring urban agglomeration...

"actually, there's a lot going on around here..."

Who are these people? What can they do?? What is their reason to exist, in a globalized networked 21st century???

Somehow they have to re-connect with some kind of source

Somehow the social / economic / cultural city system can begin to reinvent itself – things happen – possibilities emerge

Figure 15.9   Creativity: From the personal to the political

invoking other interactions besides the function of purchase. If we scale up such a small transaction up towards the urban economy or urban society, we can start to envisage how a relational creativity approach can enable the evolution of the 'relational' economy which is needed for the LCE. A similar agenda applies in the peri-urban territory, as in the example in Figure 15.9.

New Thinking: A Relational Approach

Here we step back and look for underlying concepts on creativity and innovation as evolutionary processes. This is not a new theme; but what we can contribute here is an application to cities and their global challenges under fast changing conditions, and some new insights into methods that bring together analytical and experiential approaches.

'Relational Thinking' has become a topical framework for debate in various fields in recent decades. There is a perceived 'relational turn' in human geography, which can then work with relationships between actors and ideologies in space and time (Darling 2009, Jones 2009, Murdoch 2006). A 'relational economic geography' helps to understand the extended supply chains, value chains and knowledge flows between countries and sectors (Bathelt and Gluckler 2003, Sunley 2008). In management and organization theory, 'relational governance' is a powerful way to explore the internal dynamics of actors and agents in complex institutions (Ferguson et al. 2005, Grandori 2006). This overlaps onto psychoanalytic theory and the shift from 'transactional' to 'relational' analysis (Cornell and Hagarden 2005, Hollway 2008).

Here an application of relational thinking has been developed as a practical response to the wider challenge of systemic complexity, indeterminacy, and criticality (Ravetz 2010b). The application draws on experience of 'emergence' – how complex natural and human systems such as cities, self-organize, develop and make transitions, and build their 'intelligence' and adaptive capacity (Portugali 2000). It draws from many lines of current thinking: for instance, the Institutional Analysis Framework which looks at the underlying structure of rules and norms (Ostrom 2005): another looks at sub-cultures such as the 'cultural creatives' (Ray and Anderson 2000). Looking at human systems as evolutionary processes, we can see several stages, overlapping and often fuzzy in practice. These are summarized, with examples from experiences of creative cities in Figure 15.10. Overall, we can see an evolutionary process working in some kind of cycle: at each stage there is a process of emergence of self-organizing complexity. There are approximately six main stages in this (although each is fuzzy and overlapping, and the task of mapping and analysing can start or finish at any stage that is relevant to the task).

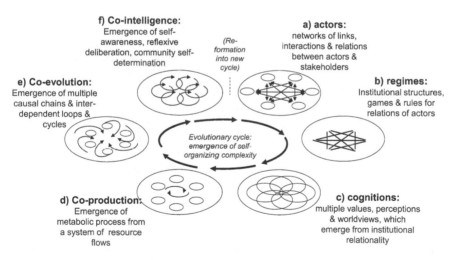

A cyclic process of emergence of increasing complexity & self-organization

f) Co-intelligence: Emergence of self-awareness, reflexive deliberation, community self-determination

(Re-formation into new cycle)

a) actors: networks of links, interactions & relations between actors & stakeholders

e) Co-evolution: Emergence of multiple causal chains & inter-dependent loops & cycles

b) regimes: Institutional structures, games & rules for relations of actors

Evolutionary cycle: emergence of self-organizing complexity

d) Co-production: Emergence of metabolic process from a system of resource flows

c) cognitions: multiple values, perceptions & worldviews, which emerge from institutional relationality

Figure 15.10 Relational thinking: Analytic approaches

1. *Actor processes*: Networks of interactions and exchanges, between actors/ stakeholders (who may be individuals, organizations, other units). For instance, urban communities are organized not only by economics or material functions: there are many kinds of interactions at many scales.

Creative processes often depend on multiple relationships, as in the example artist above, working with a wide community of critics, dealers, media, and audiences.

2. *Institutional processes*: Regimes, structures, games and other kinds of rules, which emerge from actor processes. Cities are a physical embodiment of rules, from the arrangement of streets or traffic signs, to legal or financial systems. The creative industries often revolve around rules that are more tacit, such as what is acceptable to visualize, or how to sell art to collectors.

3. *Cognitive processes*: Multiple values, perceptions, worldviews and ways of thinking, which emerge from institutional processes. Cities are often focal points for contrasting or dissonant worldviews: cultural, political, psychological, ethical and others; and each of these tends to underlie urban systems that are apparently 'rational' such as economics or governance. Cities that are more 'creative' are more likely to work with this wider range of resources and perceptions.

4. *Co-production*: Emergence of a metabolic process from a system of resource flows. In the urban example, local markets or clusters can form, where producers, sellers, landlords, intermediaries and consumers each play a role in an extended socio-cultural business model. Co-production in creative industries shows how the efforts of one actor can be amplified into a cycle of economic development, reinvestment, and renewal of confidence and aspiration.

5. *Co-evolution*: Emergence of multiple causal chains, with interdependent loops and cycles, until a tipping point and the larger system 'flips' into a rapid transition. Such transitions in cities can produce local concentrations of extreme wealth or poverty; renewed identity in some cities, or decline in others. This concept of transition also expresses the 'creative moment' (although in practice creative activities also depend on routine work).

6. *Co-intelligence*: Emergence of self-awareness, reflexive deliberation, community self-determination and other 'higher purpose' processes. Generally a system that is more interconnected, self-aware, innovative and 'intelligent', will be more durable, adaptable, resilient and sustainable. For the urban example, this suggests the mobilization of bottom-up social enterprise, as much as the mobilization of global investors for flagship projects.

We have found that this relational thinking and mapping approach can be very helpful in finding practical pathways through a jungle of complexity, uncertainty and controversy; or in dealing with 'wicked' problems that are outside the scope of rational solutions. Conflict between policy regimes, ideologies or cultures can be understood as stages in an evolutionary process. Solutions and opportunities can be explored, not only as one-off fixes, but in a context of an evolutionary journey, which builds networks of collaboration and collective intelligence. The creative process can be seen as an evolutionary process for economic and institutional systems (Beinhocker 2006). Such processea can be framed as simple actors who collectively build emergent complex systems, as in the cellular automata approach (Epstein and Axtell 1996). It can also be framed as more complex actors, that is, human actors, who collectively build emergent complex systems, with the added dimension of co-intelligence or reflexive awareness and deliberation.

Relational thinking also benefits from widening the channel from text and rational evidence. To enable more creative and holistic thinking that is more responsive to patterns of co-interaction and co-intelligence, it can use multiple channels for debate and communication, as shown next (Figure 15.11).

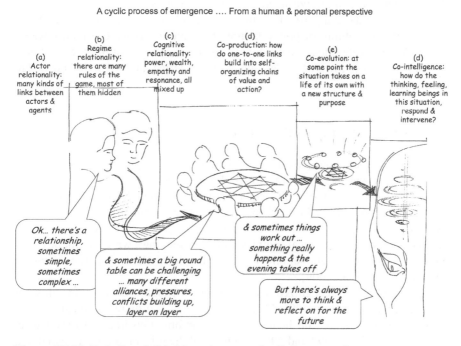

Figure 15.11 Relational thinking: An experiential approach

As above, there is an overlap of global challenges with local responses – global climate and urban systems: local creative capacity and innovation. This agenda in many ways holds the future of global bio-physical survival. But from experience we know that cities and urban policy agendas are often problematic, messy, divided, controversial, out of balance, often on the edge of chaos and creativity.

Much urban development can be seen as the intersection of power, wealth and ideology, played out on a physical and spatial basis. This agenda is also an example of the 'post-normal' – high levels of uncertainty, complexity, controversy, urgency, impacts and strong probability of policy failure (Funtowicz et al. 2002). To explore this more we can identify some of the different kinds of knowledge involved, in terms of a diversity of 'cognitive agendas':

- 'Technical agendas': Formal theoretical models, natural or social science, often quantitative and implemented in software.
- 'Systems agendas': More holistic systemic concepts, mind mapping or soft cognitive mapping.
- 'Discourse agendas': Underlying, fuzzy, qualitative issues, conflicts, cultural waves, ideologies, and so on.
- 'Policy agendas': Real-time action-focused problems, issues, risks, opportunities, strategy and tactics.

Working with such 'cognitive diversity' is an important step forward in responding to the LCE agenda, and particularly the ideological 'dissonance' and conflict that suggests the probability of policy failure. To explore this we can look towards various kinds of heterodoxy – behavioural, institutional or evolutionary economics; we can also look towards wider and more experiential forms of knowledge production.

New Methods: Multi-channel Dialogue

From the examples above we can begin to map out the scope of a 'visual thinking' approach that enhances the relational thinking process. Using a variety of interactive techniques, the result aims towards a wider and deeper response to the global urban challenges – low carbon, peri-urban development and so on – than is likely with rational text. Such visual thinking then points the way towards more holistic patterns of 'ecological complex adaptive thinking', which might be better equipped than 'linear rational thinking', for the global challenges now facing us (Waltner-Toews et al. 2009).

Visualization is recognized as a powerful enabler of new insights on complex problems (Horn 1998, Tufte 1983). There is a more technical-analytic approach which focuses on human-computer-information interfaces (Huang et al. 2010, Humphrey 1999 and 2008). In parallel there is a more experiential and creative approach, which uses the visual as one way to access the unconscious (Nachmanovitch 1990 and 2007). Through a wide variety of channels and techniques, audiences and cultural exchanges, visualization can offer to the research task:

- trans-disciplinary perspective, grounded in social experience, with open cognitive processes;
- applications on a spectrum, from systems analysis and problem mapping to experiential envisioning.

The alternative options diagram here maps out a field of possible visualization approaches with two main axes (Figure 15.12):

charting the axes in a knowledge systems context

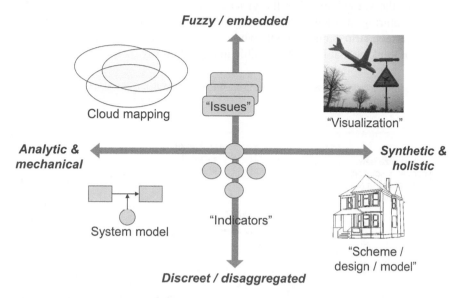

Figure 15.12 Systemic visualization approaches

- from analytic/mechanical (focusing on abstractions) to synthetical and holistic (focusing on figurative substance);
- from discrete/disaggregated (specific purpose) to fuzzy/embedded (general purpose or aesthetic communication).

This analytic approach is useful for mapping out the possibilities. But there is an alternative approach to visual 'actualization', which shows in contrast how the visualization can speak for itself, rather than as an explanation of text.

Relational Visualization

'Graphic facilitation' is now established as a valuable technique in process workshops, with a London institute and active network of practitioners (www. vizthink.co.uk). Independently, the concept of *relational visualization* has arisen from a 10-year programme of strategy and futures workshop processes, where visual material produced on site or off site was often a powerful stimulus and catalyst to creative group thinking. By extension, the visual material then became more embedded in the process itself (for examples, see Ravetz 2009c). There are broadly two strands in this:

- In the first strand is visualization *as* a process – that is, used in workshop or discussion situations – visioning, consensus building, conflict mediation, strategy forming, negotiation and bargaining.
- In the second strand is the visualization *of* a process – that is, directly capturing dialogue, debate, argument and even conflict. The classic cartoon strip is one example where a dialogue can communicate a pattern of thinking that is hardly possible in any other way.

Participative process for generating opportunities & deliberating solutions

f) Visualize new synergies & solutions with feedback in real time

a) Visualize scope of problem

b) Represent the key actors, interactions & relationships

e) Develop opportunities for added value in working together

c) Feed comments & dialogue into the actors & generate feedback

d) Explore alternative ways of thinking & worldviews

Figure 15.13 Relational visualization

We have found that a series of steps can be achieved in small groups, approximately following the 'relational mapping' cycle above, depending on the balance of process facilitation and graphic facilitation:

- Visualize a problem or 'agenda', and explore its dimensions and boundaries.
- Simulate the roles of characters and forces (either by discussion, role play or direct visualization).
- Explore relationships, arguments, conflicts and other interactions (by feeding words and dialogues into the mouths of characters on sticky notes).
- Explore alternative ways of thinking (by overlaying and playing with the comments and dialogues). This can explore what is 'interesting', not just direct positives and negatives, or costs and benefits.

- Develop opportunities for added value in working together: using cartoons, flow charts and other media, looking at the possibilities for increasing added value.
- Visualize new synergies and solutions, based on real-time feedback. This can steer towards an interim outcome – rather than trying to 'solve the problem', explore what is interesting and relevant to further discussion.

Such a method is still at the prototype stage, and will continue to learn from a great many alternative approaches in visual thinking, graphic facilitation, visioning and mediation, and so on (Figure 15.13).

Some Examples

Three examples demonstrate the range of creative action in the urban and peri-urban agendas for LCE. Each is an ongoing development in progress, so readers are advised to check the urls given, because this information will be more up to date at any point in time.

From Futuresonic to Futureeverything

The Environment 2.0 Open Lab 'unConference event' generated some innovative forms of participatory observation, mapping and mass participation on the environment, or in issues surrounding 'citizen science'. This was a highlight of the Social Technologies Summit at Futuresonic 2009, now re-titled Future Everything (www.futuresonic.com and www.futureeverything.org). Futuresonic 2009 presented fun and surprising outdoor events to discover the urban climate and wildlife of Manchester in new and unexpected ways. There were games and experiences in the streets and online, using locative and mapping technologies to develop 'new senses'.

Many urban environments are insulated from both nature and the consequences of our actions, as surely as the tarmac of the road cuts off the earth beneath. How can web technologists interested in the interface between digital and environmental footprint, and artists concerned with collaborative intervention in the environment, bring a fresh approach to 'citizen science'? The Environment 2.0 art exhibition features 30 international artists and 10 world premiers, including a public recital of the recent report by the Intergovernmental Panel on Climate Change (IPCC), an art device for striking it rich by prospecting for oil in the city centre, and an installation of ceramic plates with portraits of presidents created by exposure to smog. Each of these is an example of creative action in the new agenda of 'social urban technologies' (Figure 15.14) (Ravetz 2009b).

Multiple applications of Web 2.0 and beyond

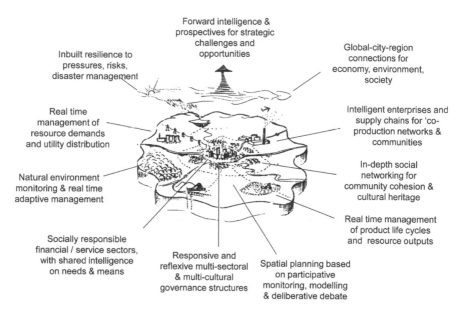

Figure 15.14 Social urban technologies

Manchester Is My Planet

In 2005 'Manchester Is My Planet' (MIMP) emerged from a unique local authority, business and academic partnership. It aimed to raise awareness and build capacity for a future low-carbon Greater Manchester through innovative green energy projects. The MIMP initiative made use of strong branding and upbeat messaging that individuals can get involved and make a difference. By March 2008 almost 20,000 people had pledged 'to play my part in reducing this city's greenhouse gas emissions by 20% before 2010, to help the UK meet its international commitment on climate change'. A new programme is now training volunteers to help communicate the message to others (details on manchesterismyplanet.com/).

Through the pledge campaign, encouraging individuals to reduce their carbon footprint, and a number of innovative green energy projects, MIMP aims to make Manchester the coolest and greenest city on earth. It is estimated that CO_2 savings of 100,000 tonnes will be achieved across the city-region by 2010.

Other projects underway include promoting the development of a Circle of Wind (turbines) around the region, a Carbon Fund to offset local emissions, local sustainable energy projects and low-carbon industrial estates. Experience is being shared with other European cities through two new EU-funded projects.

The most important feature of MIMP is that of cultural identification. An initial survey asked a random control group 'would you make a personal sacrifice to help avert climate change', and the responses were 10–20 per cent at best. A parallel survey asked a second group 'would you make a personal sacrifice to help MANCHESTER lead the fight against climate change': the responses to this were 70 per cent positive. This shows clearly how a cultural identification – in this case related to social belonging to the home city in a similar way to that of football – can turn the problem of public attitudes into an opportunity.

Peri-urban Local Food Systems

The blue–green infrastructure agenda is one that has been shown many times to increase economic investment, health and education, and social well-being. In practice it is often sidelined and fragmented between different parts of the government machine. However, there are many creative initiatives that find new ways of combining the roles of public, private, civic and community sectors, such as the community forests and country parks. A new generation of partnership organizations show new possibilities in multi-level, multisectoral networked types of strategy-making; and these in turn can help to mobilize local social enterprises. One example is the Todmorden 'Incredible Edible' scheme for local food cultivation (www.incredible-edible-todmorden.co.uk). This is a fast moving entrepreneurial scheme that in the space of three years has transformed the way that this town of 20,000 people thinks about food. It has drawn in landowners, farmers, schools and school children, markets, delivery merchants, supermarkets, health services, restaurants and tourism operators (Figure 15.15).

A visual mapping again shows a rich pattern of interconnections and relationships between social groups, which demonstrates the relational thinking dimension of such systems, which may not fit easily with conventional economic or governance models. In particular the 'creative city' agenda – here focused on social enterprise in a peri-urban small town situation – is one that has the potential to transform costs into benefits, and risks into opportunities. To explore this requires a lateral view on food and its significance in social, economic, political and cultural levels. It also suggests openness to a wider set of 'factors for success', including space for creative action and debate; public/private/ social collaboration: multi-level self-organizing governance systems; active knowledge and learning networks; innovation and enterprise culture; policy intermediation and enabling:; and so on. What begins as a set of co-interactions and co-production chains, becomes a highly evolved co-intelligence.

example of local food development scheme at www.incredibleedibletodmorden.org.uk

Key factors of success:

* Space for creative
  action & debate
* Public / private /
  social collaboration &
  investment
* Multi-level self-
  organizing
  governance systems
* Knowledge and
  learning networks
* Innovation &
  enterprise culture
* Policy intermediation
  and enabling of inter-
  mediaries

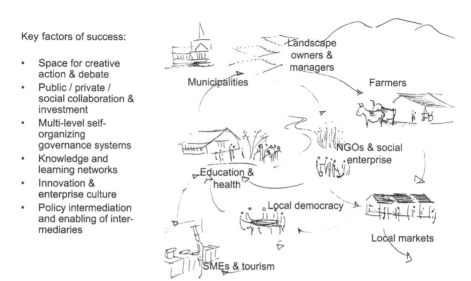

Figure 15.15 Creative local integrated development

Implications

Overall, the relational thinking approach set out here has powerful implications for the 'sustainable and creative cities' agenda:

* Urban- regional planning and the peri-urban challenge: This is a new kind of governance challenge that requires innovation on many levels. In turn the peri-urban situation can offer new spaces for creative processes, as in the local food example above.
* Low carbon urban development: This requires creative innovation on many levels. It begins with positive relationships between many types of actors, and then begins to develop extended chains of value added and co-production.
* Creative industries and policies for intervention, either in the conventional 'creative quarter', or a wider 'urban system', can be seen in their context. As cities re-invent themselves for new challenges, the physical and developmental space for creative, deliberative, participative and 'relational' processes, may be at least as important as the conventional range of studios, galleries and opera houses.

A further implication is that to evaluate policies for creativity, an apparently objective measurement of fixed factors (number of patents or web designers, etc.), may be only a partial solution. So, we need to focus on the more challenging task of looking for emergent qualities and opportunities for interactions and co-productions between different actors and social groups. This will be helped by use of multiple methods and tools – visualization and visioning, scenario testing, foresight programmes, road mapping and so on (Georghiou et al. 2007). In this task the reflexive and proactive development of 'co-intelligence' of actors and agents is essential. We look forward to a new generation of relational thinking approaches to creative and sustainable cities, which can explore new possibilities in collective intelligence and creative innovation.

Bibliography

Bathelt, H. and Gluckler, J. 2003. Toward a Relational Economic Geography. *Journal of Economic Geography*, 3(2), 117–44.

Beinhocker, E. 2006. *The Origin of Wealth: Evolution, Complexity and the Radical Remaking of Economics*. New York: Random House.

Borja, J. and Castells, M. 1997. *Local and Global: The Management of Cities in the Information Age*. London: Earthscan.

Braczyck, H., Cooke, P. and Heidenreich, M. (eds) 1998. *Regional Innovation Systems*. London: UCL Press.

Clay, C. 1994. *Real Places: An Unconventional Guide to America's Generic Landscape*. Chicago: University of Chicago Press.

Cornell, W.F. and Hargaden, H. (eds) 2005. *From Transactions to Relations: The Emergence of a Relational Tradition in Transactional Analysis*. London: Haddon Press.

Darling, J. 2009. Thinking Beyond Place: The Responsibilities of a Relational Spatial Politics. *Geography Compass*, 3(5), 1938–54.

de Roo, G. and Silva E.A. 2010. *A Planners Meeting with Complexity*. Aldershot: Ashgate.

DECC (Department of Energy and Climate Change). 2009. *The UK Low Carbon Transition Plan*. London: DECC.

Epstein, J.M. and Axtell, R.L. 1996. *Growing Artificial Societies: Social Science From the Bottom Up*. Cambridge, MA: Brookings Institution and MIT Press.

Ferguson, R.J., Paulin, M., Moslein, K., Muller, C. 2005. Relational Governance, Communication and the Performance of Biotechnology Partnerships. *Journal of Small Business and Enterprise Development*, 12(3), 395–408.

Florida, R. 2002. *The Rise of the Creative Class: And How it's Transforming Work, Leisure, Community, and Everyday Life*. New York: Basic Books.

Funtowicz, S.O., Martinez-Alier, J., Munda, G. and Ravetz, J. 2002. Multi-criteria-based Environmental Policy, in *Implementing Sustainable Development: Integrated Assessment and Participatory Decision-making Processes*, edited by H. Abaza and A. Baranzini. Cheltenham: Edward Elgar, 53–77.

Georghiou, L., Cassingena Harper, J., Keenan, M., Miles, I. and Popper, R. (eds) 2007. *International Handbook on Foresight and Science Policy: Theory and Practice*. Cheltenham: Edward Elgar.

Grandori, A. 2006. Innovation, Uncertainty and Relational Governance. *Industry and Innovation*, 13(2), 127–33.

Hansen, J. 2009. *Storms of My Grandchildren*. London: Bloomsbury Publishing.

Hall, P. and Pain, K. 2006. *The Polycentric Metropolis: Learning from Mega-City Regions in Europe*. London: Earthscan.

Hollway, W. 2008. The Importance of Relational Thinking in the Practice of Psycho-social Research: Ontology, Epistemology, Methodology and Ethics, in *Object Relations and Social Relations: The Implications of the Relational Turn in Psychoanalysis*, edited by S. Clarke, P. Hoggett and H. Hahn. London: Karnac, 137–62.

Horn, R.E. 1998. *Visual Language: Global Communication for the 21st Century,* Macro VU Press.

Huang, M.L., Nguyen, Q.V. and Zhang, K. (eds) 2010. *Visual Information Communication*. Berlin: Springer.

Humphrey, M.C. 1999. A Graphical Notation for the Design of Information Visualisations. *International Journal of Human-Computer Studies*, 50, 145–92.

—— 2008. *Creating Reusable Visualizations with the Relational Visualization Notation*: Interactive Visual Communication working paper. Available at: www.iviz.com.

Jones, M. 2009. Phase Space: Geography, Relational Thinking, and Beyond. *Progress in Human Geography*, 33, 487.

Murdoch, J. 2006. *Post-structuralist Geography: A Guide to Relational Space*. New York: Sage.

Nachmanovitch, S. 1990. *Free Play: Improvisation in Life and Art*. London: Penguin.

—— 2007. Bateson and the Arts. *Kybernetes*, 36(7/8), 1122–33.

Neuwirth, R. 2005. *Shadow Cities: A Billion Squatters, A New Urban World*. London and New York: Routledge.

Nilsson, K., Sick Nielsen, T., Pauleit, S. and Ravetz, J. 2008. A 'Plurel' Approach to Peri-urban Areas. *Town and Country Planning*, December, 519–24.

Ostrom, E. 2005. *Understanding Institutional Diversity*. Princeton, NJ: Princeton University Press.

Portugali, J. 2000. *Self-Organization and the City:* Berlin: Springer-Verlag.

Ravetz, J. 2000. *City-Region 2020: Integrated Planning for a Sustainable Environment*. London: Earthscan (Chinese language version, transl. Jian-Cheng Lin and Tian-Tian Hu, Taipei, Chan's Publishing Co. (Taiwan)).

—— 2009a. Towards the Low Carbon Economy: Transforming the Way Things Work. *Environmental Scientist* (Special issue on environmental futures), 18(2), 25–30.

—— 2009b. New Technology Applications: Risks and Opportunities for Environmental Governance. *Environmental Scientist* (Special issue on environmental futures), 18(2), 36–52.

—— 2009c. Graphic Visualization, in *Foresight for Smart Globalization*, edited by C. Bezold et al. West Virginia: Institute of Alternative Futures. Available at: www.altfutures.com/pubs/propoor/Foresight_For_Smart_Globalization_ Summary.pdf.

—— 2010a. Peri-urban Ecology: Green Infrastructure in the 21st Century Metro-scape, in *Handbook of Urban Ecology*, edited by I. Douglas. Oxford: Routledge, 599–620.

—— 2010b. Rethinking Low-carbon Strategy in the Regions: Applications of the Territorial Principle in a Networked Landscape, in *Regions and the Environment: Proceedings of the Regional Studies Association*, edited by A. Beauclair and E. Mitchell. Seaford, UK: Regional Studies Association.

—— 2010c. *'It's not the real economy, stupid': Transforming Management: Original Thinking Applied* (Manchester Business School on-line publication). available at: tm.mbs.ac.uk/features/it%E2%80%99s-not-the-real-economy-stupid/.

Ravetz, J., Fertner, C. and Sick Nielsen, T.S. 2011. The Dynamics of Peri-urbanization, in *Peri-Urban Futures: Land Use and Sustainability*, edited by S. Pauleit, S. Bell and C. Aalbers. Berlin: Springer.

Ray, P.H. and Anderson, S.R. 2000. *The Cultural Creatives: How 50 Million People are Changing the World*. New York, Three Rivers Press.

Rittel, H. and Webber, M. 1973. Dilemmas in a General Theory of Planning. *Policy Sciences*, 4, 155–69.

Roberts, P., Ravetz, J. and George, C. 2009. *Environment and City: Critical Perspectives on the Urban Environment around the World*. Oxford: Routledge.

Rockström, J.W. et al. 2009. Planetary Boundaries: Exploring the Safe Operating Space for Humanity. *Ecology and Society*, 14(2), 32–74.

Rotmans, J., Kemp, R. and van Asselt, M. 2001. More Evolution Than Revolution: Transition Management in Public Policy. *Foresight*, 3(1), 15–32.

Sandercock, L. 2003. *Cosmopolis II: Mongrel Cities of the 21st Century,* London; New York: Continuum.

Sassen, S. 1994. *Cities in a World Economy*. Thousand Oaks, CA: Pine Forge Press.

Scharmer, C.O. 2007. *Theory U: Leading from the Future as it Emerges*. California: Berrett-Koehler Publishing.

Scott, A.J. 2000. *The Cultural Economy of Cities: Essays on the Geography of Image-Producing Industries*. NY, Sage.

—— 2006. Creative Cities: Conceptual Issues and Policy Questions. *Journal of Urban Affairs*, 28(1), 1–17.

Soja, E. 2000. *Postmetropolis: Critical Studies of Cities and Regions*. Malden, MA, Oxford: Blackwell.

Sunley, P. 2008. Relational Economic Geography: A Partial Understanding or a New Paradigm? *Economic Geography*, 84(1), 1–26.

Tufte, E.R. 1983. *The Visual Display of Quantitative Information*. New York: Graphics Press.

UNDESA (United Nations Department of Economic and Social Affairs). 2007. *World Urbanization Prospects: The 2007 Revision*. New York: United Nations. Available at: www.un.org/esa/population/publications/wup2004/wup2004dh. pdf.

Waltner-Toews, D. with Kay, J. and Lister, N. 2009. *The Ecosystem Approach: Complexity, Uncertainty and Managing for Sustainability*. New York: Columbia University Press.

Watts, A. 1957. *The Way of Zen*. New York: Pantheon.

Whitaker, M.D. 2006. *Toward a Bioregional State: A Series of Letters About Political Theory and Formal Institutional Design in the Era of Sustainability*. Indiana: iUniverse Publishing.

Chapter 16

Innovative Geospatial-visualization Techniques to Collect Creative Input in Participatory and Sustainable City Planning

Eduardo Dias, Henk J. Scholten, Arda Riedijk and Rob van de Velde

Introduction

City planners have the task of designing and planning the future of the city. While sustainability is always at the top of the agenda as one of the most important design criteria, the definition of 'sustainable city' and the solutions to promote it are not objective and can depend on numerous criteria. Consequently, citizens are being called to participate in the planning process to stimulate the development of creative solutions that more closely meet the sustainability needs of the city inhabitants. One of the most important planning processes of the city where citizens can make creative contributions is the spatial planning process, where the future city is played out and we can visualize how the city will change and look like in the future. Spatial planning is strongly supported by geospatial technology where maps and geographic information play a crucial role. In order for the citizens to contribute their ideas about the development of the city, it is crucial that they have a common understanding of the geographic space and the already proposed plans and changes. To facilitate this common understanding, geospatial visualizations can help the cognitive process of translating the abstract plans into real world changes.

In this chapter we will see that geospatial-visualizations are part of an extremely broad array of institutional arrangements that contribute to a successful – or unsuccessful – integration of geospatial-visualization tools and techniques into spatial planning processes in order to enable the sustainable city and creativity.

Geospatial-visualizations are, as defined by Kraak (2003):

> visual [...] displays designed to explore [geo]data and through that exploration
> to generate hypotheses, develop problem solutions and construct knowledge.
> Maps and other linked graphics play a key role in this process.

We use the term *tools* to denote the technological instruments, know-how and equipment needed to implement geospatial-visualizations, and *techniques* to mean the methods, practices and procedures employed by organizations that are introduced or adapted in order to implement new geospatial-visualization tools. It is acknowledged that geospatial-visualizations have great potential in the planning process (Dias et al. 2003) and while the geospatial field has showed enormous developments in the last years both in functionality and sales growth, there is still a decelerate adoption by the organizations responsible for sustainable planning.

The central question of this chapter is: *Which factors determine the successful integration of geospatial-visualization tools and techniques for creative input into organizations responsible for participatory sustainable city planning?*

To answer this question, we first break it down into three sub-questions:

- What can be learned from the scientific literature about the adoption and acceptance by organizations of innovative geospatial-visualization tools and technologies?
- Which factors determine the successful implementation of geospatial-visualization tools and techniques from an organizational point of view?
- What is the state of the art regarding the use of geospatial-visualizations by governmental organizations, and, more specifically, what can be said about the development of spatial data infrastructures in Europe?

The next section starts with a vision of the future role of geographic information systems (GIS) in participatory spatial planning. We emphasize the dynamic nature of IT and defend the idea that organizations should follow these dynamic trends and keep their tools for participatory processes up to date and attractive to use. However, it may not always be possible to do this in practice. Therefore, in the third section we describe how new technologies, and specifically Geo-ICT (information and communications technology), are accepted and implemented. We will discuss several theories that explain how they are taken up within organizations and diffused through society, and the user's role in the diffusion-process. We compare several transition theories, discuss the mechanisms and operational requirements from different perspectives, and consider the user perspective. Before an organization can implement new Geo-ICT technologies it must first have a stable infrastructure for storing and sharing spatial data, a Spatial Data Infrastructure (SDI). The fourth section explains the initiatives taken by governments to set up SDIs, considering and comparing the efforts taken at both the national and international level. In the fifth section we present a number of case studies to illustrate the role of Geo-ICT for public participation. In the sixth and final section we present our overall conclusions and discuss the lessons learned.

Visioning about Geo-ICT for Public Participation

Web 2.0: Challenging the Traditional One-way Communication Protocols

GIS technology was developed in the days when information had a less dynamic character, from the 1960s until the early 1990s. De Man (2003) calls this period the 'prior information age'. In his view, information at this time was supply driven and produced in standard presentation formats. Furthermore, cartographic organizations were large and prestigious and collected and used spatial information according to their own requirements. The processes and structures for translating data into information were hierarchical, monolithic and monopolistic, and were generally in the hands of governments. As ideas about governance, accountability and transparency became widely debated within and among governmental organizations, these tendencies were gradually replaced by networking and multi-party information infrastructures. Data collection and production, as well as the presentation of data, became more creative and customized to the demands of other parties.

These trends were accompanied by increasing attention to public participation. Not surprisingly, researchers and practitioners started to develop methods and techniques for facilitating public participation that make use of geographic information. These pragmatic approaches focused on supporting various stages of participatory planning processes, such as disseminating planning-related information online, involving more stakeholders in planning, making analyses more easily understandable through the use of visualizations and weighting alternatives using graphical user interfaces (Sieber 2006). We agree with Carver (2001) that, parallel to these developments, GIS has gradually been transformed into a tool for the masses, driven to a large extent by the rise of internet-based GIS applications. In particular, since the introduction of geo-referenced services such as Google Earth, Google Maps and Virtual Earth, casual users can interact and see spatial data that were previously only available to GIS experts. Any citizen that is interested in information about spatial transformations wants to be able to find that information quickly and easily.

Modern information services are moving on from traditional one-way communication protocols. The introduction of Google Earth is a good example. In 2005, after the acquisition of Keyhole (Google 2004), Google shocked the geo-information sector with the launch of the improved Google Earth. How was it possible that this company had accomplished what so many other companies and government institutions could only dream of, namely to construct a web-based infrastructure, accessible to the whole world, for viewing high-resolution aerial photographs of every location on earth with fast and intuitive multi-dimensional geospatial-visualization interface. The introduction of Google Earth established a worldwide standard for obtaining and sharing geo-spatial data publicly. Unique to such new information and communication platforms is the combination of worldwide coverage, powerful visualization, intuitive three-dimensional (3D) interfaces and a heavily user-oriented approach that enables the user to obtain and exchange geospatial data very easily (Riedijk et al. 2006). It was just a matter of weeks before frequent internet users started to use and explore Google Earth. Of course, the first thing people do is to find their house, their school, their car, or their favourite holiday location. But the real enthusiasts soon started to experiment with the sharing function of Google Earth, which makes it possible, for example, to design routes along sites of historic, tourist or aviation interest. These files are shared among users of special forums, such as the Google Earth Community. This sharing function of Google Earth is an example of what in the ICT world is called 'Web 2.0', which stands for the second-generation internet technology. More recently, online library services have emerged in which people can upload information or data they wish to share or make available for anyone else. Examples of existing libraries for files are Flickr (photos), YouTube (videos), Scribd (documents) and Slideshare (presentations). Blogs, wikis and podcasting are other examples of Web 2.0 internet technologies called 'social software' and are characterized by user-created content. They are services that connect people through a shared interest in information. The software is open, which means that the flow of microcontent between domains, servers and machines depends on two-way access (Alexander 2006). With Web 2.0 the internet has become a place to share and create information, rather than to just collect information. Downes (2005) speaks of a new attitude, rather than a new technology:

> In a nutshell, what was happening was that the Web was shifting from being a medium, in which information was transmitted and consumed, into being a platform, in which content was created, shared, remixed, repurposed, and passed along. And what people were doing with the web was not merely reading books, listening to the radio or watching TV, but having a conversation, with a vocabulary not just of words but of images, video, multimedia and whatever they could get their hands on. And this became, and looked like, and behaved like, a network.[…] Web 2.0 is an attitude, not a technology. It's about enabling and encouraging participation through open applications and services.

Making GIS available to the masses implies that the success of GIS data and applications depends to a great extent on the ease with which an application can be used; in other words, people that do not have GIS skills should be able to use them. To respond adequately to this trend, governmental institutions need to shift from internal communication within spatial projects towards external communication. It seems that the trend called Web 2.0 is starting to create a new attitude towards communication that requires open access to all kinds of information. For organizations this means that they will have to deal with a more external orientation that serves users' needs. This growing external orientation is expected to place additional demands on geospatial-visualizations. They will no longer be used solely by experts, but by everybody affected by spatial planning projects. If organizations want to follow this trend, they will have to adopt a usability standard for the geospatial-visualizations they use. Moreover, technologies in the field of geospatial-visualization are evolving rapidly. Organizations need to be aware of this and should be flexible enough to adjust their standards and goals where necessary.

Going with the (Technological and Social) Flow

As we have seen, people are getting used to finding all the information they need on the internet. Moreover, they are getting used to giving feedback on that information instantly, also through the internet. Government organizations now have to ask themselves whether it is necessary to respond to this development and whether they are 'ready' to use new ICTs to involve citizens in spatial planning. Going back to the Google Earth example, would it not be a great advantage if people could log onto their local authority's website to consult the land use plans for their neighbourhood? The maps and plan scenarios could be viewed on a Google Earth type of interface, where citizens could leave their comments or ideas about the plan proposals using the special tool in the interface. They could also see any comments or ideas already made by their neighbours. This would be an example of participatory spatial planning in which the public enters the stages of the planning process *prior* to the legal participation procedures that allow citizens to object to ready-made plans. A survey conducted in the Netherlands showed that citizens and small businesses rarely raise questions about specific issues in existing local land use plans or other spatial plans. The type of questions they pose are: Do I need a building permit? What am I allowed to do, and are my neighbours allowed to do what they are doing? What are the consequences of the plans for my living environment in terms of noise, traffic and green space? They also ask very diverse questions about the planning procedures and want to be involved in the planning process as early as possible so that they can influence the outcome (Novio Consult Van Spaendonck 2006). This shows that there is indeed a clear demand for consulting spatial information in a user-friendly way on the internet. Some Dutch municipalities already make their local land use plans available for consultation online. In fact, in 2008 all Dutch municipalities are legally obliged to publish any new plans online in a portal called Spatial Planning Online (De Swart 2007). At the moment, though, services that allow the public to react to these plans in the early stages of their development are rare. To make proper use of the possibilities of the new possibilities such as Google Earth, Vritual Earth (Bing) government organizations must be prepared to stimulate openness and transparency. By making information and data accessible, they would be challenging the traditional relationships between experts and members of the public by positioning the public as experts on their living environment (Petts and Leach 2000).

Practice shows that technology is not a barrier at all. Current geospatial-visualization tools are becoming more flexible and accessible as the limitations of both hardware and software are constantly being pushed back: more and more data files are being made available, exchangeable and manageable, and tools are being developed to manage, combine and open up data files. Riedijk et al. (2006) have summarized the trends in geospatial-visualization technology as follows:

- large and growing volumes of information are being made available (pictures, geomorphologic data, contour maps and aerial photographs);
- tools are being developed to improve the presentation and analysis of data;
- computers are becoming faster and more powerful and are therefore able to handle the data in combination with tools;
- more and more people are familiar with the internet and its possibilities;
- governments are increasingly using the internet as a platform for sharing and exchanging information;
- internet connections are becoming more powerful for uploading and downloading data;
- tools are being developed for building 3D geo-information systems and animations;
- standardization of the interoperability of geographic information systems (GIS) (Open Geospatial Consortium, world wide standards) is making it easier to share and integrate data from different sources.

Technological developments make it possible to communicate a realistic picture of present and future spatial scenarios to groups of people involved or interested in spatial transitions. We assume that geospatial-visualization tools and techniques can support better involvement of stakeholders. However, this involvement can only be realized if the future geospatial-visualization tools and techniques meet a number of criteria. The main criteria are that the tools and techniques:

- can be used by people with little or no knowledge of spatial planning (intuitive interfaces);
- can be used on 'standard' computer systems (e.g. internet browser);
- are challenging for people to use;
- offer a transparent picture (clear terminology);
- have options for displaying the whole plan area and for zooming in to particular details and to retrieve extra information from those details;
- have options to view the area from different angles; and
- offer possibilities to analyse the effects of different scenarios.

In this section we have described the trends that might motivate government organizations to implement geospatial-visualization tools and techniques in their daily operations. There are many technical and visual requirements for successful geospatial-visualizations. To implementing geospatial-visualization tools and techniques, organizations will have to invest time, money and people. As we will see below, this is not just a matter of balancing the costs and benefits, but of 'going with the flow' of social software and geospatial-visualization tools as opposed to sticking to the old tried and tested, less complex information systems.

The acceptance and implementation of new technologies has many features, from both the organization's as well as the user's point of view. These features will be dealt with in the next section.

Diffusion of New Technologies

Conditions for Change

There is no blueprint for the adoption and use of geo-information technologies in organizations because these will vary according to organizational conditions. A key concept here is that of 'change'. But change is a sensitive issue when it is about the way government authorities interact with the public. According to Caluwé and Vermaak (2006), people are willing to change but they are not willing to *be changed*. This implies that change management is effective only when people are directly involved in procedures of change. In this section, therefore, the question of change will be examined in depth and related to the organizational changes needed to benefit fully from innovative geospatial-visualization techniques that facilitate participatory spatial planning. Many authors have conducted research into the conditions that facilitate change and some of their findings are applied to the field of geo-information. To understand the diffusion of geospatial-visualizations within governmental organizations and society we need to know which factors contribute to the adoption of new tools for public participation.

Technology Transition

Much research has been done into the process of adopting GIS in organizations. In the 1990s Scholten and Grothe (1996) carried out extensive research into the status of geo-information within 1602 Dutch governmental organizations. To answer their questions about the process of adopting new technologies, the authors used Nolan's stage model of computer development in organizations. There are four stages in this model, represented by the well-known S-curve: Initiation, Contagion, Control and Integration (Nolan 1973, see Figure 16.1). Nolan developed the curve initially for the integration of computer systems into organizations. However, his model can also be applied to the uptake of technologies in general within organizations (King and Kraemer 1984).

Nolan's growth model has four stages:

I. In the Initiation stage the information system is used by less than five people in the whole organization. This stage is characterized by limited and decentralized control and minimal planning.

II. In the Contagion stage more than five people experiment with the new system, acceptance grows and the number of applications increases. The use of the system accelerates as more people start to use it. The costs also grow at an increasing rate, in turn demanding greater management efforts.

III. In the Control stage organizational measures to control the new system are taken to ensure greater use and cost efficiency. Standards are established and documentation is prepared on the use of the system.

IV. In the final Integration stage, the applications are integrated within the organization. Planning and control of the systems has been established and the information system is adapted to meet the needs of the organization.

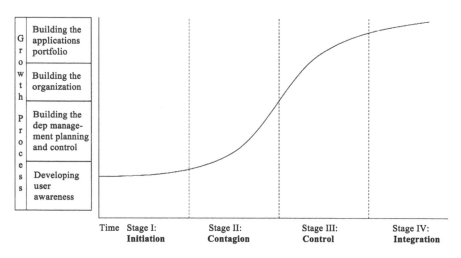

Figure 16.1   Nolan's growth curve

Source: Nolan 1973.

Similar analyses of the uptake of a new technology adoption have been made by others. Rotmans et al. (2000), for example, also distinguish four different phases, which resemble the Nolan growth curve:

1. A predevelopment or exploration phase: The developers of the technology look for opportunities to introduce this new technology.
2. The take-off phase: The change starts to occur.
3. The breakthrough phase: Embedding of the technology, characterized by visible structural changes.
4. The stabilization phase: The speed of social change decreases.

The phases described by both Nolan (1973) and Rotmans et al. (2000) can be found at a range of scales, with similar phases and transitions encountered at each scale. Rotmans et al. (2001) have made a distinction between different aggregation levels at which technological transitions can be studied. At the micro level, *niches* of individuals or companies that adopt an innovation arise. At the meso level, *regimes* of networks and communities start to adopt the innovation. And at the macro level the adoption of the innovation is widely spread along *landscapes* of conglomerates or governmental organizations. The adoption of internet technology in organizations and society is a clear example of this. The internet started as a military service to improve communication between members of the organization, but is now used by people all over the world. The speed at which the transition occurs depends on the level at which it takes place. Moreover, transitions are neither uniform nor deterministic: there are large differences in the scale of change and the period over which it occurs. It is likely that the changes within a niche follow the growth curve faster than the changes on the macro level. For example, a transformation process in which society changes in a fundamental way takes one or more generations (Rotmans et al. 2001).

Mechanisms for the Diffusion of New Technologies

Nolan's growth curve shows that the translation of innovation does not follow a linear path from research to development and implementation. Some technologies do not even complete a full path at all. The adoption of innovations tends to involve complicated mechanisms and interactions involving science, technology, learning, production, policy and demand (Edquist 1997). Geels (2002) describes how technological transitions consist of a change from one socio-technical configuration to another, involving the substitution of a technology as well as changes in elements such as user practices, regulation, industrial networks, infrastructure and symbolic meaning. With this he indicates that a change is also necessary within the workflow procedures and infrastructure. Just substituting a technology will not lead to a successful transition.

One way to approach the diffusion of an innovative, emerging technology in society is to look at it as a Technology Innovation System (TIS). This system does not follow a growth curve like those of Nolan and Rothman, but considers the boundary conditions for the successful integration of a new technology within organizations or society. TIS was first defined in 1991 by Carlsson and Stanckiewicz (1991) as 'a network or networks of agents interacting in a specific technology area under a particular institutional infrastructure to generate, diffuse and utilise technology'. These agents may be firms, research and development (R&D) infrastructures, educational institutions or policy-making bodies (Carlsson, Jacobsson 1997). The system can be applied within each separate niche to study the characteristics of the system associated with a specific emerging technology. The TIS must have a number of functions that have to be used before a system becomes successful. The functions of innovative systems are the activities that contribute to the goal of the innovation system (Hekkert et al. 2006). It is assumed that as more functions are served, and the better they are served, the better the performance of the TIS will be and thus the better the development, diffusion and implementation of innovations will be (Edquist 2001). The system functions described below can be seen as critical success factors that contribute to successful adoption of innovations in society:

- *Entrepreneurial activities*: These are essential for the introduction of innovative systems.
- *Knowledge development*: Mechanisms of learning are essential and new knowledge has to be developed. This can be done by experimentation, R&D, learning by doing and imitation initiatives.
- *Knowledge diffusion through networks*: The network determines the structure of the innovation system Because it provides the channels for direct contact between the organizations and the market.
- *Guidance of the search*: There needs to be a focus on further investment. This function indicates the activities that can positively affect the visibility and clarity of specific needs and wishes of technology users.
- *Market formation*: New technologies have to compete with existing embedded technologies. This function involves the creation of a niche or a temporary competitive advantage, such as preferential tax treatment.
- *Resource mobilisation*: Financial and human input are necessary for all activities within the system. Actors in the system will always complain about insufficient resources.
- *Advocacy coalitions*: The technology has to become part of an incumbent regime. The regime can act as a catalyst by placing the technology on the agenda, lobbying for resources, and creating a favourable tax regime.

The Users' Perspective

In the previous sections we described the mechanisms that play a role in introducing new technologies into society and organizations from the organizational and technological perspective. However, the opinions and attitudes of the users also have to be taken into account. These users can be government officials, planning professionals, policy-makers, journalists or citizens interested in a particular geo-related topic. This section deals with the users' perspective and what drives the users to start using the new technology in the first place.

There are several theories that attempt to predict the adoption and acceptance of a new technology by potential users. One of the first theories was the Theory of Reasoned Action (TRA) developed by Fishbein and Ajzen (1975, 1980). This theory tries to explain the relation between attitude and behaviour. Put simply, a positive attitude towards a new technology does not automatically lead to its use – there has to be a link between attitude and behaviour. Fishbein and Ajzen analysed the determinants of behaviour, such as social norms, time and context elements and found that a person's behaviour is to a large extent influenced by how they think other people would view them if they performed the behaviour (Manstead 1996). In addition, if people are to translate their positive attitude towards a new technology into actually using it, they must have time, knowledge, money, and the necessary technical infrastructure.

A more commonly used extension of the TRA is the Technology Acceptance Model (TAM) devised by Davis (1989). The TAM uses two main factors to predict the acceptance of a system: the Perceived Ease of Use (PEoU) and the Perceived Usefulness (PU) of the system. Both factors are explicatively stated as being 'perceived' because both the ease of use and the usefulness of the technology may be different for different users. The diagram in Figure 16.2 illustrates the basic TAM model. Many varieties and elaborations have been made of the model to make it applicable to specific user groups.

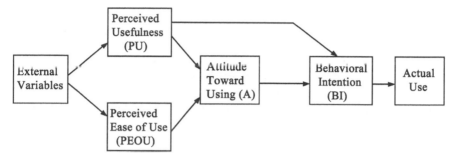

Figure 16.2   Diagram of the Technology Acceptance Model
Source: Davis 1989.

From the TAM we may logically conclude that when people find a new technology useful and easy to use, the right conditions for acceptance are created. The authors were involved in a project where an online 'geo-discussion' tool is gradually gaining acceptance among the professional stakeholders as they become more aware of the benefits of participatory approaches. But more importantly, they already have a strong affinity with the use of maps and the use of internet. They also consider the geographical datasets to be a major requirement for future development in the field of participatory spatial planning (Hoogerwerf 2005). This example shows that a shared positive attitude towards new technologies leads to good initiatives. Another good example is the introduction of the mobile telephone. Because mobile telephones are as easy to use as standard telephones, the perceived ease of use factor will not be much different from that for standard telephones. As it turned out, this new technology was perceived to be very useful because it opened up a host of new possibilities for staying in touch with work, friends and family outside the home. This perceived usefulness has led to widespread adoption of the mobile telephone.

In this section we have seen that for the acceptance and implementation of our vision as described in the second section, a simple cost–benefit analysis will not be enough for the successful implementation of new Geo-ICT technologies. From both the organizational and the users' viewpoint, we need to create the right preconditions. One such precondition, the accessibility of all spatial data, is of crucial importance and is explained in the next section.

Access to Spatial Data

The Role of Standardization in GIS

Before organizations can start using geospatial-visualizations in their day-to-day work they have to fulfil some basic requirements. One such requirement is the establishment of an SDI. According to the 'GSDI Cookbook' (Nebert 2004), this term is often used to denote the relevant basic collection of technologies, policies and institutional arrangements that facilitate access to spatial data. The SDI provides a basis for spatial data discovery, evaluation and application for users and providers within all levels of government, the commercial sector, the non-profit sector, academia and by citizens in general. For example, in a country where an SDI is a common good, it is much easier for an organization to access the spatial data necessary to make effective visualizations and to let citizens interact with these data and images.

Defining SDI

The first forms of what were later called SDIs started to emerge from the mid 1980s (Masser 1999). They were developed mainly because geographic information was expensive and an SDI made it possible to exchange spatial data, avoiding the need to gather and manage the same spatial data many times (Rajabifard et al. 2003). Masser (1998) notes that the higher purpose of SDI initiatives was to promote economic development, stimulate better government and guarantee a sustainable environment.

According to Loenen (2006) the definition of SDI has developed over the years from being a pure technical device into being a combination of information, content and people. The SDI framework continuously supports the effective and efficient construction, processing and use of relevant geographic information within or between organizations. Crompvoets (2006) describes the following core elements of an SDI:

- *Access networks*: The access network technically facilitates the use of data by people. It seeks access to relevant data sources and spatial information services by anyone, anywhere. The best example of an access network is a 'national clearinghouse'. Crompvoets defines a spatial data clearinghouse as 'an electronic facility for searching, viewing, transferring, ordering, advertising, and/or disseminating spatial data from numerous sources via the internet and, as appropriate, providing complementary services. Such a clearinghouse usually consists of a number of servers that contain information (metadata) about available digital data.' Other names used for such access networks are 'catalogue services', 'spatial data directories' and 'data information systems' or 'metadata information systems'.

- *People*: Increased use and awareness of spatial information leads to an increasing number of people using GIS. Three groups or people are involved in a clearinghouse environment: data suppliers, managers (service administrators) and end-users. The power of a clearinghouse is that an extensive number of data suppliers can disseminate their products via the facility.
- *Policy*: The policy and administrative component of the SDI definition is critical for the construction, maintenance, access and application of standards and datasets for SDI implementation. In general, policies and guidelines for SDI concern the following topics: spatial data access and pricing, funding, spatial data transfer, custodianship, metadata and standards.
- *Technical standards*: Standards are essential to ensure interoperability between the datasets and access mechanisms defined by an SDI. They can be applied at many different levels within an SDI. Standards for metadata, for example, tell the user about the content, quality, source and lineage of the data. Widely used metadata standards are those of the Federal Geographic Data Committee (FGDC), the European Committee for Standardisation (CEN/TC 287) and the ISO Standard for Geographic Information (ISO 19115).
- *Data*: Data within an SDI should be compatible in terms of format, reference system, projection, resolution and quality. Most clearinghouses provide access to standardized metadata. Data can be transmitted via email or Web Feature Servers and the variety and numbers of datasets that can be accessed vary considerably. For example, the US Federal Clearinghouse allows the user to access to more than 139,000 datasets, while the 25 European clearinghouses together give access to only 10,000 datasets.

SDI Moves onto the Political Agenda

The national SDI in the USA was launched by Executive Order 12906, which was issued by President Clinton in 1994. Europe has also taken steps to create a European SDI. The European Commission launched the GINIE Programme (2001–04) with the goal of realizing a 'Geographic Information Network In Europe' (Craglia et al. 2002). This culminated in the adoption by the European Commission in July 2004 of a proposal to draft a directive on the establishment of a European SDI. The INSPIRE Directive (INfrastructure for SPatial Information in Europe) was approved on 23 November 2006 by the European Parliament and the Council and provides the legal framework for realizing a European Spatial Data Infrastructure. It obliges European member states to make practical arrangements to ensure the interoperability and where possible harmonization of their spatial datasets and cartographic services. The directive will initially be implemented in support of environmental policies and will later be enforced in the agriculture, transport and energy sectors. The member states themselves are responsible for making their national geo-information infrastructures compatible with the provisions of the directive (Europa Decentraal 2006). INSPIRE was initiated to solve the problems of data gaps, missing documentation and incompatible spatial datasets and services arising from the use of different standards, and to remove the barriers to the sharing and reuse of spatial data. In its proposal for the INSPIRE Directive, the Commission stated that action at the Community level is necessary for the following reasons (Commission of the European Communities 2004):

- 'Few Member States have developed a framework for establishing a national infrastructure for spatial information that addresses operational, organizational and legal issues. Where steps have been taken, they have often been restricted to specific regions or specific sectors.'
- 'In most Member States where a framework has been adopted, not all problems have been addressed or initiatives are not compatible.'
- 'Without a harmonised framework at Community level, the formulation, implementation, monitoring and evaluation of national and Community policies that directly or indirectly affect the environment will be hindered by the barriers to exploiting the cross-border spatial data needed for policies which address problems with a cross-border spatial dimension.'

Implementing SDI: Barriers and Successes

How much progress has been made worldwide with implementing national SDIs? This question was answered in a doctoral thesis by Crompvoets (2006). From 2000 to 2004 Crompvoets assessed the worldwide status and development of national spatial data clearinghouses and identified the critical internal and external factors for the successful development of national clearinghouses.

According to Crompvoets, the challenges of making larger amounts of spatial data more accessible are more likely to be organizational than technical in nature. From an analysis of the results obtained by Kok and Loenen (2005), Loenen (2006) identifies six critical organizational aspects that determine the development of SDI: leadership, vision, communication channels, the power of a GI community to reorganize, awareness, and sustainable resources. Rajabifard and Williamson (2001) mention six key factors for speeding up SDI development. Three of these factors are related to organizational development: (i) awareness of the variety of applications in geo-information and SDI; (ii) involvement and support from politicians; and (iii) cooperation between diverse stakeholders. Furthermore, Craglia et al. (2002) and Rajabifard et al. (2003) note that, in addition to strong leadership, awareness of the added value of geo-information in relation to multi-level policy-making is an important stimulating factor.

The limited cooperation and coordination between public sector organizations on defining clear data exchange policies also seems to be an important constraining factor (Nebert 2004). According to Hoffmann (2003), this has to do with four competing ethics within government:

- *Open government*: Information produced by the government is public and should therefore be inexpensive and easy to access.
- *Individual privacy*: The privacy of citizens is paramount and data cannot be made public.
- *Security*: Security of the state is a major factor and data that compromise that security cannot be made public.
- *Fiscal responsibility*: Government should be entrepreneurial in its approach to data that have a market value.

Another important constraining factor, as described in the 'GSDI Cookbook', is the fact that most of the motivation to employ geographic information and tools is still internal to institutions with the aim of serving their primary needs. Outreach and education are not emphasized because existing systems primarily serve their own clientele without concern for the needs of other potential users. Furthermore, transparency remains a major problem. Organizations do not seem to know what exactly is available, where different types of data are available and who is in charge of producing this information. Instead of being able to draw on an organized spatial data infrastructure, sharing data is largely a matter of good or bad luck (Nebert 2004).

Besides these internal organizational aspects, Crompvoets (2006) found that societal conditions have a strong impact on the establishment of national clearinghouses. The standard of living is a critical factor. Countries with a high standard of living have a sound investment climate and policies that promote openness and high-quality (technological) infrastructures and services. They also have a good education infrastructure and supportive legal and regulatory instruments. Societal conditions therefore have an important role in the success or otherwise of national clearinghouses.

Applications of Geo-ICT for Public Participation

This section presents four example applications for facilitating public participation in spatial planning process. The first two examples: Virtual Landscape and Urban Strategy are for communication with the stakeholders where an expert shows the effects of the future changes in a presentation type of session. The other two examples make use of an interactive, multi-touch device (the Microsoft surface) where the citizens can explore the spatial data of the area and contribute (draw) ideas to the proposed plans in an interactive way. From these experiences, it became clear that ease of use and usefulness play a very important role, but it also was clear that in interactive spatial planning, it is crucial to have access to background spatial datasets in order to understand the context, limitations or potential benefits of the project. The schetgids application (last example) can access, connect and interface with Open Geospatial Consortium (OGC) standards and therefore in a straightforward way display all contextual data. These can be spatial limitations for a new road due to Natura2000 reservations, or the positive potential social impact of a new library based on displaying the youth inhabitants of a certain area via the geo-demographic dataset from the census bureau.

Virtual Landscape

The 'Virtual Landscape' prototype integrates different geospatial datasets and the stakeholders are able to fly over the landscape and to 'zoom in' to access detailed and different geo-referenced data. The rationale behind this development is to help the cognitive understanding of landscape change projects where the stakeholder group is numerous and heterogeneous by nature. The different stakeholders have different sensibilities, different interests and concerns about the project, therefore it is fundamental to display the correct information in an appropriate way to assure that all have the same understanding of the goals and consequences of the project. 3D visualization provides an effective way of presenting large amounts of complex information to a wide audience, including those with no GIS or mapping experience. The system was designed taking into consideration cognitive principles and is able to integrate high-quality mapping of the current situation, 3D representations of the future and (geo)multimedia (regarding real-world information). The people involved can understand the proposed plans

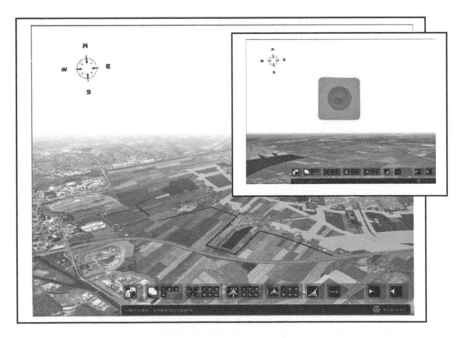

**Figure 16.3   Virtual Landscape interface and floating 3D icon representing a
geo-referenced audio comment**

and proposed changes. This new approach was built based on a geo-information infrastructure that supports open plan process and participation and is able to integrate all available sets of data. Once the participants understand the project, they can leave geo-referenced comments (audio or text) (see Figure 16.3). The intuitive interface allows the collection of creative ideas that can be aggregated and taken into account for the sustainable criteria of the city developments.

Urban Strategy

Urban Strategy (TNO 2010) is an interactive, decision support system developed by TNO that contains modeling expertise in local environments, involving aspects such as air quality, noise, accessibility, safety, parks and gardens, and water. Similar to the previous example, Urban Strategy can display an overview of the city environment in 3D. The unique distinction from other visualization tools is that changes can be implemented in an interactive way (e.g. extra residential area, road closure, clean trucks only in environmental zone) and the effects of these on the quality of the surroundings are shown directly using state-of-the-art calculation models. The current models include: Traffic (intensity and accessibility); Air quality; Noise (level, nuisance, sleep disturbance); External safety (group and place related risk); Groundwater flows and levels; Accessibility of parks and gardens; Shade and shadow; Sustainability (e.g. climate effects); and Cost/benefits of demolition/new development. This application is meant for collaborative planning with the stakeholders, who can see the immediately the result of their ideas into the environment and know whether they are improving the sustainability of the city.

Geodan's TouchTable: Leerdam

In January 2010, the company Geodan implemented a participatory spatial planning tool for the municipality of Leerdam, the Netherlands. The tool, called Geodan's touchtable, was able to connect to predefined contextual layers (available as web services) via the local spatial data infrastructure. These spatial layers provided diverse information to facilitate the collaborative discussion such as census data (geo-demographics and the spatial distributions of different age groups), location of special institutions (e.g. schools, hospitals) and, location of the new planned buildings within the municipality. During three days the citizens were invited to visualize the planned constructions and provide their input, creative critics and ideas about the sustainable development of their city (Figure 16.4).

The main distinctions from the previous two examples are: (i) the citizens (even without any mapping or geo-spatial expertise) were able to draw their opinions in the spatial planning tool via a very intuitive and touch interface; (ii) the access to the SDI played a crucial role in the discussions because the users were able to easily turn layers on and off that were relevant for the discussion; (iii) there was no quantitative modeling of the impacts; the goal was a extensive and cross-demographic qualitative collection of the opinions of the local inhabitants.

Figure 16.4   Citizens contributing their creative ideas and opinions about the new plans in Leerdam, the Netherlands

This participative session was facilitated using the Geodan spatial planning interactive interface, implemented on a multi-touch device, the Microsoft Surface, which connects to the local SDI for contextual information with spatial layers such as topographic maps, aerial imagery, census data and geo-referenced 360 photographs (right photo) and allows the participants to draw on digital map with touch and gestures (left photo).

Conclusions

This chapter assumes a fundamental link between a city's quest for sustainability and the participation of all the stakeholders in the spatial planning processes and decision-making (and give special attention to the inclusion of the local inhabitants, which can provide creative and off-the-box solutions to the local problems). In order to participate in the spatial planning process, the first step is to understand the proposed changes. The use of geospatial-visualizations helps the cognitive process and allows all the stakeholders to have a common and clear understanding of the plans. Nevertheless, the adoption of geospatial-visualization, seems to lag behind all the developments in the field. This chapter elaborates on the factors that have an impact on the adoption of geospatial-visualizations in organizations responsible for the participatory planning processes.

We started the chapter by introducing the Web 2.0 trends in society. We learned that for a Web 2.0 technology to be successful, it must be highly user-centred, giving users the power to mould the available information to meet their needs and questions. Our vision is that this Web 2.0 'attitude' can be exploited to encourage active, environmentally conscious citizens to become more involved in the development of their spatial environment. Governments should therefore be willing to invest time, money and people to find out what Web 2.0 could mean for their future participatory processes. The question then is how we can embed these technologies in society. We looked in more detail at the processes that can influence the adoption and acceptance of new technologies in organizations and in society.

We have found answers to our first two research questions:

- What can be learned from the scientific literature about the adoption and acceptance by organizations of innovative geospatial-visualization tools and technologies?
- Which factors determine the successful implementation of geospatial-visualization tools and techniques from an organizational point of view?

In answering the first question we found that the TAM and Theory of Reasoned Action can best predict the adoption and acceptance of a new technology. When a technology is perceived to be useful and easy to apply, acceptance of the technology will straightforward. Perceived usefulness is a particularly important consideration when creating new technologies for public participation. Users must feel that the new technology is indeed useful, otherwise it will not be used.

The answer to the second question can be derived from the TIS concept described in the third section. The TIS concept states the functions that can determine the successful implementation of a new technology within an organization. Factors such as resources, training, legitimacy, entrepreneurial activities, expectations and research have to be considered if public participation techniques are to be successfully introduced. In summary, we can say that the adoption and acceptance of new technologies depends on a combination of factors that together contribute to successful implementation.

The third question to be answered was:

- What is the state of the art regarding the use of geospatial-visualizations by governmental organizations, and, more specifically, what can be said about the development of spatial data infrastructures in Europe?

We acknowledged that a crucial requirement for the implementation of geospatial-visualizations in organizations is to have an organized, standardized system for consulting and sharing geo-information. By studying the state of the art of SDI development in Europe we can say much about progress in Europe in the field of geo-information technologies. We have seen that SDI is still in the developmental phase and has not yet reached the integration stage described by Nolan. However, in recent years SDI has become an important asset in the European political agenda owing to the INSPIRE Directive. Here we can see that political commitment is a major stimulator of innovation. Furthermore, locally based projects are likely to have a large impact on knowledge dissemination across Europe. Likewise, there are many initiatives in society that will ultimately lead to a sense of urgency for a broader uptake of participatory processes in planning procedures using new communication and visualization tools and techniques.

The main question to be answered in this chapter was: Which factors determine the successful integration of geospatial-visualization tools and techniques into organizations responsible for participatory spatial planning? We drew on empirical and theoretical examples to show that successful integration of geospatial-visualization tools and techniques depends on a combination of organizational factors and societal factors. Organizational factors include leadership, vision, communication channels, the power of a GI community to reorganize, awareness and sustainable resources. Societal factors are trends that occur externally but have a direct influence on the operational activities of organizations. These factors include legal initiatives, developments in internet technology and growing awareness among citizens of planning-related issues.

These organizational and societal factors are the success factors for organizational change. As every country has its own characteristic organizational culture, planning culture, technological culture, economic culture and political culture, it is difficult to conclude with a set of recommendations that are valid for every country. Therefore, on the basis of the information provided, we finish this chapter with an open-ended conclusion: the climate for change is very positive. There seem to be few technological barriers to the development and use of geospatial-visualization tools and techniques in spatial planning procedures, and many encouraging private and public initiatives. Moreover, many national planning cultures in Europe embrace participatory approaches, or have the potential to do so. Consequently, our recommendation to local organizations is to focus on awareness raising and knowledge dissemination within the organization to highlight the benefits of using geospatial-visualization tools and techniques, and in doing so to emphasize the need for a SDI. The tools, practices, approaches and commitment presented in this book provide an excellent starting point.

Bibliography

Alexander, B. 2006. Web 2.0. A New Wave of Innovation for Teaching and Learning? *EDUCAUSE review*, March/April, 32–44.

Caluwé, L. de and Vermaak, H. 2006. *Leren veranderen: een handboek voor de veranderkundige*, 2nd edition. Alphen aan de Rijn: Samsom.

Carlsson, B. and Jacobsson, S. 1997. Diversity Creation and Technological Systems: A Technology Policy Perspective, in *Systems of Innovation – Technologies, Institutions and Organizations*, edited by C. Edquist. London: Pinter, 266–94.

Carlsson, B. and Stankiewicz, R. 1991. On the Nature, Function and Composition of Technological Systems. *Journal of Evolutionary Economics*, 1, 93–118.

Carver, S. 2001. *The Future of Participatory Approaches using Geographic Information: Developing a Research Agenda for the 21st Century*, position paper presented to the ESF-NSF Meeting on Access and Participatory Approaches in Using Geographic Information, Spoleto Italy, 5–9 December.

Commission of the European Communities. 2004. *Proposal for a directive of the European Parliament and of the council establishing an infrastructure for spatial information in the Community (INSPIRE)*, (COM 2004 516 Final) Brussels, 23 July.

Craglia, M. et al. 2002. *GINIE; Geographic Information in the Wider Europe*. Sheffield: University of Sheffield.

Crompvoets, J.W.H.C. 2006. *National Spatial Data Clearinghouses: Worldwide Development and Impact*. Doctoral thesis, Wageningen UR, Netherlands.

Davis, F.D. 1989. Perceived Usefulness, Perceived Ease of Use and User Acceptance of Information Technology. *MIS Quarterly* (September), 319–40.

De Man, W.H.E. 2003. Cultural and Institutional Conditions for Using Geographic Information: Access and Participation. *URISA Journal*, 15(2), 29–33.

De Swart, N. 2007. *Onderzoek naar functionaliteit RO-Online website* Draft Internal Report Geonovum.

Dias, E.S. et al. 2003. *Virtual Landscape Bridging the Gap between Spatial Perception and Spatial Information*, paper presented to the 21st International Cartographic Conference (ICC), Durban, South Africa, 10–16 August.

Downes, S. 2005. Feature: E-learning 2.0. *eLearn Magazine*, 10.

Edquist, C. 1997. *Systems of Innovation: Technologies, Institutions and Organisations*. London: Pinter.

—— 2001. *The systems of Innovation Approach and Innovation Policy: An Account of the State of the Art*. Aalborg: DRUID.

Europa Decentraal. 2006. *Overeenstemming bereikt over INSPIRE-richtlij*. Available at: www.europadecentraal.nl.

Fishbein, M. and Ajzen, I. 1975. *Belief, Attitude, Intention, and Behavior: An Introduction to Theory and Research*. Reading, MA: Addison-Wesley.

—— 1980. *Understanding Attitudes and Predicting Social Behaviour*. London: Prentice Hall.

Geels, F.W. 2002. Technological Transitions as Evolutionary Reconfiguration Processes: A Multi-Level Perspective and a Case-study. *Research Policy*, 31(8–9), 1257–74.

Google. 2004. *Press Release: Google Acquires Keyhole Corp.* Available at: www. google.com/press/pressrel/keyhole.html.

Hekkert, M.P. et al. 2006. *Functions of Innovation Systems: A New Approach for Analysing Technological Change.* Technological Forecasting and Social Change.

Hoffmann, M.C. 2003. The Ethics of Public Data Dissemination: Finding the 'Public' in Public Data, in *Proceedings of the Second Annual Conference of PPGIS, Portland, OR, 21–23 July.* Park Ridge, IL: USA, 273–82.

Hoogerwerf, T. 2005. *Workvisit Poland.* Internal report for PSPE project, Wageningen University.

King, J.L. and Kraemer, K.L. 1984. Evolution and Organizational Information Systems: An Assessment of Nolan's Stage Model. *Communications of the ACM*, May, 456–75.

Kok, B. and Loenen, B. van 2005. How to Assess the Success of National Spatial Data Infrastructures? *Computers, Environment en Urban Systems*, 29, 699–717.

Kraak, M.-J. 2003. Geovisualization Illustrated. *ISPRS Journal of Photogrammetry and Remote Sensing*, 57, 390–9.

Loenen, B. van 2006. *Developing Geographical Information Infrastructures: The Role of Information Policies*, PhD thesis, Delft University, Netherlands.

Manstead, A.S.R. 1996. Attitudes and Behaviour, in *Applied Social Psychology*, edited by G. Semin and K. Fiedler. London: Sage Publications.

Masser, I. 1998. The First Generation of National Geographic Information Strategies. In *Selected Conference Papers: 3rd Global Spatial Data Infrastructure Conference, Canberra, Australia.* London: Taylor and Francis.

—— 1999. All Shapes and Sizes: The First Generation of National Spatial Data Infrastructures. *International Journal of Geographic Information Science*, 13, 67–84.

Nebert, D.D. 2004. Developing Spatial Data Infrastructures: The SDI Cookbook, Version 2.0. *Global Spatial Data Infrastructures* (24 January). Available at: www.gsdi.org.

Nolan, R. 1973. Managing the Computer Resource: A Stage Hypothesis. *Comm. ACM*, 16(7), 399–405.

Novio Consult Van Spaendonck. 2006. *Klik en Klaar? Ruimtelijke informatie voor burgers op internet.* Report, 16 March.

Petts, J. and Leach, B. 2000. *Evaluating Methods for Public Participation: Literature Review.* R and D Technical Report (E135), Environment Agency, Bristol.

Rajabifard, A. and Williamson, I.P. 2001. Spatial Data Infrastructures: Concept, SDI Hierarchy and Future Directions, in *Proceedings of GEOMATICS'80 Conference, Tehran, Iran.* Available at: www.geom.unimelb.edu.au/research/publications/IPW/4_01Raj_Iran.pdf.

Rajabifard, A. et al. 2003. National GDI Initiatives, in *Developing Spatial Data Infrastructures; from concept to reality*. London: Taylor and Francis, 101–5.

Riedijk, A., Van de Velde, R.J., Pleizier, I.D., Hoogerwerf, T., van Lammeren, R., Baltussen, W., Jansen, J., Wynia, P., van Uum, J. and van Wilgenburg, R. 2006. *Virtual Netherlands: Geo-visualisations for Interactive Spatial Planning and Decision-making: From Wow to Impact. Definition Study*. Amsterdam: Vrije Universiteit.

Rotmans, J., Kemp, R. and Asselt, V.M. 2001. More Evolution than Revolution: Transition Management in Public Policy. *Forsight/The Journal of Future Studies, Strategic Thinking and Policy*, 3(1), 15–31.

Rotmans, J. et al. 2000. *Transities and Transitiemanagement: de casus van een emissiearme energievoorziening*. Assen (The Netherlands): Koninklijke van Gorcum.

Scholten, H.J. and Grothe, M. 1996. *GIS in de publieke sector: een inventarisatie naar gebruik van geo-informatie en GIS bij de Nederlandse overheid*. Utrecht: Koninklijk Nederlands Aardrijkskundig Genootschap, Vakgroep Ruimtelijke Economie Vrije Universiteit Amsterdam.

Sieber, R. 2006. Public Participation Geographic Information Systems: A Literature Review and Framework. *Annals of the Association of American Geographers*, 96(3), 491–507.

TNO. 2010. *Urban Strategy*. Available at: www.tno.nl/urbanstrategy.

Chapter 17

Creative Urban Transport

Robert U. Ayres

Introduction

The idea of creative cities is a little strange. People can be individually creative – they can paint pictures or write novels. In the context of urban sustainability, however, the term *creative* must be understood in the sense of a non-traditional, yet effective, response to the challenges of rapid urbanization. Those challenges are all too evident in the mega-cities of the developing countries. They range from lack of employment and educational opportunities, lack of fresh water and sewage treatment to overcrowding and congestion. Even crime and drug culture can be blamed on the stresses of urbanization.

This chapter deals with the transportation dimension of the larger problem. At the end of the day, barring enormous investments in transportation infrastructure (for which there is no plausible source of funding in most developing countries) the possible responses in mega-cities are remarkably limited, and most of them (not quite all) are essentially technological. Moreover, inaction is not an option. People living in the favelas and barrios of the world need a way to get to the central cities or the industrial zones where there are jobs. Moreover, the favelas and barrios need food and they need ways to get rid of human and other wastes, other than dumping everything into the rivers or streams from which people downstream are getting drinking and washing water.

This chapter cannot deal with all of the problems of rapid urbanization, nor is there any clear connection between transportation and most of the ills listed above. For this reason, I will concentrate on the people-mover problem, which applies to cities of all sizes and degrees of wealth, albeit with differences of detail.

Cities are located where they are today for a variety of historical reasons, most of which are as unimportant today as city walls, except for the continuing importance of transportation, especially ports and harbours. As water transportation developed, safe harbours and navigable rivers became critical. But canals extended the river systems and heavy or bulky goods moved on water until the mid-nineteenth century when railways made canals much less important, though not entirely obsolete. But railways began and ended at disconnected stub-end terminals in the biggest metropolitan cities, such as Paris, London, Berlin, Milan, Vienna, New York and Chicago. There was a time, a century and a half ago, or so, when virtually all goods and most intercity passengers and commuters were transported by rail, when most railways used coal as a fuel for their steam-powered locomotives.

Those locomotives were noisy, smoky and thermodynamically inefficient. But coal was so cheap and thermodynamic theory was so undeveloped at the time that nobody bothered to redesign the engines.

The steam-railway was never an integrated urban system, and it never served the internal needs of big cities themselves except as a faucet (tap) serves a bathtub. Though cities were expanding rapidly, internal transportation in the nineteenth century was entirely on foot, on horse, by cart or horse-drawn carriage. The horses themselves became a major element of traffic and a source of pollution, and the stables and other facilities required to house and feed them occupied quite a lot of space.

The electrified surface and underground rail transit systems that were built starting in the 1880s and 1890s were a partial solution, but only along the busiest routes. The surface trams were in direct competition with other traffic, which made them slow, while the underground systems were disruptive during construction, extremely expensive and took years to complete.

About the same time that electric trams became practical, the internal combustion engine was successfully applied to the (formerly) horse-drawn carriage by Daimler-Benz and others. The primitive engines of the 1890s used natural gasoline – a by-product of petroleum refining – as a fuel. In those years steam engines and electric motors were also applied to the automobile. It took a number of years for the new form of automotive transportation to develop beyond the 'toys for (rich) boys' stage. But thanks to Henry Ford and a number of other industrial pioneers, mainly in the United States, automobiles became reasonably efficient, reliable, user-friendly and – most important – mass-producible. By 1911 the demand for motor fuel exceeded, for the first time, demand for illuminating oil (kerosene), which had been the primary product of the petroleum industry since the discovery of oil in Pennsylvania in the 1850s.

The US was the centre of the global petroleum industry and the world's dominant producer in the first half of the twentieth century (until 1970), and gasoline was cheap. The 'cracking' technology introduced starting in 1911 also made more gasoline available from each barrel of crude oil. In the United States the automobile market was already well-established by 1920 and horses were rapidly disappearing from American city streets. Urban tramlines were mostly torn up – largely thanks to a 1930s campaign led by General Motors (GM), opening the market for intra-urban travel to diesel-engine powered buses (supplied by GM).

Not so, in Europe, and the rest of the world, thanks to the economic havoc of the First World War and its aftermath. Petroleum had to be imported, either from the United States, Rumania or the Caucasus. Governments wanted to minimize imports or tax them. Also, most European and Asian towns and cities were built around ancient cores with narrow streets. Hence, European automobile manufacturers built small cars, towns and cities kept their electric trams and governments did not encourage automobiles or the creation of an automobile industry. That only changed when Hitler's Germany created the Volkswagen (peoples' car) and started building autobahn's in the 1930s, as an employment stimulus. However, before the

Second World War, the European automobile industry was comparatively dwarfed by the US industry, and there was virtually no auto industry in the rest of the world.

Since then, the situation has changed radically, mainly on account of petroleum. After the Second World War, the United States was still a petroleum exporter, and an automobile exporter. By the end of the 1960s that had changed. In the early 1970s the Arab oil embargo and the first major price spike, signalled the coming change, but the US auto industry was selling big inefficient cars to a domestic market dominated by suburbanites used to wide streets, easy parking and cheap gasoline. The rest of the industrialized world (Europe and Japan) had always depended on expensive (and heavily taxed) imported petroleum, cities with narrow streets and dense populations and relatively few suburbs in the US sense. No wonder, in a world of higher oil prices, the US auto industry became less and less competitive.

It is important to distinguish between goods transport and passenger transport. Virtually every city in the world depends on gasoline or diesel oil for the bulk of its goods transport, at least from the harbour or rail terminal to the factories and shops. Some cities in the developing countries still depend on animals or human muscles to perform this function, but that is not a permanent or satisfactory solution for the same reasons horses and bullocks became *non grata* in Europe and America. The good news, however, is that small delivery vehicles are more amenable to electrification than passenger cars. In fact, there is quite a long history of electric parcel delivery vans and milk trucks in western cities, and the main obstacle in developing countries is probably the unreliable electricity supply.

On the other hand, passenger transport in cities outside of North America is somewhat less dependent on the internal combustion engine, at least where investments in alternative options such as trams and metro-rail have been made in the past. But even in a city such as London, Paris, Berlin or New York where it is easy to get along without owning a car, the streets are congested. Cities without extensive rail systems in China, India, Africa and Latin America depend much more heavily on buses, trucks and variants on the theme.

Clearly, the global addiction to oil is unsustainable in the long run. This, in turn, means that oil consumption *must* be cut dramatically, not just because of the need to cut greenhouse gas (GHG) emissions that are causing climate-change, but also for reasons of economic stability and international security. And the place where oil consumption must be cut most urgently is, of course, the transportation sector.

The technological possibilities are very limited: they are (i) to increase drastically, conventional fuel efficiency, (ii) to shift to alternative fuels (biofuels, hydrogen), (iii) to adopt electric vehicles (EVs) or electric hybrids for city driving or (iv) to develop truly alternative modes of transportation. The first option consists mainly in weight reduction, which is desirable but has limited potential in countries where vehicles are already very small in relation to load factors. The second option primarily involves bio-fuels, which also have very limited potential because they ultimately compete with food production. In the remainder of this chapter, only the last two options will be considered, not necessarily in the original order. However, before discussing technologies, it is appropriate to consider the policy options that could hasten the needed technological changes. One policy option, in particular, is city oriented, namely congestion charges, or some equivalent, discussed below.

This chapter does not attempt to identify problems – or solutions, if there are any – specialized to the developing countries. In fact, the transport problems in Mumbai or Cairo or Lagos or Mexico City are not different in kind. They differ mainly in the greater degree of congestion and the lesser availability of funding for capital-intensive projects. Thus, it is probably still true that heavily overloaded trucks are still a primary mode of urban (and rural) transport in South Asia and much of Africa. It is true that some modes, such as tricycle taxis, water taxis and rickshaws are found only in the more congested parts of Asia. Water taxis are important in river cities such as Cairo, Lagos, Bangkok, Calcutta, Canton and Shanghai. It is also true that cities that are less dependent on automotive technology than the cities of the industrial countries, may adapt more readily to the end-of-oil.

There are those who argue new institutions rather than new technology are the way forward. If the point being made is that new technology is not 'the answer' to congestion and pollution, and still less for poverty and lack of opportunity, I would have to agree. The introduction of congestion charges, as discussed next, is a desirable and perhaps inevitable institutional change. The creation of pedestrian-only zones, widely applied in Europe, is another example. A variety of tax schemes, on fuel, vehicles, engine power or parking are potentially useful. Many of these would also be classified as institutional. On the other hand, it is sad and frustrating to observe China and India trying to accelerate economic growth by developing domestic automobile industries, and building freeways, with virtually no thought for the consequences for cities.

Congestion Charges

Ideally, any large city should also impose a road-pricing or congestion charge of some sort on private vehicles (including trucks) entering the central area. A wide variety of such schemes have been implemented around the world since Singapore introduced the first congestion charge in 1975. The first version was controlled by police. Now it is entirely electronically controlled and variable pricing schemes (according to traffic conditions and time of day) are being considered. Singapore also imposes high fuel taxes, high parking fees and even vehicle quotas to keep the number of vehicles down. At the same time, Singapore has invested heavily in public transportation. As a result, only 30 per cent of families there now own a car.

Bergen, in Norway, introduced a charge in 1986; today there are six such schemes operating in Norway with a national coordinating scheme called AutoPASS. Stockholm has had a congestion charge in effect for the central city since 2007. Durham was the first in the UK (2002). London has had a congestion charge in place for the central districts since 2003; it was extended in 2007. A similar scheme is proposed for Manchester.

Valetta, in Malta, has imposed an innovative congestion charge scheme for the central city since 2007. It includes variable pricing, variable exemptions and a variety of electronic billing schemes. Austria and Germany now have vehicle pollution charges for trucks. Austria has a comparatively simple scheme, operational since 2004. The German scheme, operational since 2005 is more elaborate, depending on the Galileo satellite and other advanced technologies. It charges trucks on a per kilometre basis depending on their emission levels and the number of axles. Milan introduced a vehicle pollution charge in 2008 for all vehicles, but with a lot of discounts and exemptions. New York seems to be the only US city to consider the possibility seriously. In April 2007 Mayor Bloomberg proposed a congestion charge of $8 per day for the southern half of Manhattan only, with exemptions for people using the peripheral highways and staying only for the weekend. It was rejected by the New York State legislature, largely due to complaints from suburbanites.

The funds from such a charge, as well as parking and other fees, should be used to subsidize any revenue shortfall in the public mass transportation systems, with some preference for bus rapid transit (BRT). The funds transfers from vehicle congestion charges to public transport users, from bicycles to buses, should be revenue-neutral in financial terms, if possible.

To be sure, none of the above examples applies (yet) to most of the mega-cities in developing countries, probably because congestion charges are mainly applicable to automobiles. However, China has been encouraging development of the auto industry, and has recently surpassed the United States as the world's largest market for cars. The large Chinese cities, especially Hong Kong and Beijing, could soon be candidates for the Singapore solution.

Bus Rapid Transit

The simplest way to simultaneously save money and energy currently used for transportation, is to reduce the amount of driving we have to do The simplest way to do this is to utilize capital equipment more efficiently. And the simplest way to do this is to provide more and (much) better bus service on existing routes and to expand the area coverage by the routes. Underground rail transit is energy efficient in densely populated cities, but very costly to create or extend once the city above is already in being. To be sure, the world's largest cities will go on building metro-rail systems, and there may be some integration of the underground systems with regional rail systems on the surface (as in the Paris area), but this will have at most a very modest – and long delayed – impact on energy use. The most cost-effective solution in the near term can be called Bus Rapid Transit, or BRT (Levinson 2003, Vincent and Callaghan Jeram 2006). As compared to metro-rail, it is much less expensive – as little as one twentieth per passenger km travelled. Mexico City has a new BRT system along its major artery (*Insurgentes*) that cost altogether no more than two underground metro stations (Anonymous 2006).

BRT began in Curitiba, Brazil in 1966. Thanks to support from the World Bank, there were (in 2008) 49 BRT systems operating around the world, many of them in developing countries. Of these, 16 are in Brazil (but only three are in the US). Another 26 systems were being planned, mostly in Asia and Latin America. Curitiba's pioneer BRT system carries 2,190,000 passengers per day, and it is used regularly by over a third of the population. BRT is an integrated system, not just an upgrade of existing bus lines. Tickets are prepaid. It operates along a designated corridor between purpose-built stations. For optimum performance, it requires dedicated (confined) bus lanes, and station stops spaced several blocks apart, ideally in the centre of the corridor between two opposite traffic lanes. It usually involves articulated vehicles (similar to a tram) but current vehicle designs require no rails or overhead wires.

Bogota, the capital of Colombia provides a newer example. It has a population of 7 million. Before the *Transmilenio* BRT project began it had 21,000 microbuses in operation, averaging 14 years old. They provided 70 per cent of the city's transportation needs, but they caused great congestion, pollution and many accidents. Phase I of the new system (now under construction) consists of three trunk corridors, seven entrance routes to the system, four transfer stations, four stations that link with other systems, 53 ordinary stations, four full service garages for the buses and a smaller repair facility (Ardia 2006). *Transmilenio* now moves 40,000 passengers per hour each way during peak hours. When the system is complete in 2015 it will have 6000 buses in operation, carrying five million people a day on 22 lines. Incidentally, the construction project also includes parks and cycle paths as part of a 'city for people' concept.

A BRT system called *Trolebus* was built in Quito, Equador, starting operation in 1995 (Levinson 2003). The system is still operational, but its effectiveness has been severely degraded by a political decision by the city to allow taxis and ordinary cars to use the bus lanes. As a consequence of this decision the average speed of the buses has been cut from 17 mph to 12 mph, with consequent reductions in mobility service. This experience illustrates two important lessons: (i) a BRT system with confined bus lanes is capable of much 40 per cent to 50 per cent higher average speeds than otherwise, thus cutting travel time by at least a third, and attracting more passengers; and (ii) it is important to organize the system from the start as an independent legal entity – preferably a trust – to protect it from political meddling by special interests.

So much for feasibility and utility. However, the BRT systems in place or under construction (2008) are mostly in Latin America, with a scattering in other continents; six in North America, 11 in Europe, five in Asia (two in China) and four in Australasia. Planned systems are still mostly (15) in Latin America, plus five in India, another in China and five in Africa. The World Bank and United Nations Development Programme (UNDP) have provided some of the financing for the Latin American and African systems, driven in large part by the transport catastrophe created in developing countries by rapid, unplanned urbanization. It might be of marginal interest to note that in 1940 Sao Paulo and Pittsburgh had about the same population, three quarters of a million each. Today Sao Paulo's population is approaching 20 million while Pittsburgh's population has actually declined by 50 per cent because everybody has moved to the suburbs.

Two questions now arise: Does the BRT fit the needs of the industrialized world, where most of the biggest cities have invested in underground metro rail systems, and are not growing so much as spreading out. And could it constitute a cost-effective solution for a gas-guzzling city such as Houston or St. Louis or Los Angeles?

The answer depends on costs, and the cost depends on finding a way to recapture some of the land now devoted to multi-lane feeder highways and expressways that are now used almost exclusively by private cars and trucks. One articulated BRT vehicle of current design can carry 160 people, equivalent to 100 cars averaging 1.6 passengers per car. Suppose that bus were allowed to travel on a reserved lane, stopping only at stations from 1 to 3 or 4 km apart in the outer suburbs. Such a reserved lane might be shared during rush hours, for instance, by cars carrying at least three people. Suppose passengers used prepaid electronic charge cards, allowing entrance to the bus stations (located in the centre of the freeway) and used again to exit the BRT terminus, thus relieving the driver of any responsibility for collecting money or making change. Suppose each trip consists of an entrance and an exit and fares are computed automatically and deducted from the balance on the e-card. If the balance goes negative during a trip the passenger must make up the difference (in an 'add-fare' machine) before departing. In the suburbs, with stops several kilometres apart, such a bus could achieve average speeds in the range of 50 or even 60 kph. It would arrive at the central city as fast, or faster than, a typical automobile commuter in rush-hour, and there would be no need to find (or pay for) a parking place.

The electronic fare-card system described above is nothing new. Exactly the same system has been in operation on the Washington, DC metro-rail system since it opened in 1976. Essentially the same system is now used by the Bay Area Rapid Transit (BART) system around San Francisco and more recently in the London underground system ('Oyster cards') and the Paris metro-rail (Navigo), and throughout Europe in various forms. As for finance, the Washington, DC metro-rail system receives about 57 per cent of its revenues from fares, the remainder being provided by the counties it serves, according to a formula based on population and population density (Wikipedia 2008a). Something similar would be needed for any area-wide BRT. The formula for cost sharing, as well as the ownership arrangement (to prevent political meddling as in Quito) are both important elements of the original planning exercise. The capital costs of creating this system would depend a lot on the geography, the existing highway network and the cost of land, but any place where a rail system can even be contemplated should first consider a surface BRT network.

BRT users along any reasonably well-travelled route would save money, perhaps quite a lot. The overall energy savings and pollution reduction could be quite large, but again depending on the specific details. Incidentally, it should be pointed out that future buses need *not* be noisy and need not spew black smoke, as many existing buses do. In fact they need not depend on diesel engines, even though virtually all buses today do so. In fact, buses are excellent candidates for either compressed natural gas or battery-assisted fuel cell operation.

Bicycles

Bicycling is good exercise and it is the national sport of France. The increasingly obese population of the United States would be healthier if a lot more people walked or rode bicycles instead of driving cars (Pucher et al. 2010). In fact, there are many more bicycles in the world than automobiles. As of a few years ago there were more than 450 million ordinary human-powered bicycles in China, for instance, as compared to a few tens of millions of private cars.

But in China the bicycles are being squeezed off the roads by cars. The newly affluent businessmen and functionaries who are entitled to official cars in China are far more influential than the cyclists. Automobile sales in China have skyrocketed in recent years, to exceed the US level. This development has been supported by a belief in high places that it is necessary for China to develop its own automobile industry in order to industrialize. Thanks to their disproportionate influence the car-owners are driving the government of China to build freeways where rice paddies used to be. These expensive highways certainly cut travel times between major centres of activity, including airports, for the lucky few, but simultaneously they spill automobile traffic onto all the connecting streets and roads. This increases congestion, pollutes the air and makes it much harder for local cyclists to compete for road space. (The conversion of rice paddies to highways also drives up the price of land and rice.)

In some European countries such as the Netherlands and Scandinavia, bicycles are now the dominant mode of transportation in towns and cities. Nearly half of all urban trips in the Netherlands were by bicycle as long ago as 1987, when gasoline prices were very low. People in those countries use bicycles to go to school, to university classes, to shops and offices, sports facilities, and to the nearest railroad station (which is never very far). Copenhagen is much the same. In all these places cyclists will find convenient bicycle parking spaces. The climate in northern Europe is by no means sunny and warm for much of the year, but it doesn't stop the cyclists. To be sure, the bicycle solution is most suitable for the young, especially the students, and those commuters and shoppers who don't mind a little damp in case it is raining. But older folks ride too. To be sure, bicycles are not really suitable for commuters wearing business suits (though some do and others change clothes at the office), for mothers with little children in tow, shoppers with large or awkward packages, and even less so for the elderly and infirm. But the fact that many person-trips in Amsterdam and other Dutch cities do not require automobiles saves energy, cuts emissions and leaves more room for the buses and taxis (and private cars) to move.

In big cities there are enough people able to make use of a bicycle for short trips to make a significant dent on urban traffic congestion, if (and only if) the automobile-oriented traffic authorities would cooperate. The first step must be to prohibit on-street parking by cars in downtown areas absolutely, except for delivery vehicles during non-rush hours. Segregated bicycle lanes, often in parks or along rivers or canals, are features of most European cities, if only a few in North America. Segregation is inexpensive; it is only necessary to paint stripes on the road and signs. Starting in Amsterdam in the 1960s, several large European cities have experimented with free 'bicycle sharing'. Most of the early plans were inadequately thought through or too small in scale to succeed, But the schemes are getting more sophisticated.

Paris is the most exciting example. On-street parking has been banned on most streets for many years, and the no-parking zones are spreading. All new buildings must have underground parking facilities, and parking is never free. The city has recently set aside special lanes on major boulevards, reserved for buses, taxis and bicycles. (The idea is that professional drivers can share road space with cyclists without endangering them.) In July 2007 the city inaugurated a programme called 'Velib' (a contraction of '*velo libre*', or 'free bike') with an initial endowment of 10,600 bicycles, of uniform design (Wikipedia 2008b). They were paid for by Cyclocity, which is a subsidiary of the big advertising firm JCDecaux and allocated to 750 reserved parking racks around the central city. The number of bicycles was increased to 20,600 within the next year, and the number of parking stations increased to 1450. The bikes are activated by an electronic credit card and electronically monitored. There is a small annual fee to belong to the 'Velib club'. The first half hour is free of charge, with a nominal hourly rental fee thereafter. Vandalism has been a problem, but even that problem seems to be gradually getting under control.

In Lyons, France, where the system has now been in operation since 2005, each bike was used on average 12 times each day, and 95 per cent of the trips are free. Most uses of the trips are point-to-point, between one reserved parking place and another. It is difficult to obtain good data on the impact of these shared-bicycles on other modes of travel. Hopefully they are replacing at least some private auto trips, although probably not very many (because very few central city residents use cars for short trips inside the city.) The best current estimate is about 10 per cent in Lyons. The newly available shared bicycles mainly replace walking trips, or bus and metro trips, reducing congestion on those modes. The main benefit to users is speed. The bicycles are faster between most pairs of destinations than either bus or metro. So far there is very little evidence of energy conservation or cost saving, but the savings will grow as the system is widely accepted.

However, the shared bicycle programmes in Paris, Vienna and elsewhere may be stepping stones to a much more significant future programme combining arterial BRT routes with shared EVs aimed at reducing commuter trips from within the city or the inner suburbs, and later from the outer suburbs, where most commuters in Europe and America travel by private car.

At present the average time/distance for an average human-powered urban bicycle is less than half-an-hour or 5 km, enough for many commuters, and enough for many others to get to a bus or tram station or a train station. Bicycle-parking facilities at railroad stations (as in the Netherlands) would sharply increase the utility of this option. Obviously a great many young men and women can go much faster and further, at least where the terrain is flat, as in most cities.

There are a number of municipal policies that can (and have) influenced the relative popularity of bicycles *vis-à-vis* motor vehicles. These include speed limits, on-street parking regulation, one-way streets, reserved parking for bicycles, and others. For a useful review, see (Pucher, Altshuler, and Womack 1981, Pucher and Lefevre 1996).

e-bikes and e-scooters

The next step beyond the human-powered 'push-bike' is the battery-powered 'e-bike'. Such bikes are capable of an average speed of 15 to 35 km/hr, depending on traffic. By the end of 2007 there were already at lest 30 million of these e-bikes in China, out of a total bicycle population of 450 million (Weinert et al. 2007). The market for e-bikes in China was 40,000 units in 1998, but it has exploded to an estimated 16 to 18 million units in 2006, produced by over 2000 firms, mostly small and local.[1] There are two types, 'bicycle-style' (perhaps with pedals for supplementary muscle power) and 'scooter style'. The former carry batteries with a capacity ranging from 0.4 to 0.6 kWh, while the latter carry batteries with about twice as much capacity (around 1 kWh) and are capable of higher speeds. Indeed, it is now a subject of debate whether scooter-type e-bikes should be allowed in bicycle-only lanes.

The dominant battery type today is the valve-regulated lead-acid (VRLA), which accounts for 95 per cent of Chinese production, and almost all of domestic consumption, since the other battery types, especially the bikes with nickel hydride and lithium-ion batteries, are mostly exported. The lead-acid batteries have a lifetime of only about two years. However the more advanced battery types are lighter, more powerful and have a much longer expected lifetime. They are rapidly becoming more popular and optimists expect them to account for 20 per cent of Chinese e-bike output by 2010. In 10 years or so the lead-acid batteries for bikes are likely to go the way of the dodo.

Electric bicycles and scooters are still rare in Europe and even rarer in the United States, but they have the potential for changing commuter behaviour radically, even in a sprawling American city such as Los Angeles. All it takes is serious effort on the part of municipal authorities to discourage cars – especially

---

1   These local firms don't really manufacture anything; they convert ordinary bikes into e-bikes by adding a battery pack and replacing a conventional wheel by a wheel with a motor. The battery packs and motorized wheels are made by a much smaller number of suppliers.

through congestion charges and parking restrictions – and to make it easier for bicycles to use the roads and to find safe parking facilities at bus stations, tram stations, railroad stations, and so on. The changes cannot take place overnight, but over a period of (say) 20 years they can make a significant difference.

Motorcycles

Obviously gasoline powered motorcycles are already a feature of the highway. They by-pass the cars stalled in traffic jams, and they can be parked almost anywhere, or at least in a lot more places than cars. However, many of us over the age of 30 who don't ride ourselves, see them as noisy, brutish machines – called 'hogs' by the afficionados – driven much too fast by crazy young men such as Jimmy Dean, Marlon Brando and Evel Knievel, Hell's Angels and so on. That image is not very accurate today, but it persists. Another image is the motorcycle races frequently televised on weekends, which feature daredevil riders leaning so far into the high-speed turns that their knees virtually scrape the pavement, and every so often one of those daredevils crashes spectacularly, in a welter of flying bits and pieces followed by a fire from which the driver is usually (but not always) dragged alive. Exciting to watch, but not suitable for commuting to school or work, at least in a business suit.

But a much more accurate picture is the one seen by commuters stuck in the frequent traffic jams during rush hours on the main auto-routes into large European cities. The car-bound folks curse in frustration at the steady stream of motorcycles that pass freely between the stalled lanes of cars, without even slowing down. These are not recreational drivers; they are people going to or coming from work, consuming much less gasoline and producing less carbon dioxide (but much more noise) at much less cost per km travelled, than the four-wheel vehicles they pass by. The two-wheelers are also easier to park. Why doesn't everybody do it? Well, increasing numbers or people are doing it, because it is so practical. But there are problems and barriers.

One barrier for many is simple fear. Powerful motorcycles are known to be dangerous. Accidents are much more frequent than for four-wheelers. Remember the opening scene from *Lawrence of Arabia* where T.E. Lawrence dies in a crash? Quite a lot of people seem to have chosen to drive big, heavy SUVs – rather than small, more fuel efficient cars – precisely because (encouraged by auto company marketing) they believe big cars are safer. Such people are not likely to switch to motorcycles any time soon. Then, there is the need for a special licence to ride a high-powered BMW, Suzuki, Honda or Harley-Davidson motorcycle on the highway. To obtain that licence you need special training, registration and insurance. In Europe the training is fairly expensive, yet millions take it. In the United States the motorcycle safety foundation (MSF) offers free or inexpensive classes. In many states, proof of passing this course is enough to get a waiver of the state road test and written test.

The other problem for the gasoline powered two-wheelers is noise and pollution. These vehicles are supposed to be equipped with mufflers, but the mufflers often are not functional, and there is still no legal constraint on emissions. As a result SUVs produce a lot more carbon-dioxide (because they consume more fuel) but they are 95 per cent cleaner than motorcycles with respect to other pollutants, such as carbon monoxide, unburned hydrocarbons, nitrogen oxides and particulates (soot). To put emissions control on a motorcycle adds significantly to weight and cost, and detracts from performance. Up to now, many motorcycle 'gear-heads' simply disconnect the unwanted equipment.

But as the number of motorcycles on the roads increases, the pressure to eliminate this loophole in the anti-pollution and anti-noise laws will grow. And the likely answer is the electric motorcycle. Of course, the electric bikes now on the market are several times more expensive than comparable gasoline-powered bikes (comparable if you don't care about noise and pollution).[2] But they have no tailpipe emissions whatsoever, and operating costs (electricity) are considerably lower than gasoline-powered vehicles. Even if the electric power is generated by burning coal or natural gas, the electric version will be about twice as efficient (in life-cycle terms) as the gasoline-powered version. Of course, the other 'solution' to the noise and pollution problem is to reduce the power and go slower. In fact, scooters and mopeds are really a substitute for private cars in the mega-cities of the developing world. However, the majority of moped owners are probably dreaming of moving up to a four-wheeler as soon as circumstances allow.

The average middle-aged suburban commuter in Europe or the United States – male or female – is not likely to buy and use either a powerful 200 kph gasoline-powered motorcycle or an $11,000 electric scooter to ride to work on. But the costs of the e-scooter are going to drop radically in the next few years, both because the lithium-ion battery technology is still being improved rapidly, and because there will soon be cheap imports from China. Moreover, some of those cheap imports, even if they are limited to 20 or 30 km/hr, will be able to use the bike paths or reserved lanes that many cities are going to build into their traffic plans for the coming decades.

---

2   It is already available from a firm called Vectrix, in Newport, RI. Vectrix will offer a scooter, weighing 200 kg, capable of 65 mph, using nickel-hydride batteries (same as the Toyota Prius) with an expected lifetime of 10 years. It will cost $11,000. Two new electric motorcycles, using lithium-ion batteries, with even longer expected lifetimes, are scheduled to appear on the market in 2008. One, the Enertia, from Brammo Motorsports of Ashland, OR, weighs 125 kg. It costs $12,000. The batteries are made by Valence Technologies of Austin, TX. The other, from Zero Motorcycles, of Scotts Valley, CA, is an off-road motorcycle using lithium-ion batteries from A123 systems of Watertown, MA. It weighs just 54 kg, partly because the battery charger is not integrated into the bike. For sports fans, it is said to be capable of a 20 metre jump. It costs $7000.

It is highly likely that electric bikes and scooters are soon going to become extremely popular in Europe and (initially) in California and other western states of the United States. This development will, in combination with some other changes (including some developments described below, offer the simplest way out of the suburban sprawl transport dilemma.

Electric Four-wheel Vehicles

The hybrids, such as the Prius pioneered by Toyota (1997) and the Honda Insight (1999) have been a success story in the marketplace (the Prius has sold over a million) and a lot of technological development has occurred along the way. They are actually electric only for purposes of starting and acceleration. The nickel metal-hydride batteries used in these cars are now mass producible and they last nearly as long as the car. One replacement, at most, is likely to be needed. However, hybrids available today are still made of steel, with small gasoline engines assisted by batteries. This combination allows higher fuel efficiency and cuts emissions, but it requires 30 per cent more components than a conventional internal combustion engine (ICE) car, and will never be cheaper to build or sell.

The next generation of so-called 'plug-in' hybrids was introduced in 2009, and several others are being unveiled in current year (2010) or the next year. The mass-producible lithium-ion battery will almost certainly soon replace the nickel metal-hydride battery of yesterday. The problem is that the plug-in hybrids currently in production or planned use a standard 'platform' shared by other models, and based on the mass-produced steel body and frame. This problem also applies to the all-electric Chevrolet 'Volt', which has been widely publicized, though less so at the time of writing given GM's financial troubles and uncertain future. A different approach is Daimler's SmartForTwo (2010), which is an adaptation of the Smart city car introduced over a decade ago in Europe.

Today (2010) the THINK Citycar, based on Amory Lovins's original 'hypercar' concept (Lovins 1996, Lovins et al. 1996) is the only EV to be fully certified according to US and EU vehicle safety standards. The initial price is €21,000 or about $30,000. That is almost certainly not high enough to be profitable, nor low enough to be widely attractive. But the costs and prices will come down as production increases and as li-ion batteries permitting greater range and shorter recharging times eventually replace the NiCad batteries currently used. The THINK Citycar is not the only adaptation of Lovins's hypercar concept. A startup company, Fisker Automotive, will soon be manufacturing a plug-in hybrid called the 'Fisker Karma' with a power train and lithium battery pack from Quantum Technologies. Others in this category include the Kewet Buddy (2009), Reva Citycar (2009), Microvett Doblo EV (2009), Microvett Fiorino EV (2009). Some startups will fail (as usual) but their time in the sun seems to be coming.

Lithium-ion Batteries

Lithium-ion batteries have anodes and cathodes both consisting of metal alloys or compounds with open three-dimensional structures such that small lithium ions can move freely. A number of such materials have been tried, including lithium cobalt oxide, manganese spinel, and others. One of the most interesting electrode compounds discovered to date is lithium-iron phosphate ($LiFePO_4$) alloy. This material has been commercialized for mobile phones, digital clocks, laptop computers and other uses by several companies.

The early lithium batteries have to be charged and discharged slowly. However, under certain (rare) conditions, lithium-ion batteries can discharge rapidly enough to get very hot and even cause a fire. (There have been a few incidents in small electronic devices.) Before being adopted for widespread use in cars, these safety concerns will have to be resolved. But a number of battery producers claim to have solved this problem or to be close to solving it. For instance, involving Massachusetts Institute of Technology (MIT) and the State University of New York (SUNY), Stony Brook, claims a similar improvement in li-ion battery performance due to modifications in the micro-structure of their proprietary lithium-nickel manganese electrodes (Bullis 2006a). Altair Nanotechnologies, of Reno Nev. has developed a new lithium-electrode material permitting both faster charging – in as little as six to eight minutes – and three times longer battery life (Bullis 2006b).

A Silicon Valley startup, Tesla Motors, has been manufacturing a high-performance electric sports car with a proprietary lithium-ion battery since 2004. It has a range of 220 miles and an acceleration 0 to 60 mph in 3.9 seconds (but not both on a single charge).[3] Tesla has contracted with Daimler-Benz to provide 1,000 lithium-ion battery packs for its 'SmartForTwo' EV, due to be introduced in the United States in 2011. Daimler also has its own lithium battery production plans. The advanced plug-in hybrid vehicles will boast very low operating costs, but still at significantly higher prices. Over the next two decades manufacturing costs should decline significantly, thanks to technological improvements in the batteries and economies of scale and experience.

The Trouble with Lithium?

The big problem for mass production of all-electric cars or plug-in hybrids is mass-production of high-performance batteries, almost certainly of the rechargeable lithium-ion type that is now widely used for laptop computers and other such devices. Lithium availability has been a concern to some authors, since crustal content is low (estimates range from 20 ppm to 65 ppm) and there are no good ores with a high lithium content (e.g. Råde and Andersson 2001, Tahil 2006). Current

---

3　Tesla now has a competitor with an even more powerful electric sports car, Shelby Supercars, whose Ultimate Aero EV recently was clocked at a world's record 208 mph.

output of lithium is largely from brines and low-grade ores such as spodumene and petalite, hectorite clay or brines. The largest known recoverable deposit is in the Salar y Uyuni area of Bolivia (a dry salt flat) The US Geological Service (USGS) estimates the Bolivian lithium reserves to be 5.4 million tons, with another 3 million tons in salt flats in Chile a possible 2 million in Argentina and 1 million tons from a similar deposit in Tibet (China) and another in Brazil. US reserves are only 410,000 tons. There are probably other salt flats in Africa, Australia and the Middle East that have not yet been tested for lithium content.

The current price of lithium carbonate (used in rechargeable batteries) is only about $2.80 per pound, which would amount to about $31 for a single 9 kWh rechargeable lithium-ion battery for an EV, in current dollars. Cumulative demand up to the year 2100, assuming very rapid growth rates for batteries (15% pa for 10 years, 10% pa for the next 10 years and 3% pa until 2050), plus other uses, could be as much as 16 million tons (Yaksic Beckdorf and Tilton 2009). This seems like quite a lot compared to current output. Nevertheless, there is probably enough lithium available from terrestrial sources – probably some of it undiscovered – to keep going that long.

Will that be the end? Not at all. In the first place, lithium-ion batteries can be recycled. But that may not even be necessary. The major global lithium resource is the ocean (0.17 ppm), which contains 224 *billion* tons of the metal in the form of lithium chloride. The recovery technology is straightforward (electrolysis of brine). Studies were done in the 1970s on the potential availability of lithium for tritium production for a possible future nuclear fusion power system. Adjusting for inflation, it can be estimated with reasonable confidence that lithium is potentially available from the ocean at between $4 and $5 per pound (current dollars) of carbonate equivalent (Yaksic Beckdorf and Tilton 2009). If lithium recovery were to be combined with sea-water desalination plants, possibly combined with solar concentrators to provide the electric power, the costs might be even lower. Desalination plants purify saline water, by removing the salt. The latter could – in turn – be the feedstock for a lithium recovery plant as well as other activities such as chlorine and caustic soda production. There is no lithium scarcity problem.

The first users might well be car-sharing firms using EVs. These possibilities will be discussed later in the section on public transport.

Car-sharing

Another part of the solution of the problem of low-density suburbia will be car-sharing. This idea was first tried out in Switzerland (1987), mainly in connection with the railway system. People arriving in Zurich or Basel by train were given a new option, an inexpensive short-term rental for local travel. It soon spread to Germany. A survey conducted by a researcher at the University of California found 18 programmes in the US with a total of 234,483 members, sharing a total of 5261 vehicles. These numbers increased by 75 per cent and 45 per cent respectively between 1 January 2007 and 1 January 2008, so this innovation has clearly passed beyond infancy to adolescence (Wikipedia 2008c). There are now at least 600 cities worldwide – possibly as many as 1000 as this is written – with car-sharing clubs, and a membership of 350,000. That number has doubled since 2006 (Britton 2009).Within five years membership in car-sharing clubs could climb to several million people and in 10 years it could reach 100 million. Yet the untapped potential is still enormous. Only 5 per cent of the population of Zurich and only 1 per cent of the Swiss (where it all began) belongs to a car-sharing club, and a far lower fraction in other countries.

The standard car-sharing model is similar to the Velib bicycle-sharing model now operating in Paris, except that the vehicles are necessarily more expensive and parking facilities for them are also more expensive. For this reason, car-sharing sites now tend to be concentrated in a central location (such as a railway station) or perhaps a few such central-city locations. The system operates like the bicycle-sharing programme in other respects. There is an annual membership fee, and cars are rented by the hour at modest rates – but with no free period. The weakness of the system today is that cars must be reserved in advance and picked up (and parked) in specific locations. The annual fee for Zipcar is $50, and the hourly rate in the United States starts at $11 in New York. Elsewhere costs are a little less. The cost of a four-hour trip in the United States averages around $30. For multi-day trips, rates are comparable to standard car-rental. A number of car-sharing ventures have failed, but the trend is up, nevertheless, as the statistics cited above indicate. The difference between success and failure is partly attributable to the business model and the available capitalization, partly in the degree of encouragement by municipal authorities, and partly the price of fuel.

A recent development of some interest is exemplified by the City of Philadelphia, which has contracted with car-sharing firms since 2004, resulting in the elimination of 330 vehicles from the city fleet. Zipcar, the largest car-sharing firm, claims that each Zipcar takes 15 other vehicles off the road; that each member drives 4000 miles per year less than before joining the programme and that each member saves an average of $435 per month by using the service. If true, this is very definitely a double-dividend and negative cost solution, and it is available today with existing technology.

Traditional car-rental firms are getting into the car-sharing business, probably because, thanks to the economic downturn they are having difficulty selling their used vehicles. Avis began with a partnership with some Paris parking garages, but their programme has not (yet) prospered. U-haul recently started car-sharing programmes in several college towns, Berkeley, California, Madison, WI and Portland, OR. Enterprise Rent-a-Car started in St. Louis and recently expanded into half a dozen universities. The most important new entry is 'Connect by Hertz', launched with some glitz on 17 December 2008 in New York, London and Paris. Hertz is starting with 10 locations in Manhattan, as compared to Zipcar's 300 locations in the New York area. The entry of Hertz, the largest car-rental firm, into this field is probably a milestone.

Whereas central city customers tend to take the cars outside the city (where buses are rare and taxis even rarer) the new trend is to locate car-sharing facilities in the suburbs. To compete with private cars, such centres need to be located within a few kilometres of lots of residents and reachable by bicycle (or e-cycle). Many existing automobile dealerships, auto rental offices and used-car lots, facing declines in traditional business, will find it makes sense to double as car-sharing centres. Others will co-locate with shopping malls and suburban railroad stations, small airports, and so on. A great many customers, especially commuters, will reserve vehicles for their use on a regular schedule. There will be a charge for the reservation service at some times of day (as there is for a reserved seat on a train). But walk-in clients will be able to rent a shared car with minimal effort, probably no more than a credit card with a PIN number. The on-board GPS guidance system and map in the car will direct the user to a vacant parking place as near as possible to their desired end-location. Charges will be based on distance travelled, zone in which the trip takes place (in case there are congestion charges), time of day and location of final parking place with respect to the nearest car-share facility (in case of users who want to leave the vehicle outside the 'allowed' region).

All of this computation will be based on the route recorded by the GPS system, linked to the central dispatching and locational system via integral 'hands free' cell-phones or (in the case of purpose customized car-sharing EVs) built-in WiFi communications. The cell-phones will permit users to contact the central dispatching centre. All of these technologies are currently available, and all that is needed is some integration and packaging. The leader at present seems to be Eileo, an on-line fleet management system for car-sharing firms. It currently provides real-time vehicle monitoring and tracking, GPS mapping, maintenance management, booking logistics, and customer relationship management in a single software package.

Car-sharing firms, like car-rental firms, mostly now use standard vehicles. However, since the mode of usage is different – a shared car may make several short trips each day – it will (like a taxi) accumulate mileage much faster than a conventional vehicle. Typically a rental car or taxi is put out to pasture (i.e. sold second-hand) after two years intensive usage and something like 120,000 miles on the speedometer. At that point is will already be ready for its third set of tyres, and will be showing distinct signs of mechanical wear in other areas. Yet many of the components of such a car are still nearly new. The car-rental firms are finding such cars well-suited to the car-sharing business, after being fitted with the electronic package described above, and they have the capability to provide regular and rigorous maintenance.

One of the most interesting new entrants into the car-sharing field is a firm named MoveAbout. It was launched at the headquarters of DNV in Oslo in December 2008, with five vehicles and four more corporate customers scheduled to go online in February 2009. The Oslo parking firm Entra Eiendom has made parking spaces with recharge facilities available at each of its stations. There will be four EVs at each Entra station. Another five vehicles will begin operations at

Lindholmen Science Park in Gothenburg, Sweden, in February 2009.[4] MoveAbout plans to rent electric city cars to corporations for a fixed monthly fee and offers 24/7 access, a web-based booking system, individual access contactless cards for users and full service, including roadside assistance. The fleet management will utilize the Eileo system, mentioned above. MoveAbout's vehicle of choice is the Norwegian TH!NK (discussed above), although other electric city cars are now or will soon be available. It may take 10 years or more to get to the smaller suburban towns but car-sharing is on a roll. For the next two decades most clients will also maintain their own private cars, but they will use them less and less. Like global oil production, private car ownership in the industrial countries is about to peak and begin a long decline. In the so-called BRIC countries (Brazil, Russia, India and China) the peak will occur later, but it is coming. This is bad news for GM, Toyota and VW, but good news from an environmental standpoint.

In late 2007 a Silicon Valley 'whiz kid' Shai Agassi (who had sold his software company to SAP for $400 million) raised $200 million in venture capital for an audacious project. In early 2008 he formed a partnership with Renault-Nissan and the state of Israel, to create a nation-wide EV rental system, with the intention of drastically reducing the need for petroleum. Israel is a good place to test such a system since 90 per cent of the population drives less than 70 km per day. In 2008 Better Place had 50 EVs in Israel with 1000 charging locations and three battery-exchange stations. By the end of 2009 he was aiming to have 500 cars, 10,000 charging spots and 100 battery change stations in Israel. Each car will have a GPS system built in, to guide users to recharging stations when needed. Customers will buy mobility kilometres, analogous to mobile phone companies that sell minutes of conversation. Another similarity with the mobile phone industry is that Better Place will give a car free to a consumer who will sign a service contract for four to six years. Better Place already has agreements with Denmark, with negotiations under way with Australia and other countries with suitable densely populated urban areas (Orsato 2009: Chapter 7). The system is designed to fill the gap between the standard car-rental operation, and the Velib shared bicycle system operating in Lyons and Paris. In the former case clients rent a car for a day or several days and (normally) return it to the same rental station, for a daily fee that is independent of use, plus a charge for fuel. In the Velib case customers rent the bicycle for a few minutes, (the first 30 minutes are free) and return it to any of hundreds of parking stations around the city.

---

4   It is interesting to note that Gothenburg is also the headquarters city for both Volvo and SAAB.

Goods Delivery

There are various ways to increase the energy efficiency of individual truck vehicles, which need not be listed here in detail since they are generally similar to the options for car design, e.g. lighter weight, use of carbon-fibre composites, hybrids, and so forth. Since energy costs are a significant cost element for truckers, it can be assumed that most of the low-cost options will be adopted routinely, especially if cap-and-trade legislation is adopted setting a minimum price on carbon emissions. Diesel-hybrid engine designs may be introduced in the future. Radical improvements in heavy vehicle efficiency are not to be expected, however.

A more complex way to save energy would be to develop a technical means of cargo-sharing among shippers to increase the average vehicle load. This is what the Postal Service, DHL Federal Express and UPS do already for small parcels. There may be a role for medium sized containers, spanning the size range between a large suitcase (for example) and the large containers that are now standard for shipments of industrial goods and household moves, by rail or ship. However, whereas Fedex and UPS pick up and deliver door-to-door, security and loading/unloading problems arise for smaller shippers at the interface when one shipper has to transfer goods to another. For the moment, these problems seem to trump some of the potential advantages of load consolidation, although some consolidation is obviously taking place through mergers and partnerships.

Conclusions: The Urban Transportation System in 2050

It is a truism that forecasting – especially of the future – is very difficult. However the range of future possibilities is not infinite. A number of mega-trends that are now under way and that are effectively irreversible, will constrain the range of possibilities. One of them is population growth. Two or three billion more people will be living in 2050 than today, all of them in so-called developing countries. Another mega-trend is urbanization: future population growth will be in the cities, and the rural villages will depopulate, if anything. A third continuing trend is electrification; as liquid fuels for internal combustion engines become scarcer, electrical alternatives will be necessary. A fourth is the build-up of greenhouse gases in the atmosphere. A fifth is 'peak oil': the decreasing availability (and rising price) of liquid petroleum-based fuels.

From this last fact alone, it follows that the enormous industrial superstructure built around the internal combustion engine over the last 150 years and, ultimately, based on cheap liquid hydrocarbon fuels, is going to have to change radically. The details are unclear, although there is some light in the tunnel, but the fact that radical change is necessary cannot be denied. The most obvious option, and the one undoubtedly preferred by the powerful auto industry, will be to use natural gas or gas from coal so produce a synthetic liquid fuel can be burned in a standard engine. If this scenario happens the cars of the next 30 or 40 years will not look or perform much different from today's car.

However, that solution is only temporary at best, because global natural gas output is also approaching a peak, almost certainly before 2050. Moreover, only a few countries will have enough gas to export by 2050 (Russia and Iran among them) and the more fundamental changes that need to be made will be significantly delayed.

The next scenario depends upon a breakthrough in bio-fuels from cellulose. Either ethanol or methanol can be combined with gasoline, as an additive or supplement, thus extending the life of the existing ICE-based liquid fuel system. The current option, ethanol from sugar cane or corn, is not sustainable for long in a world with growing population and limited arable land and fresh water. The idea of switching down the road to ethanol or methanol from cellulose, supplied by specialized cellulose plantations (bamboo, switch grass, eucalyptus trees) is appealing to some tropical countries, such as Brazil. But there is no practical way to restrict such plantations to marginal or poor land where food crops will not grow.

On the contrary, there is every economic incentive to use the best and most productive land for bio-fuels if motor vehicle owners are allowed to 'bid' without restriction. Thus, the rising demand for automotive fuel as well as paper and lumber, will inevitably compete with food and feed crops, and raise prices for basic foodstuffs, unless effective regulations to prevent it are put in place. It is true that there is some scope for integrated production of cellulose plus food and feed, using plant wastes for fuel rather than ploughing them back into the land, But that could have a bad effect on long-term soil fertility. In short, the bio-fuel option is a dead end, though far from recognizing that fact. The question at issue for governments is whether the prospect of short-term profits to a powerful alliance between agro-business and motor vehicle owners will trump long-term considerations.

The long-term solution has to be largely based on electrification, which in much of the world is likely to be heavily dependent on nuclear power because of the scarcity of alternatives. Electrification includes trams and rail-based systems, but most of all it means EVs, whether with four wheels, three wheels or two wheels. The usual objection is that EVs have short range and limited speed and power, and that the battery packs are expensive. This is true, to a degree, but less so as the price of hydrocarbons increases. Indeed, some of these objections are already being answered, as battery technology has progressed in leaps and bounds in recent years. But the crucial point is that most trips are local and do not require a vehicle with a very long range. Another objection is that electrification of the vehicle fleet will require a lot more power plants. The answer to that objection is that a large rechargeable battery capacity in vehicles can smooth the daily demand fluctuations and is therefore actually synergistic with intermittent power sources such as wind and sunshine.

As more and more people live in cities, the need for cars to get to work will necessarily decline, if only because of congestion. The ideal solution is for people to live where they work, which can be achieved to a limited extent by clever city planning, and also, to some extent by more widespread encouragement of 'telework'. Congestion charges, parking charges and insurance costs will also discourage private cars, as is already the case in Singapore and London. The large steel-body privately owned family car will not disappear immediately, but more and more of its functions can, and will, be taken over by more specialized modes: shared bicycles (as in Paris), scooters and e-cycles, buses and bus-rapid-transit, car-sharing based on fleets of small light EVs, and rental fleets.

These changes will not be painless, especially for the auto industry. Companies that do not see what is coming will not survive, and those that do survive will have to depend much less on profits from production and sales of motor vehicles, and much more on a new business model, selling transport services (km travelled). The coming decades will be 'interesting times' for them, as the Chinese used to say.

Bibliography

Anonymous. 2006. *Mexico City Metrobus*. Mexico City: BRT Policy Center. Available at: www.gobrt.org/MexicoCityMetrobus.html.

Ardia, A. 2006. *P Mexico City Metrobus*. Available at: go.worldbank.org/TWEUFJ5SV0.

Britton, E. 2009. *Car-sharing Database*. New Mobility. Available at: www.carshare.newmobility.org.

Bullis, K. 2006a. Battery Breakthrough. *Technology Review*, 21 February. Available at: www.technologyreview.com/business/16384/.

—— 2006b. The Lithium-ion Car. *Technology Review*, 24 March. Available at: www.technologyreview.com/business/16624/.

Levinson, H.S. 2003. *Bus Rapid Transit on City Streets: How Does it Work?*, paper presented to the Second Urban Street Symposium, Anaheim CA.

Lovins, A.B. 1996. *Hypercars: The Next Industrial Revolution*, paper presented to the 13th International Electric Vehicle Symposium (EVS 13), Osaka, Japan, 14 October.

Lovins, A.B. et al. 1996. *Hypercars: Materials, Manufacturing, and Policy Implications*. Snowmass, CO: The Hypercar Center, Rocky Mountain Institute.

Orsato, R.J. 2009. *Sustainability Strategies: When Does it Pay to be Green?* Basingstoke: Palgrave.

Pucher, J.R. and Lefevre, C. 1996. *The Urban Transport Crisis in Europe and North America*. London: Macmillan.

Pucher, J.R., Altshuler, A. and Womack, J.P. 1981. *The Urban Transportation System: Politics and Policy Innovation*. Cambridge, MA: MIT Press.

Pucher, J.R. et al. 2010. Walking and Cycling to Health: RecentEvidence from City, State and International Comparisons. *American Journal of Public Health,* 100(10), 1986–92.

Råde, I. and Andersson, B.A. 2001. Requirement for Metals of Electric Vehicle Batteries. *Journal of Power Sources,* 93(1–2): 55–71.

Tahil, W. 2006. *The Trouble with Lithium.* Meridian International Research. Available at: www.meridian-int-res.com/Projects/Lithium_Problem_2.pdf.

Vincent, W. and Callaghan Jeram, L. 2006. The Potential for Bus Rapid Transit to Reduce Transportation-Related CO_2 Emissions. *Journal of Public Transportation* BRT Special Edition, 9(3), 219–37.

Weinert, J., Burke, A. and Wei, X. 2007. Lead-acid and Lithium-ion Batteries for the Chinese Electric Bike Market and Implications on Future Technology Advancement. *Journal of Power Sources,* 172(2): 938–45.

Wikipedia. 2008a. *Washington Metro.* Available at: en.wikipedia.org/wiki/ Washington_Metro.

—— 2008b. *Velib.* Available at: March.httm://www.technologyreview.com/read article.aspx?id=16384&ch=biztech.

—— 2008c. *Carsharing.* Available at: en.wikipedia.org/wiki/Carsharing.

Yaksic Beckdorf, A. and Tilton, J.E. 2009. Using the Cumulative Availability Curve to Assess the Threat of Mineral Depletion: The Case of Lithium. *Resources Policy,* 34(4), 185–94.

Chapter 18

Sustainable Cities with Creativity: Promoting Creative Urban Initiatives – Theory and Practice in Japan[1]

Emiko Kakiuchi

Introduction

In the face of an aging and declining population, Japan is being forced to change its socio-economic structure from one based on growth to a more sustainable model. Globalization is also having a great impact on the whole of Japanese society, including on people's daily lives. Devolution and coordination among the private, government and 'public' sectors are essential in reorganizing society in Japan.

Several theoretical frameworks have been discussed concerning the development of cities in Japan. From 1980 to 1990, the 'global city' concept was introduced in Japan, whereby the megacity can play an important role in dealing with globalization, serving as an incubator for innovation, and in developing talent through training (Sassen 1991). This model focuses strongly on networks of highly specialized services such as accounting, finance, advertising, telecommunications and other management functions. It was suggested that global cities with a concentration of these functions, such as New York, London and Tokyo, would be the leading players.

Alternatively, the concept of the 'sustainable city' was introduced in the late 1990s as an anti-global city concept, and several measures to create sustainable cities were recommended to the Japanese Government by the OECD in 2001.[2] These recommendations strongly influenced arguments and trends concerning urban development policy. The Japanese government has taken some actions, including by introducing a series of legal and financial measures such as enactment

1 This work is supported by KAKENHI (Grant-in-Aid for Scientific Research) 19530233.

2 OECD (2001) 'Urban Policy in Japan'. Their recommendations are: revitalization of urban centres and managing urban growth in suburbs to achieve sustainable cities, achieving appropriate land use patterns in urban areas, restructuring regulations, expanding investment for cities, securing financial measures for improvement, reconciling private rights and the public interest, and re-evaluating the role of national government, taking a comprehensive approach.

of the Landscape Law, and other steps to revitalize cities. In particular, the concept of the 'compact city' was derived from the sustainable city concept, aiming at a liveable city for the handicapped, aged and pedestrians through human-scale development. Due to motorization (and lack of public transportation), suburban sprawl and large roadside shopping centres led to urban decay. The inner cities, which are not convenient for cars and not easy to develop because of complicated land rights frameworks, have deteriorated, and streets with shuttered shops are seen in many places. In order to solve these problems, many cities are now following the compact city model, aiming at a more pedestrian/bicycle oriented environment.

In the twenty-first century, the concept of the creative city was introduced, focusing on the importance of creative talents and creative industries based on individual creativity, skill and talent. They are expected to create economic wealth and jobs through developing intellectual property regimes.[3]

In the most recent decade, content businesses, which are normally considered creative industries in Japan, have comprised less than 3 per cent of total GDP.[4] On the other hand, in almost all industries, even those that are not necessarily classified as 'creative industries', creativity and innovation have become essential. For example, the Toyota Motor Company, which is categorized as a manufacturing company, now emphasizes not only the function of its automobiles, but also design and comfort to meet consumers' demands. Furthermore, production processes and know-how are protected by patents, and rapid industrial structural change has been brought about through creativity and innovation. Thus, it is not easy to define which industries are creative and which are not.

In Japan, the importance of 'creativity' has long been discussed in relation to education, culture, and science and technology in particular.[5] Since Japan achieved rapid economic growth in the 1960s and 1970s based in part on uniform education that produced a well-trained workforce that allowed Japan to catch up economically with advanced western countries, more attention has been paid to the creation of human capital with diversified character and talent. In the course of discussion of this issue, the number of patents, Nobel prize winners, research papers published in international journals, and so on, are used as measures of creativity in Japan. At the same time, in the field of business management, creativity has been analysed as one of the important elements for innovation, which allows Japanese companies to adapt their operations to the ever-changing business environment, by redefining

---

3   See World Intellectual Property Organization, n.d.

4   According to the Digital Content White Paper (Digital Content Association of Japan 2009), the market size of contents business such as books, images, music, broadcasting, internet services, game software and related services in 2008 was about 13.8 trillion Yen, while nominal GDP of 2008 was 494.2 trillion Yen (Cabinet Office n.d.).

5   For example, the Central Council for Education issued a report in 1971 referring to the importance of developing creativity through formal and informal education (Ministry of Education, Science and Culture, The Central Education Council 1971).

the value of their goods and services, adding new values to their products, and creating new markets and clients. In this discussion, tacit knowledge and corporate culture shared by corporate members are considered to play an important role (Nonaka and Takeuchi 1995). The newly introduced 'creative' city concept, at least as it is taken in Japan, seems to basically be an extension of this philosophy, but with the slight difference of paying more attention to arts and culture (in a narrow sense) as well as science and technology.

In this chapter, considering the above theoretical models, I will introduce best practice of sustainable cities with creativity in Japan. The concept of 'city' will be interpreted not by size but by coherence as a 'community'. Sustainability will be broadly defined as 'improving the quality of life in a city without leaving a burden for future generations', focusing mainly on cultural components.[6] 'Creativity' as used in this chapter is not confined to creativity as defined by World Intellectual Property Organization (WIPO), but is interpreted in a broader sense to include all of the efforts and wisdom of the community members applied to solving problems that threaten to cause the deterioration of a community. Then, the chapter discusses three topics with a strong relationship to culture: the definition of creativity; its potential role in fostering sustainable cities; and the implementation of policies that encourage creativity, with reference to several cases in Japan. First, socio-economic trends will be outlined in brief, and several aspects of urban development will be discussed. Each case study will then be examined by considering quantitative data, and the relationship between creativity and urban development will be analysed.

Socio-economic Change in Japan

Macroeconomic Trends

Japan has modernized itself for the past 150 years. After the Second World War, Japan's new constitution renounced war, and Japan placed a strong focus on economic development. In the 1960s and 1970s, in particular, when Japan experienced rapid economic growth, serious social problems such as disorderly development, public nuisances, over concentration, and depopulation of rural areas became part of the political agenda.[7]

After the two oil shocks of the 1970s, Japan still maintained reasonable economic growth through an export drive, which led to the 'Plaza Accord' agreement to devalue the US dollar relative to the Japanese Yen and other currencies in 1985. Accordingly, credit was relaxed in Japan, which caused excessive property and stock speculation in the late 1980s. This led to soaring

---

6   Urban 21 conference, see Regional Environmental Center, 2001.

7   The National Income Doubling Plan and the Comprehensive National Development Plan were put into effect in 1960 and 1962, and the Shinkan-sen (bullet train) service was inaugurated, and the Olympic Games were held in Tokyo in 1964.

land prices and spurred haphazard and excessive development in what became
known as the 'bubble economy'. After the bubble burst in 1989, national and local
governments, carrying the burden of huge debts, were squeezed for resources
for proper development of urban infrastructure, and the economy was stagnant
for a period commonly characterized as the 'lost decade'.[8] Due to the financial
crunch as well as local needs, the devolution process has been ongoing.[9] However,
even now, concentration of many economic and commercial functions in the
Tokyo Metropolitan area and along the Pacific coast continues, and disparity and
inequality among cities have been widening.[10] Japan's population growth stopped
in 2004, and is now declining. In sum, by the early twenty-first century, the society
and economy of Japan has matured.

In the course of the above developments, several cities in Japan that had served
mainly as sites for industrial production lost many valuable functions such as
the incubation of local culture, identity and social coherence. Since the bubble
economy collapsed, they have also lost their role as sites for industrial production
due to the movement of production facilities to low-cost countries.

In the 1980s and 1990s, 'culture' and 'region' increasingly became key words
for all aspects of life, and people gradually recognized cultural properties as an
assets for regional development. Many local governments took action to preserve
the historic atmosphere of their city and utilize historic sites. People began to
recognize the importance of cultural assets as resources for tourism, commerce
and industry.[11] The national demand for a better quality of life has increased.
People are seeking their own identity in their community, and cultural and artistic
activities have been integrated into local communities. In other words, culture is
considered to be an important component of having a good quality of life.

Demand for Non-material Satisfaction and Quality of Life

Data from a national survey of values is shown in Figure 18.1. As shown by
the graph in Figure 18.1, Japanese people increasingly consider non-material
satisfaction to be more important than material satisfaction.

---

8   The outstanding amount of government bonds is 685,213 billion Yen as of 2008,
and the outstanding amount of local authority bonds is 139,059 billion Yen as of 2006.

9   The Decentralization Promotion Law, enacted in 1995, was the third significant
reform of the local government system, following the Meiji Restoration in 1868 and the
reforms after the Second World War.

10   'Greater Tokyo' includes an area located within about a 70 km radius of the centre
of Tokyo (including portions of surrounding prefectures such as Kanagawa, Yamanashi,
Saitama, Ibaraki and Chiba prefectures), and has a population of more than 34 million.
Also the urban agglomeration of Tokyo as defined based on population density by the UN
has a population of 26 million, while the total population of Japan is about 120 million as
of 2008.

11   In 1992 a new law was enacted to promote regional industries and businesses
utilizing traditional performing arts and other related activities.

Figure 18.1 National Survey of values

Based on these trends, the Japanese government has undertaken several initiatives. In 2001, the Fundamental Law for the Promotion of Culture and Arts was enacted. This law is the first of its kind; it reflects a broad social consensus on the importance of culture. The law makes provision for the support of cultural activities developed not only by the national and local governments, but also by non-governmental organizations (NGOs), private companies and individuals.

Another important law is the Landscape Law, which was enacted in 2004, aiming to promote pleasant and beautiful scenery in cities and villages. This is the first law that refers to the importance of the 'beauty' of cities and villages in Japan and stipulates that the national government is responsible for extending public support through zoning, and if necessary, through restriction of the private rights of land owners. This law also calls for the active participation of citizens.

In 2005, Japan's '21st Century Vision' was released. This document calls for Japan to become a 'culturally creative nation'. Japan is now seeking 'cultural creativity' for the future in terms of socio-economic development. However the 'content business' in Japan (audio-visual, TV broadcasting, music, radio, game software, books, newspapers, etc.) comprises less than 3 per cent of GDP, although this proportion is growing.

Alternatively, as a source of local development, more emphasis is being accorded to promoting cultural tourism and local traditional industry with a view to harmonizing economic development and local sustainability. In 2006, the Fundamental Law for the Promotion of Tourism Nation was enacted to strengthen strategic measures to attract tourists from all over the world, utilizing local cultural assets such as historic sites, beautiful scenery, monuments, landscapes, hot springs

and traditional industry. The National Land Council recommended effective land
use by utilizing culture for local sustainable development in 2008.[12]

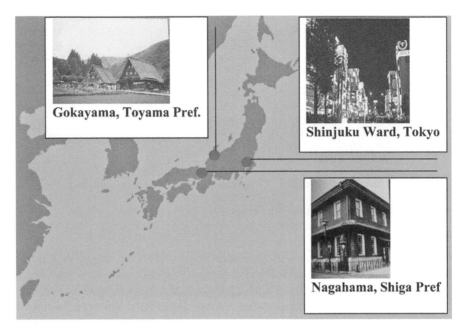

**Figure 18.2   The cases considered in this chapter: Nagahama, Gokayama,
and Shinjuku Ward Cases**

**The Case of the City of Nagahama, Shiga Prefecture: Best Practice
in Revitalizing the Inner City**[13]

Outline

Nagahama, a city of 80,000 inhabitants in Shiga prefecture, is located on the
largest lake in Japan, Lake Biwa. Hideyoshi Toyotomi, a famous feudal lord in the
sixteenth century, built a castle there. Nagahama has a financial capability indicator
(defined as the capability of an area to raise income for necessary expenditure) of
0.59, which is slightly higher than the national average. Of the working population,
most residents are engaged in the service sector (57.1%) and the industrial sector[14]

---

12   See Ministry of Land, Infrastructure, Transport and Tourism, National and
Regional Planning Bureau 2008.

13   This section is mostly based on the DVD produced for the World Bank by the
author (see National Graduate Institute For Policy Studies, Cultural Policy Program, n.d.).

14   Chemistry, machinery and textile factories are located in the city.

(38.1%).[15] The average per capita income of the city was estimated to be 3.09 million Yen, slightly lower than the national average in 2006.[16]

As the establishment of his castle led to the foundation of the city of Nagahama, Lord Hideyoshi Toyotomi is a symbol of the city. The Wheeled Float Festival, for example, which is held every spring, celebrates the birth of Hideyoshi's son. However in the 1970s and 1980s, despite its long prosperous history and special characteristics, the city faced urban decay and depopulation due to motorization, suburban sprawl, and the development of a large-scale shopping complex located on the outskirts of the city that damaged the business of the inner city shops.

Four hundred years after the founding of the original castle, the Nagahama Castle Historical Museum was built in 1983. It was funded by donations from the public, ranging from elderly citizens who donated some money from their pensions to children who contributed from their pocket money. This popular support became the driving force of the regeneration of this region. Various events were held to celebrate the opening of the newly built castle museum, and these events generated momentum for further efforts.

The concept of a 'museum city' was introduced then. This concept was composed of two ideas: first, to live in beautiful surroundings in a modern society utilizing tradition; and, second, to learn about the passion and wisdom of ancestors and pass their spirit onto the next generation, including as a practical measure, renovating the streets in the inner city area.

Kurokabe Initiative

The building of the Nagahama branch of the former Hundred Thirty Bank, which was nicknamed the Kurokabe (Black Wall) Bank, was built in the Meiji era, over 100 years ago. In the late 1980s, a developer bought the building and made plans to demolish it. However, strong opposition from local residents led to the creation of a company, Kurokabe, Inc., in April 1988. Kurokabe, Inc. was founded with capital amounting to 40 million Yen from the city government and 90 million Yen from eight regional companies. This company, which is a 'third sector' joint venture between the government and private industry, purchased the bank building in order to preserve it, and made further plans for the preservation and regeneration of the inner city. The goal of this company is to promote economic activity in businesses that do not compete with pre-existing shops and industries in the city. In addition, the company aims to promote the image of the city of Nagahama. Thus Kurokabe, Inc. decided to focus on the craft of glass blowing,

---

15   Data as of 2005, provided by City of Nagahama.

16   Japan Planning Systems (2008). In this indicator estimate, the figures are based on the city, town or village tax revenues surveyed by the Ministry of Internal Affairs and Communications. The national average is estimated at 3.39 million Yen per person in 2006.

which is new to the region. This is compatible with the arts festivals held in Nagahama and helps to create an international aura.

Kurokabe, Inc. is a joint venture between the government and private industry, but the participants from the private sector actually run the company. Kurokabe, Inc. chose artistic glassware as the nucleus of its business and has been trying to integrate this with the old townscape of the city. By taking this approach, Kurokabe, Inc. is not only compatible with existing local industries but is also becoming one of the leading new industries in the city. The Kurokabe glass crafts building, and other shops, reflect the atmosphere of the Meiji period in the late nineteenth century (Figures 18.3 and 18.4).

Figure 18.3   **Figure 18.4**
Exterior of the Kurokabe building   **Interior of the Kurokabe building**

Figures 18.3–18.13 are provided by the Kurokabe Inc.

More than 10 shops in Kurokabe combine to form one of the centres for artistic glassware in Japan, whereas glassware shops in other cities in the country tend to be isolated shops featuring the works of individual artists, or souvenir shops or retail outlets for industrial glassware. Keeping the atmosphere of the houses of old style merchants, the Kurokabe shops feature glassware culture. Some are decorated with stained glass on the walls, and some use glass tableware. In addition, other shops selling different items combine to create an attractive inner city in Nagahama with the glassware cultural industry as the essential core. This charm helps to attract tourists to the city of Nagahama.

Some locations before and after renovation are shown in Figures 18.5–18.7. Decayed buildings have been remodelled to create a souvenir shop, using the traditional style. The interior is decorated to coordinate with the exterior.

Figure 18.5
A Nagahama location
before renovation

Figure 18.6
A Nagahama location
after renovation

Figure 18.7
The interior of a
renovated Nagahama
location

Figures 18.8–18.10 show an example of a traditional house and its conversion into a glassware museum.

Figure 18.8
A traditional Nagahama
house before renovation

Figure 18.9
A traditional Nagahama
house after renovation

Figure 18.10
The interior of a
traditional Nagahama
house after renovation

Glassware is not as common as pottery in everyday life in Japan. The glassware factory is trying to create distinctive Nagahama styles that reflect its history and traditional design (Figures 18.11–18.13).

Figures 18.11–18.13   Kurokabe collection

Evaluation

In 1998, Kurokabe, Inc. created an independently operated centre for regional regeneration. This centre functions as a bridge between local governments, shops and citizens, providing information and coordinating services for various events. The centre has achieved the regeneration of the inner city shopping mall and the main street, Hokkoku Kaido Street. This street runs more than one kilometre from the north to the south of central Nagahama and features a historic townscape, maintained in cooperation with local residents who invested in restoration of old buildings and houses. The beautiful townscape and artistic glassware have been attracting more and more visitors, leading to a virtuous cycle.

The trend in the number of visitors to Nagahama has been increasing constantly and now amounts to 5 million per year. Among them, visitors to Kurokabe, Inc. comprise a major part, amounting to 2 million in 2002 (Figure 18.14).

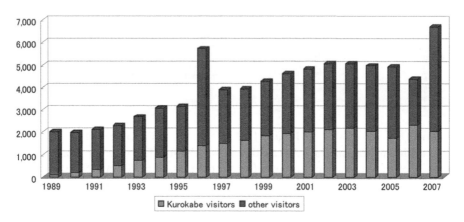

Figure 18.14 Number of tourists visiting Nagahama, 1989–2002
Source: City of Nagahama.

Public Opinion

In order to estimate the benefits and identify the beneficiaries of these cultural initiatives, Table 18.1 shows the result of a citizens' survey conducted in Nagahama in 1995. According to this survey, public opinion regarding the cultural initiatives is quite positive. In addition to the preservation of the historic townscape, the respondents also think that Kurokabe created a lively image of the city and made it a new tourist spot. It also upgraded regional cultural standards and contributed to the local economy.

Table 18.1   Public opinion (multiple answers)

Good to preserve the old townscape	55%
Created a new tourist spot	61%
Created a good lively image	64%
Upgraded regional cultural standard	41%
Brought economic impact to the region	28%
Too noisy	4%
Tourism spoiled quiet atmosphere of the city	8%

Economic Impact Estimates

An economic impact study of tourists' consumption was conducted in 1995. At that time, there were 1.2 million visitors annually to Kurokabe, in comparison to more than 2 million in 2002. According to the surveys, one third of the respondents said that the main purpose of their travel was to visit Kurokabe square. Based on this figure, direct consumption was estimated to be at least 1,500 million Yen, together with an indirect effect of 1,930 million Yen and an induced effect of 380 million Yen. Visitors spent at least 300 million Yen at shops, other than Kurokabe, and 460 million Yen at nearby hotels. In 2002 another visitor's survey was conducted in Nagahama. According to the survey, more than 60 per cent of visitors said they came to Nagahama to visit Kurokabe. Therefore, the present economic impact of visitors to Kurokabe is even larger than the 1995 estimate.

Table 18.2   Consumption data (unit: million Yen)

Direct consumption (1,500 million Yen)	Kurokabe sales	690
	Nearby shops	300
	Parking	60
	Hotels	460
Indirect effect (1,930 million Yen) and **Induced effect** (380 million Yen)	Hotels	520
	Glass production	430
	Commerce	330
	Restaurants	160
	Transportation services	114

Conjoint Analysis[17] of the Cultural Facilities and Policy Implications

According to our survey of tourists conducted in 2003, only 10 per cent of visitors stayed overnight, and half of them stayed in Nagahama for only half a day.[18] If people stay longer in the city, they will spend more money. To encourage visitors to stay longer, it is necessary to know why they are coming and what they are seeking.

Table 18.3   Expenditure during visits to Nagahama (unit: Yen)

Sojourn time (number of samples)	Amount of expenditure during a visit to Nagahama				
	Craftwork (e.g. glass sculpture)	Food products	Eating in restaurants	Other	Totals
1–2 hours (110)	3130.0	743.6	2027.3	518.2	6419.1
3–4 hours (195)	3153.1	1592.3	2130.8	1166.7	8042.9
Half-day (190)	3763.2	1523.7	3021.1	978.9	9286.8
All day (54)	3781.5	1653.7	3163.0	8111.1	16709.3
Overnight stay	3921.9	2218.8	6453.1	5703.1	18296.9

Generally speaking, the charms of Nagahama come from its cultural institutions, such as museums and temples. Considering the importance of cultural tourism, we decided to conduct a conjoint analysis of the cultural institutions in the city centre of Nagahama. We found that the age of the visitors varied greatly, but that the average age is still young, and the income of all age groups is consistently higher than average. Most visitors come from outside the city, especially other prefectures. A total of 39 per cent of visitors come to Nagahama because they are interested in traditional landscapes and history; 64 per cent come to Nagahama because they like viewing, and shopping for, artistic glassware. Roughly half of the visitors were repeaters.

There are seven cultural institutions within 30 minutes' walk of the central railway station (Figure 18.15). Considering the location of cultural institutions in

---

17   Conjoint analysis, a multi-attribute compositional model, is a statistical technique widely used in marketing, operations research and product management. It aims to determine what combination of a limited number of attributes is most preferred by respondents.

18   Kakiuchi and Iwamoto (2008) (in Japanese).

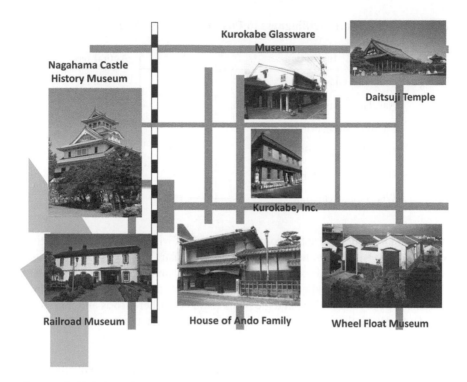

Figure 18.15 Map of Nagahama cultural institutions

the city centre, we conducted a conjoint analysis on visitors' willingness to pay (WTP) for these sites. Based on the WTP of tourists, the most attractive institution is the Kurokabe glassware gallery, and the Wheel Float Museum is the least attractive. Although the city of Nagahama spent 2 billion Yen constructing the Wheel Float Museum, the tourists' WTP is estimated to be only one tenth of that for the Kurokabe glassware gallery.

To promote tourism, it might be appropriate to prepare a special discount pass that would allow the holder to enter any institution listed on the pass, with the combination of the most popular glassware gallery and other institutions.[19]

In summary, the inner city of Nagahama has been revitalized through a series of urban initiatives, most of which originated and were promoted by citizens and the private sector, in close cooperation with local government. The Kurokabe project in particular, successfully created a glassware market which is new to the region and still growing. The core competence of the Kurokabe project comes from the artistic creativity which attracts a large number of visitors. The Kurokabe visitors

---

19   The city of Nagahama sells Nagahama Roman Passports ('Roman' is a loan word meaning romantic in Japanese) that allow the holder to enter all of the registered cultural institutions.

Table 18.4　Visitors' Willingness to Pay (WTP) (unit: Yen)

Institution	WTP	Actual fee
Kurokabe Glassware Museum	1,556	600
Nagahama Castle History Museum	716	400
Railroad Museum	557	300
Daitsuji-temple	315	300
House of Ando Family	241	200
Wheel Float Museum	184	600

are arts connoisseurs and active consumers, many of who return for a second time. Similar attempts to promote artistic glassware industries have been emerging in other cities, including glassware studios in Noto, Ishikawa prefecture, and the Tsukiyono glassware ateliers in Niigata prefecture. Facing these competitors, Nagahama needs to develop strategic plans to make further efforts and investment in order to ensure that its artistic creation remains sustainable.

The Case of the Gokayama Historic Village: The Survival of a Marginal Village[20, 21]

Outline

These historic hamlets shown in Figures 18.16–18.21 are located in a mountainous area in Toyama prefecture, in the central part of Japan. They are well known for their traditional wooden houses, which are one of the representative forms of Japanese architecture. Harmonizing with the surrounding harsh natural environment, the steep thatched roofs of the houses, which are made of natural materials from nearby forests, can endure the heavy snows of this region. These houses were maintained by *yui* (community cooperation), and the necessary skills to maintain them were passed from generation to generation. However, due to the rapid economic growth and changes in social structure from the 1950s through to the 1970s, most of these houses and historic hamlets vanished, (with the exception of the Shirakawa-go hamlet in Gifu prefecture and the two smaller

---

20　This section is mainly based on Kakiuchi (2005 and in press).

21　A marginal hamlet is defined as a village in which 50 per cent of the population is aged over 65, and where maintaining community life – mutual help or holding a funeral service – is no longer feasible.

Figure 18.16
Ainokura hamlet

Figure 18.17
Ainokura Hamlet

Figure 18.18
Suganuma hamlet

Figure 18.19
Suganuma

Figure 18.20
Inside a traditional house

Figure 18.21
Repairing a thatched roof

Figures 18.16–18.21 are provided by the Industrial Economic Dept, Nanto City.

hamlets of Gokayama: Ainokura and Suganuma).[22] These three hamlets are now designated as 'districts of groups of important historic buildings' under Japan's Law for Protection of Cultural Properties (hereafter referred to as 'the Law'), and under the UNESCO World Heritage List. The two hamlets of Gokayama were also designated as 'national historic sites' under the Law in the 1970s; since then changes in both the interior and exterior of the houses have been strictly restricted.

Both hamlets are located in 'designated depopulated villages'.[23] The annual expenditure of these villages was roughly 2.4–3.0 billion Yen in 2000.[24] During 2000–03, Taira village had around 1400 residents with a financial capability indicator of 0.15, and Kamitaira village had less than 1000 residents with a financial capability indicator of 0.42. Three villages, including these two, comprised the

---

22   Ainokura hamlet in Taira village and the Suganuma hamlets in Kamitaira village are small preservation areas with less than 100 residents. These areas contain houses, temples, shrines and surrounding forests.

23   Under the Law of Special Measures on Promoting Independence of Depopulated Areas (in effect since 2000), areas that are suffering from underdevelopment of social and production infrastructure can be designated eligible for special financial support from the national government.

24   The statistics cited in this paragraph were provided by the former Taira village.

Gokayama region, until 2004 after which they were all merged into the city of Nanto. The per capita income of both villages was estimated as relatively low: 2.75 million Yen per person in 2006, although almost no one received public assistance (welfare).[25] The house ownership rate is very high in these villages, and among the population in the work force, roughly half were engaged in the service sector and 40 per cent were engaged in the industrial sector.

Located in the former Taira village, Ainokura hamlet occupies 18 hectares where 20 thatched roof houses are located and preserved. The population of Ainokura hamlet was around 90 people in 27 households at the time of their inclusion in the UNESCO World Heritage List, and has gradually decreased since then. The hamlet residents are either families who are engaged in tourism-related business, or production process/construction workers who are employed by companies outside the hamlet, including a large electric power company. Most of them are also engaged in agricultural production, but mainly for self-consumption.

Located in the former Kamitaira village, Suganuma hamlet occupies 4.4 hectares and has nine thatched-roof houses. This hamlet, much smaller than Ainokura hamlet, has only 40 residents with eight households. Due to aging, some of the residents are pensioners, and others are engaged mainly in construction work.

These hamlets now face a serious challenge. Depopulation, rural decay due to nationwide economic development, aging and the increase in tourism-related family businesses have weakened various functions needed for maintenance of the hamlets such as making paddy fields, managing forests and re-roofing historic houses. In order to maintain this cultural heritage, the hamlets must be made sustainable and liveable. Therefore, creating jobs and upgrading living conditions is essential.

Tourism-related Issues

As the Gokayama hamlets are designated as national historic sites under the Law, the Japanese government has extended public subsidies for their preservation. At the same time, the Law requires public access as much as possible to designated historic sites. The number of visitors to the Gokayama region has been growing (see Figure 18.22).[26] In 1970 when the hamlets were designated, 100,000 travellers visited the Gokayama region annually. Since 1995, the hamlets have seen a dramatic increase in the number of tourists (more than 900,000) due to their inclusion in the UNESCO World Heritage List, but the number of tourists has stabilized at the level of 700,000 per year more recently. Due to the national highway connection in 1999, it has become much easier for tourists to visit Gokayama now, and most of these tourists visit these hamlets.

---

25   Japan Planning Systems (2008). In this indicator estimate, the figures are based on the city, town or village tax surveyed by the Ministry of Internal Affairs and Communications.

26   The figures quoted in this paragraph were provided by the Gakayama tourism association.

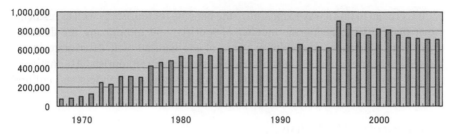

Figure 18.22 The number of visitors to Gokayama

At the time of inclusion of these hamlets in the UNESCO World Heritage List, the Gokayama region set up a committee to plan a comprehensive development strategy that included construction of highways connecting the historic hamlets to large cities, creation of an eco-museum and/or 'World Heritage Theme Park', inviting the venture businesses of artists' ateliers and attracting more tourists.[27] Even though the partial completion of an improved highway has brought about an increase in transient tourists, the number of tourists who stay overnight in the region has been quite low, averaging around 5–6 per cent of total tourist numbers. Furthermore, the other plans have not been realized and the economic and industrial structure has not changed greatly.

The heavy pressure from day-trippers has created difficulties for the residents: their daily lives are exposed to outsiders' eyes all the time, there is increased waste generated by tourists, and there is congestion in the high season. The hamlets have also had to ask tourists to visit only during the daytime (9:00 to 17:00). Still, residents recognize the importance of tourism for maintaining the hamlets. Existing public support does not fully cover the costs of maintaining their cultural heritage, and further support, including additional job creation and upgraded living standards, is necessary to facilitate the daily life of the hamlets and to create new systems to replace traditional community cooperation, in order to preserve the hamlets and their environment.

Ainokura established a foundation for collecting fees from tourists in 1996. The foundation also works towards maintaining the natural environment surrounding the hamlet. At present, the foundation has funds for only a few jobs, most of them part-time. The other Gokayama hamlet, Suganuma, has built a museum and accommodation at a nearby site and is trying to maintain the living heritage as it was in the past. Visitors are free to enter the hamlet. Today, the Gokayama region tries to connect these hamlets and other tourist spots nearby, Shirakawa-go hamlet in Gifu prefecture in particular, to create a cluster of agricultural landscapes. Also, through a partnership between government and private initiatives, the local government seeks to promote local production through tourism.[28]

---

27 Gokayama Regional Committee for Development (1996: 25–30).
28 Gokayama Tourism Association (2006).

It is crucial to find a way to maximize the benefits for tourists as well as residents rather than just increasing the number of tourists, and minimizing the demerits of tourism. Otherwise, the cultural value of the hamlets will deteriorate in the long run, which might result in a decrease in the number of tourists. Alternatively, tourists could take more responsibility for maintaining the hamlets and cultural values they enjoy, rather than having a free ride.

Implication of CVM[29] Analysis

In our survey conducted in 2001, nationwide accrual of the benefit of cultural heritage on a large scale was observed. The most important and fundamental value of the cultural landscape of the hamlets is the bequest value. The cultural landscape provides large benefits to society as a whole, and public involvement can be justified on this basis. These benefits of the hamlets derived from non-use values such as bequest value and existence value. Furthermore, there is a strong consensus among tourists, as well as from a national survey that the hamlets cannot be preserved by the residents alone.

Table 18.5   Estimate of WTP

	Nationwide survey	Tourist survey Ainokura	Tourist survey Suganuma
Mean WTP (Yen)	10,344.9	14,766.3	25,179.5
Median WTP (Yen)	1,885.3	3,038.6	3,249.5
Estimated number of households (unit: thousands) Estimated number of tourists (unit: persons)	4,637.6 —	— 56,742	— 38,694
Response rate (%)	26.9	72.0	70.0
Mean TWTP * (million Yen)	479,757.0	837.9	974.3
Median TWTP* (million Yen)	87,430.8	172.4	125.7
Mean TWTP** (million Yen)	129,054.6	603.3	682.0
Median TWTP**(million Yen)	23,518.9	124.1	88.0

Note: Cf. * a conventional model, multiplying WTP by population; ** a multiplying model by response rate of the survey, assuming that a non-respondent has zero WTP.

---

29   The contingent valuation method (CVM) is a method directly asking people in a survey; questionnaire or interview, about their willingness to pay (WTP) for specific services.

Table 18.6   Variables and coefficients

Variables	Nationwide survey	Tourist survey
Constant	++	++
In (T) bid amount	++	++
Spontaneous visit to the hamlets	—	++
Recognized as beautiful landscape	—	++
Recognized as mountain hamlets	—	++
Recognized as designated cultural property	—	+
Recognition of cultural value	—	++
Recognition of bequest value	++	++
Recognition of existence value	++	—
National government should support the hamlets	—	++
Gender: female	—	—
Age (years)	++	—
Income (unit:10,000 Yen)	++	—
Volunteer experience in the last one year	—	++

Note: Cf. + rejected with 5% significance level, ++ rejected with 1% significance level. Confidence interval is set at 90%.

These heritage values, not only aesthetic but also cultural and bequest values, are the essential core that attract tourists to the hamlets. Recognizing these values, tourists are visiting the hamlets and are willing to make donations to preserve the hamlets, as well as do volunteer work. In order to facilitate these initiatives, it is recommended that the local residents provide appropriate information and create good relationships with tourists, focusing on education, participation and awareness. These initiatives might improve the quality of culture-based tourism and lead to greater satisfaction among tourists who are sensitive to the various values of the hamlets.

Furthermore, considering the carrying capacity of the cultural landscape of the hamlets, it might be desirable to limit the number of visitors to avoid congestion and overuse of the resources, in order to maximize the benefits for those visitors who are most concerned about the hamlets.

Without a healthy economy and community, cultural heritage values cannot be maintained. Considering culture as an important component of tourism, tourism should be organized to contribute to local societies and economies in a sustainable way. Toward this end, the initiatives and involvement of local residents are crucial, and well-organized management is required.

Kabukicho Renaissance – The Case of Shinjuku Ward, Tokyo

Outline

During the Edo period (sixteenth to nineteenth centuries), the former city of Edo (now Tokyo) had a population of more than 1 million. Since the Meiji Restoration in 1868, when Japan opened the country to international society, the capital city has grown further and is sprawling, although the city was greatly damaged by the Kanto Great Earthquake of 1923, and by bombing during the Second World War. The total population of the prefecture of Tokyo is approximately 13 million (12,886,838) as of 2008, the largest among the 47 prefectures of Japan.[30] With the largest population and also the largest population density (13,000/km²), Tokyo still attracts more people from the surrounding prefectures for work since Tokyo functions as the centre of Japan in terms of politics, economy, information and communication, education, culture and entertainment.

Located in the central part of the Tokyo Metropolitan area, composed of terrace land and valleys, Shinjuku Ward is the fourth largest ward in terms of business concentration and worker numbers. Developed as a post station during the Edo era, it had thrived with large shopping centres, cinemas and theatres, and restaurants. The Great Tokyo Air Raids during the Second World War devastated most parts of Shinjuku ward and in 1947 it began to rebuild under the new post-war regime.[31] Shinjuku ward is characterized by skyscrapers, shopping and entertainment districts, and parks and gardens (Figures 18.23, 18.24 and 18.25, photos are provided by the Shinjuku Ward Office). Shinjuku's total population is 310,000 as of 2007, of which 30,000 are foreign passport holders. The daytime

---

30   There are several possible categorizations of Tokyo. *Greater Tokyo* includes an area located within about a 70 km radius of the centre of Tokyo (including portions of Kanagawa, Yamanashi, Saitama, Ibaraki and Chiba prefectures), and has a population of more than 34 million. Also the urban agglomeration of Tokyo as defined based on population density by the UN has a population of 26 million. In this chapter, however, we define *Tokyo* as the territory under the Tokyo prefectural government (Tokyo-to). Note that certain outlying islands quite far from Tokyo proper are included in *Tokyo-to*, but both as a fraction of the land area and population the proportion is negligible.

31   Shinjuku ward is one of 23 *special wards* (*tokubetsu ku* in Japanese) in the central part of Tokyo prefecture. The administrative status of these special wards is unique; although as quasi-municipalities they are autonomous local governments in most respects, certain public services, such as water supply, sewage disposal, and fire prevention services, are collectively provided by the Tokyo prefectural government, which levies some of the taxes needed to support these services.

population of 770,000 includes workers in the business districts, but not shoppers.[32] Most residents are engaged in the service sector (84.4%) and the industrial sector (11.6%) as of 2005. The financial capability indicator of Shinjuku ward is 0.65. The average income is estimated as 5.1 million Yen per person in 2006, which is much higher than the national average.[33]

Kabukicho Renaissance Project

Around 3.5 million passengers use the main railway/subway/bus station in Shinjuku every day, and the areas surrounding the station are the ward's foremost business, shopping and amusement districts. Located just to the east of the main station, the Kabukicho district that was developed after the Second World Ward is the busiest in Shinjuku ward (Figures 18.23–18.25, and Table 18.7). In the late 1940s and 1950s, the residents of Kabukicho aimed to build an entertainment district that would provide amusement services for people of all ages.[34] Part of the district has been occupied by theatres since then.[35] However, this original plan was not as successful as expected, and economic development and soaring land prices led to the concentration of the sex industry and related industries of high profitability. This caused safety and security problems, as well as community decay.

Figure 18.23
Shinjuku night view

Figure 18.24
Kabukicho

Figure 18.25
Shinjuku Gyoen Garden

Figures 18.23–18.25 are provided by the Shinjuku Ward, Tokyo.

Since 2004, under the leadership of the new mayor, Hiroko Nakayama, many initiatives have been undertaken to improve security, enhance the environment and revive the district. With close cooperation among the ward council, local

---

32   The major industries are information services, restaurants, printing and publishing, retail and real estate.

33   Japan Planning Systems (2008). In this indicator estimate, figures are based on the city, town or village tax revenues as surveyed by the Ministry of Internal Affairs and Communications.

34   Kabukicho is named after the traditional stage art *Kabuki*, even though Kabukicho has never been a significant site for kabuki performances.

35   Due to dilapidation, the Shinjuku Koma Theatre was closed in 2008.

Table 18.7   Statistical data

Residents	1,970
Foreign passport holders	809
Visitors to Kabukicho per day	300,000
Buildings in Kabukicho	632
Restaurants	3,967
Cinemas	15
Hotels	83
Businesses offering food and entertainment	1,429
Sex-related businesses (restaurants and shops)	312
Bars, saloons and other drinking places	1,610

business owners, city authorities, the police and fire departments, and other related institutions, there have been crackdowns on organized crime groups, sex-industry related illegal businesses and illegal gambling. This cooperation also includes measures to clean up the environment and the streets, such as the prevention of illegal parking, removal of obstructing bicycles and illegal signboards, and renovation of the district and its buildings.

However, without legitimate business activities replacing the illegal ones, other illegal businesses will fill the vacant spaces, which can lead to a vicious cycle. In general, it is difficult for ordinary businesses to operate in a district with such high land prices, especially when there is no sound and sustainable development of the community, given that it has an economy based on illegal sex-related business.

In 2005, the Kabukicho renaissance project started to replace sex-related businesses with entertainment related businesses such as planning, production and presentation of popular culture. The project aims to build a dynamic, attractive and safe and secure district by developing a new type of culture, and space for amenities and beauty. The Kabukicho Renaissance Promotion Committee was formed and it adopted a Charter stating this vision.

Kihei Project[36]

As one of its main projects, the Kihei Project, which aims to mobilize private entrepreneurs to play a central role in realizing the district's vision in the future,

---

36   Named after Kihei Suzuki, a former president of the Kabukicho Town Council who proposed rebuilding the district as an entertainment city from the ashes of the Second World War.

was started. In actual implementation, the committee provides support to artists, entertainers, producers and creators of 'content-oriented' products such as food, design, stage performances, audio-visual products, and so on. The most symbolic action was a collaboration with Yoshimoto Kyogo Co. Ltd.

Yoshimoto Kogyo Co., Ltd is a major entertainment company, headquartered in Osaka since 1912.[37] They opened a live comedy theatre in Shinjuku Ward in 2001. This theatre became a trendy spot and attracted large audiences. The company thereafter recognized the further potential of the Tokyo market and sought more space for operations in Shinjuku ward.

On the other hand, due to the decline in the birthrate, many elementary schools have closed in Shinjuku ward. The Yotsuya 5th elementary school (Figure 18.26) in Kabukicho district was founded after the Kanto Great earthquake in 1929.[38] The building became well known for its outstanding design, but 70 years after its foundation, it needed costly repair and earthquake-strengthening measures as well. Residents near the building requested that the city renovate the school and utilize the building for the community. To accommodate this local demand, the city decided to lend the building to Yoshimoto Kogyo as a part of the Kihei Project. Yoshimoto renovated the interior of the building, while preserving its exterior and atmosphere, and transformed it into their Tokyo office headquarters. Yoshimoto sells tickets to its events in collaboration with local shop owners and the town council, and organizes events for the local community.

Figure 18.26 Yotsuya 5th Elementary School
Provided by the Shinjuku Ward, Tokyo.

---

37   A major component of the Japanese entertainment world, Yoshimoto Kogyo Group started as a traditional theatre. It has grown to be an influential and powerful company, employing many talented performers in the field of manzai and owarai (comedy), and also producing and promoting TV shows and live stage performances.

38   The building of this elementary school is included in the *DOCOMOMO Japan 100* of excellent buildings.

Only 400 employees work in the headquarters building of Yoshimoto Kogyo. However, including visitors to the office, it makes a large economic contribution to the community. Also their theatre in Shinjuku Ward attracts a diverse audience, including many young females. This audience has significantly altered the atmosphere of the Kabukicho districts, where the major visitors were previously customers of sex-related businesses.

Public Opinion

In the central part of the Tokyo metropolitan area, the most symbolic district, Kabukicho, is changing through cooperation of many types of stakeholders: national and local governments, private institutions, the business sector and residents. The projects have just started, but some changes can be observed. According to public opinion surveys in the three years, as shown in Figures 18.27–18.30, the image and cleanliness has certainly become more appreciable year by year. Also this change is compatible with the vitality of the Kabukicho district.

It is not easy to replace profitable sex-related businesses with others. However, there is no future for a community with sex-related businesses as the dominant economic activities. Utilizing the accumulation of various service sectors suchas restaurants, retail and theatres, Shinjuku is trying to transform its centre into a more sustainable and liveable community.

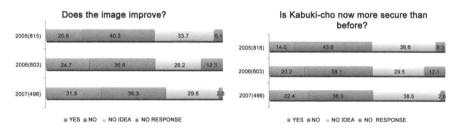

Figure 18.27 Kabukicho image   **Figure 18.28 Kabukicho security**

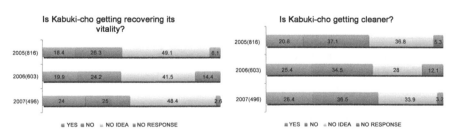

Figure 18.29 Kabukicho cleanliness   **Figure 18.30 Kabukicho vitality**

Figures 18.27–18.30 are provided by the Shinjuku Ward, Tokyo.

Conclusion

Summing Up

In the twenty-first century in Japan, the role of creativity, which is a relatively new concept in urban development, has increasingly gained attention. In this chapter, I have introduced three cases of best practice attempts to create a sustainable community, through the creative use of cultural assets such as historic buildings, artistic values and cultural landscapes.[39]

There are many other successful cases beyond these. In terms of cultural branding (the collection of images and ideas), Kyoto has been the foremost successful case, based on its world heritage temples and shrines. Also Kanazawa, Ishikawa prefecture, located in central Japan along the Japan Sea, escaped devastation during the Second World War, and enjoys a rich historic heritage. It has been moving to remake itself as a hub of many business functions in the region and as a liveable town, and has also had success in reinventing itself as a tourist spot.

The cases introduced in this chapter are quite different from one other in terms of their socio-economic conditions. However, in comparing a marginal village in central Japan, a local city and a special ward in the capital city of Tokyo, there are common features to consider. First, they are all trying to harmonize economic wealth and social benefits, which are sometimes not compatible. Nagahama is trying to revitalize its stagnant inner city to stimulate a new artistic glassware production by utilizing all its cultural historic assets, in order to preserve the historic atmosphere of the city. Gokayama, a marginal hamlet is surviving by gaining more support from tourists in order to preserve the hamlets and surrounding environment as they are now. These tourists recognize that the historic hamlets' values include non-market assets. Shinjuku ward, one of the busiest and most expensive parts of Japan, is also trying to harmonize lively economic activities with the amenities of the city.

Second, another common feature to be observed is the style of governance. All of the stakeholders in each community are working together to create a better quality of life. These stakeholders are not limited to government, but also include private land owners, residents, business people, and even tourists and visitors.

Third, in Japan, it is essential for the national and/or local governments to be involved. Although initiatives should be taken by residents and private institutions, it is the responsibility of the government to coordinate and integrate all the stakeholders with different interests. However, the governmental role should be carefully designed to create a platform to gather wisdom, harness the efforts of community members and other supporters and not to dis-incentivize private initiatives.

---

39   Cultural landscape is defined by the Law for Protection of Cultural Properties in Japan as landscape developed though interaction between human activities and natural environment.

Culture and Creativity for Sustainable Development

Florida's model of the creative city (Florida 2005), which strongly emphasizes the attractiveness of the creative city to highly skilled talent from outside the city, suggests one possible approach for how cities can deal with globalization in the increasingly competitive knowledge-based new economy. However, as many researchers have pointed out (Greffe and Pflieger 2005), this approach is applicable only to the largest cities, so-called megacities in particular.

In this sense Tokyo, the capital city, is the most creative city in Japan, not only in terms of knowledge-based industries such as IT, legal and accounting services, but also in terms of the number of artists, creators who are engaged in information-related industries, and cultural activities (Kakiuchi in press). The city of Tokyo has the great advantage of face-to-face communication among its large population of highly skilled professionals, which is an important force driving technological and business breakthroughs.

Florida's model notwithstanding, it is still not easy to establish causal relationships between creativity and economic prosperity. In general, economic prosperity and a large market will attract many talented professionals, which leads to a 'virtuous cycle' of further concentration of talented professionals and further increase in economic wealth (Kakiuchi 2010), as is the case of Tokyo, which presently enjoys both economic prosperity and creativity. The case of smaller cities in Japan, though, is quite different from the megacity of Tokyo.

As defined above, the creative city concept serves better as an instrument to mobilize community members in problem-solving efforts rather than as a goal. In Japan, for the past six decades, too much emphasis has been placed on economic growth and development, but it is now widely recognized that economic aspects are important but not necessarily sufficient to secure quality of life. At the same time, many cities, facing globalization, are losing competitiveness in terms of commodity production, and looking for alternative economic development models. Attaining a balance between quality of life and economic vitality, and finally increasing social welfare for all residents, is one of the most serious and difficult tasks facing cities.

In order to maximize social welfare and attain sustainable development, creativity can make an important contribution to smaller cities in Japan. As studies of industry suggest (Nonaka and Takeuchi 1995), tacit knowledge and culture shared by community members might be an infrastructure for creativity and innovation. 'Culture' should not be viewed as limited to arts and heritage in a narrow sense, but also value systems, ways of thinking, atmosphere and even regional history. This idea is not new, and many cities are now trying to attain an appropriate balance between economic development and quality of life through the efforts of their citizens. This alternative to the creative city approach presented in this chapter may also be applicable to tackling other issues, such as natural environment conservation, waste control and transportation problems, although this remains a subject for further work.

Bibliography

Cabinet Office, Government of Japan. (ed.) 2005. *Japan's 21st Century Vision*. Tokyo: National Printing Bureau. Available at: www5.cao.go.jp/keizai-shimon/english/publication/vision.html (in Japanese).

Cabinet Office. n.d. *Gross Domestic Product Account (Production and Expenditure Approach)*. Available at: www.esri.cao.go.jp/en/sna/kakuhou/kekka/h21_kaku/23annual_report_e.html.

Digital Content Association of Japan. 2009. *Digital Content White Paper 2009*. Tokyo: Digital Content Association of Japan.

Florida, R. 2005. *Cities and the Creative Class*. New York: Routledge.

Gokayama Regional Committee for Development. 1996. *Towards New Villages at the Inclusion of the UNESCO World Heritage List*. Gokayama: Taira Village and Kamitaira Village.

Gokayama Tourism Association. 2006. *Symposium Abstract*. Tokyama: GTA.

Greffe, X. and Pflieger, S. 2005. *Culture and Local Development*. Paris: OECD.

Japan Planning Systems. 2008. *Personal Income Indicator 2008*. Tokyo: JPA.

Kakiuchi, E. 2005. *Evaluating the Cultural Landscape*. Tokyo: Suiyo-Sha.

—— 2008. The Possible Model for Culture-based Tourism Development in Japan: Implication of CVM Survey of the World Heritage of Gokayama, Toyama Prefecture, Japan, in *Tourism and Community Development Asian Practices*. Madrid: UNWTO.

—— 2010. Reconsidering the Concept of the 'Creative City': Theory and Reality in Japan, *Research on Education and Media*, 2(2), 121–42.

—— in press. The Case of Tokyo, in I. Shirley and C. Neill (eds), *Development Patterns: Asian and Pacific Cities*. London: Routledge.

Kakiuchi, E. and Iwamoto, H. 2008. Promotion of Cultural Tourism in the City of Nagahama, Shiga Prefecture, Japan: Based on the Conjoint Analysis of Tourists' Preference for Cultural Facilities, *Urban Planning*, 30(4), 52–60.

Ministry of Education, Science and Culture, The Central Education Council. 1971. *A Report on Fundamental Policies and Measures for the Future Expansion and Development of School Education*. Tokyo: Ministry of Education, Science and Culture (in Japanese). Available at: www.mext.go.jp/b_menu/shingi/12/chuuou/toushin/710601.htm.

Ministry of Land, Infrastructure, Transport and Tourism, National and Regional Planning Bureau. 2008. *National Spatial Stratey – National Plan*. Tokyo: Nikkei Printing Inc. Available at: http://www.mlit.go.jp/en/kokudoseisaku/index.html.

National Graduate Institute For Policy Studies, Cultural Policy Program. n.d. *Heritage Conservation and Sustainable Development in Japan, The Case of Kurokabe Square, Nagahama, Shiga*. Available at: www3.grips.ac.jp/~culturalpolicy/rsc/aud/aud_kakiuchi_English.asx.

Nonaka, I. and Takeuchi, H. 1995. *The Knowledge-creating Company: How Japanese Companies Create the Dynamics of Innovation.* Oxford: Oxford University Press.

OECD. 2001. *OECD Economic Surveys: Japan 2001.* Paris: OECD.

Regional Environmental Centre. 2001. Sustainable Cities: Environmentally Sustainable Urban Development. Available at: archive.rec.org/REC/Programs/ SustainableCities/.

Sassen, S. 1991. *The Global City, New York, London, Tokyo.* New Jersey: Princeton University Press.

World Intellectual Property Organization. n.d. *Creative Industries.* Available at: www.wipo.int/ip-development/en/creative_industry/.

Index